ADP-Ribosylation in Pathogens

ADP-Ribosylation in Pathogens

Editors

Anthony K L Leung
Anthony Fehr
Rachy Abraham

MDPI

Basel • Beijing • Wuhan • Barcelona • Belgrade • Novi Sad • Cluj • Manchester

Editors

Anthony K L Leung
Department of Biochemistry
and Molecular Biology
Johns Hopkins University
Baltimore
United States

Anthony Fehr
Molecular Biosciences
University of Kansas
Lawrence
United States

Rachy Abraham
Dept. of Biochemistry and
Molecular Biology
Johns Hopkins University
Baltimore
United States

Editorial Office
MDPI
St. Alban-Anlage 66
4052 Basel, Switzerland

This is a reprint of articles from the Special Issue published online in the open access journal *Pathogens* (ISSN 2076-0817) (available at: www.mdpi.com/journal/pathogens/special_issues/ADPRibosylation_Pathogens).

For citation purposes, cite each article independently as indicated on the article page online and as indicated below:

Lastname, A.A.; Lastname, B.B. Article Title. *Journal Name* **Year**, *Volume Number*, Page Range.

ISBN 978-3-7258-0986-8 (Hbk)
ISBN 978-3-7258-0985-1 (PDF)
doi.org/10.3390/books978-3-7258-0985-1

Contents

About the Editors

Anthony K L Leung

Dr. Anthony K. L. Leung, currently a tenured professor at Johns Hopkins University, began his academic journey at the University of Oxford, where he completed his undergraduate studies. He went on to earn his Ph.D. from the University of Dundee under the mentorship of Dr. Angus I. Lamond. His completed his postdoctoral training at MIT under the guidance of Nobel Laureate Dr. Phillip A. Sharp and Dr. Paul Chang. In 2011, Anthony embarked on establishing his own lab. His scientific interests primarily revolve around exploring RNA and ADP-ribosylation, with a passion for pioneering new technology. His work extensively covers various aspects of RNA research, including RNA granules, RNA viruses, and coding and non-coding RNA metabolism. His career has been distinguished by numerous accolades, including the Top 5 Agilent Early Career Professor Award in 2013, the Inaugural Johns Hopkins Catalyst Award in 2015, the Research Scholar Award from the American Cancer Society in 2016, being a Top 10 Finalist for the American Society of Cell Biology–Gibco Emerging Leader Prize in 2016, and receiving the Shikani/El-Hibri Prize for Discovery and Innovation in 2019.

His leadership in the field is recognized by over 120 seminar invitations, chairing of conference scientific sessions, and successful funding from the National Institutes of Health, the Department of Defense, the American Cancer Society, and various private foundations. He also directly mentored a group of 60 talented junior scientists from diverse backgrounds and led the postdoctoral training efforts at the departmental and school levels. Dr. Leung's team also has a strong outreach presence by engaging and educating the public about how basic science promotes human health.

Anthony Fehr

Dr. Anthony Fehr is currently a tenured professor at the University of Kansas. He graduated with a B.S. degree from the University of Nebraska–Lincoln and then obtained his PhD from Washington University in St. Louis, where he studied the molecular biology of human cytomegalovirus under the mentorship of Dr. Dong Yu. He completed his postdoctoral training in Dr. Stanley Perlman's lab at the University of Iowa, studying the innate immune response to coronaviruses. It was there that Dr. Fehr began his work on how ADP-ribosylation impacts coronaviruses. In 2018, Dr. Fehr started his lab at the University of Kansas, where he has continued his efforts to decipher how ADP-ribosylation restricts coronavirus replication and pathogenesis. His work encompasses both basic research on viral macrodomains and host ADP-ribosyltransferases and translational efforts to develop macrodomain inhibitors and live-attenuated coronavirus vaccines.

Dr. Fehr's leadership in the virology and ADP-ribosylation fields is recognized by more than three dozen invited seminars or meeting talks, almost fifty authored manuscripts, multiple invitations to chair meeting sessions for the American Society for Virology, and several NIH awards related to his work on coronaviruses and ADP-ribosylation, including an NIGMS R35 Maximizing Investigator's Research Award (MIRA). In addition, Dr. Fehr provided a substantial level of community service, particularly during the COVID-19 pandemic, where he gave several talks about coronaviruses to the general public and conduted nearly 50 interviews to various media outlets to help educate the genral public about coronaviruses.

Rachy Abraham

Rachy Abraham, Ph.D., currently works as an assistant scientist at Johns Hopkins University. She earned her Ph.D. from the Rajiv Gandhi Centre for Biotechnology (RGCB), which is affiliated with the University of Kerala. Dr. Abraham pursued her Fulbright-Nehru Doctoral and Professional Fellowship at Hopkins during her graduate studies under the guidance of renowned Alphavirus expert Prof. Diane Griffin. Subsequently, she continued her research as a postdoctoral fellow in the same laboratory.

Dr. Abraham's research is dedicated to unraveling the intricacies of RNA viral infections, with a specific focus on alphaviruses. Through her endeavors, Dr. Abraham has contributed to the elucidation of the fundamental molecular mechanisms underlying the infection and replication cycles of alphaviruses. A pivotal aspect of her research involves the identification and characterization of the enzymatic activity of the viral macrodomain of alphaviruses. This macrodomain has emerged as a crucial factor influencing viral replication and virulence.

Preface

ADP-ribosylation was discovered over 60 years ago, and much of the research focused on bacterial toxins that used ADP-ribosylation to kill host cells. However, in the last 20 years, the field has identified many cases where host-mediated ADP-ribosylation can either promote or repress pathogenic infections. It is clear that ADP-ribosylation is a key component of the innate immune response in eukaryotes and bacteria, and many of the mechanistic details of how ADP-ribosylation impacts pathogens have yet to be uncovered. In this Special Issue, we have focused on publishing unique reviews and primary research on the topic of pathogens and ADP-ribosylation. We hope that you enjoy this collection.

Anthony K L Leung, Anthony Fehr, and Rachy Abraham
Editors

pathogens MDPI

Perspective

The Conserved Macrodomain Is a Potential Therapeutic Target for Coronaviruses and Alphaviruses

Anthony K. L. Leung [1,2,3,4,*], Diane E. Griffin [5,*], Jürgen Bosch [6,7] and Anthony R. Fehr [8,*]

1 Department of Biochemistry and Molecular Biology, Bloomberg School of Public Health, Johns Hopkins University, Baltimore, MD 21205, USA
2 Department of Oncology, School of Medicine, Johns Hopkins University, Baltimore, MD 21205, USA
3 McKusick-Nathans Department of Genetic Medicine, School of Medicine, Johns Hopkins University, Baltimore, MD 21205, USA
4 Department of Molecular Biology and Genetics, School of Medicine, Johns Hopkins University, Baltimore, MD 21205, USA
5 W. Harry Feinstone Department of Molecular Microbiology and Immunology, Bloomberg School of Public Health, Johns Hopkins University, Baltimore, MD 21205, USA
6 Center for Global Health and Diseases, Case Western Reserve University, Cleveland, OH 44106, USA; jxb745@case.edu
7 InterRayBio, LLC, Cleveland, OH 44106, USA
8 Department of Molecular Biosciences, University of Kansas, Lawrence, KS 66045, USA
* Correspondence: anthony.leung@jhu.edu (A.K.L.L.); dgriffi6@jhu.edu (D.E.G.); arfehr@ku.edu (A.R.F.); Tel.: +1-(410)-5028939 (A.K.L.L.); +1-(410)-955-3459 (D.E.G.); +1-(785)-864-6626 (A.R.F.)

Abstract: Emerging and re-emerging viral diseases pose continuous public health threats, and effective control requires a combination of non-pharmacologic interventions, treatment with antivirals, and prevention with vaccines. The COVID-19 pandemic has demonstrated that the world was least prepared to provide effective treatments. This lack of preparedness has been due, in large part, to a lack of investment in developing a diverse portfolio of antiviral agents, particularly those ready to combat viruses of pandemic potential. Here, we focus on a drug target called macrodomain that is critical for the replication and pathogenesis of alphaviruses and coronaviruses. Some mutations in alphavirus and coronaviral macrodomains are not tolerated for virus replication. In addition, the coronavirus macrodomain suppresses host interferon responses. Therefore, macrodomain inhibitors have the potential to block virus replication and restore the host's protective interferon response. Viral macrodomains offer an attractive antiviral target for developing direct acting antivirals because they are highly conserved and have a structurally well-defined (druggable) binding pocket. Given that this target is distinct from the existing RNA polymerase and protease targets, a macrodomain inhibitor may complement current approaches, pre-empt the threat of resistance and offer opportunities to develop combination therapies for combating COVID-19 and future viral threats.

Keywords: coronavirus; alphavirus; SARS-CoV-2; macrodomain; ADP-ribosylation; ADP-ribosylhydrolase; therapeutics

check for updates

Citation: Leung, A.K.L.; Griffin, D.E.; Bosch, J.; Fehr, A.R. The Conserved Macrodomain Is a Potential Therapeutic Target for Coronaviruses and Alphaviruses. *Pathogens* **2022**, *11*, 94. https://doi.org/10.3390/pathogens11010094

Academic Editor: Luis Martinez-Sobrido

Received: 22 December 2021
Accepted: 11 January 2022
Published: 14 January 2022

1. Introduction

Emerging and re-emerging viral diseases pose continuous public health threats that have increased with globalization, population growth, urbanization, and climate change. These disease-causing viruses range from zoonotic and sexually transmitted viruses such as human immunodeficiency virus (HIV), to vector-borne viruses such as Zika virus and chikungunya virus (CHIKV), to respiratory viruses such as H1N1 influenza virus and severe acute respiratory syndrome coronavirus 2 (SARS-CoV-2). The rapid mutation of these RNA viruses can facilitate infection of new hosts, improve transmission, and increase virulence with sudden spread into new regions. The medical, social, and political consequences of these emerging infections are global but often disproportionately affect resource-poor countries, resulting in unequal social, health system, and economic burdens on these populations.

Effective control requires a combination of non-pharmacologic interventions, treatment with antiviral drugs and antibodies, and prevention with vaccines. Experience with the COVID-19 pandemic and other recently emergent viral infections has exposed a lack of effective treatments capable of protecting humans from the devastating effects of highly deadly and contagious viral pathogens. This gap in preparedness has been due, in large part, to a lack of investment in development of antiviral agents for classes of viruses that were only "potential threats" and, if they did emerge, often caused only acute disease and were quickly contained. When SARS-CoV-2 emerged, this dearth of therapeutics was tragically exposed. Therefore, a multi-pronged investment in development of multiple classes of direct-acting antiviral drugs effective against emergent RNA viruses is crucial to prepare for the next epidemic/pandemic.

Coronaviruses, once thought to be mere contributors to the common cold, have now caused three notable epidemics of life-threatening disease in the last two decades: SARS-CoV in 2002–2003, Middle East respiratory syndrome coronavirus (MERS-CoV) in 2012–2015, and the current pandemic of SARS-CoV-2. All of these viruses are capable of causing respiratory infection that can result in severe disease and even death in a high proportion of those infected [1]. Currently, SARS-CoV-2 is the major public health challenge worldwide, with more than 5 million deaths, accompanied by prolonged destabilizing consequences for the global economy, life-expectancy, and education. The COVID-19-induced global recession resulted in an economic contraction of 3.5% in 2020, with the most pronounced downturn in the poorest countries. Furthermore, disruptions to health care systems have increased deaths from other causes and hindered delivery of routine health care services, such as immunizations. Another class of viruses with pandemic potential are the mosquito-borne alphaviruses. Although previously geographically restricted, these viruses continue to expand into new regions to cause epidemics of rash, arthritis, and encephalitis [2,3]. The past 50 years have seen a dramatic emergence/re-emergence of epidemic arboviral diseases, with 3.6 billion people (nearly half the world population) living in urban areas with exposure to efficient *Aedes aegypti* mosquito vectors [4]. In addition to incapacitating acute disease, a high proportion of individuals infected with arthritis-causing alphaviruses, such as CHIKV, Sindbis virus (SINV), and Mayaro virus (MAYV), develop chronic joint pain [5,6]. Although infectious virus is cleared promptly after most acute RNA virus infections, including those caused by alphaviruses and coronaviruses, viral RNA often persists, accompanied by ongoing immune stimulation that may contribute to prolonged symptoms [5,7–13]. Therefore, lack of effective treatment results not only in deaths due to acute disease but also in prolonged disability that could likely be prevented with antiviral treatment [14,15]. Availability of effective antiviral drugs for both coronavirus and alphavirus infections would decrease hospitalizations, deaths, and long-term disability, with benefits for individuals and society. A common feature of these virus families that offers a potential target for antiviral drug development is the highly conserved macrodomain.

2. Macrodomains Represent a Unique Target for Pathogens of Pandemic Potential

Few antivirals are available for treatment of SARS-CoV-2 [16], and no treatments are available for infection with any alphavirus. Antiviral drug development for coronaviruses has mostly focused on nucleoside analogs as inhibitors of RNA polymerase function (e.g., remdesivir or molnupiravir) and peptide analogs that inhibit viral proteases (e.g., nirmatrelvir). Likewise, preliminary evaluation of drugs that inhibit alphavirus replication have most commonly targeted the proteases or polymerase [17,18]. However, there remains a substantial need for developing additional antivirals with novel targets. Identifying distinct targets that require different antiviral mechanisms may complement current approaches, pre-empt the threat of resistance, and offer opportunities to develop combination therapy. Both viral families contain a highly conserved macrodomain that is critical for viral replication and virulence, making it an attractive therapeutic target [19,20]. Macrodomain inhibitors, if successfully developed, could be significant weapons in the future antiviral armamentarium to combat these pathogenic viruses with epidemic potential. Here, we summarize the role of the macrodomain in viral replication and virulence and review the current efforts to develop macrodomain inhibitors as direct-acting antivirals.

3. The Biochemistry of Macrodomains

ADP-ribosylation is a post-translational modification catalyzed by ADP-ribosyltransferases (ARTs, also known as PARPs [21]) that transfer an ADP-ribose moiety from NAD^+ onto target proteins [22]. The ADP-ribose molecule is transferred as a single unit of mono-ADP-ribose (MAR) or as consecutively attached single units of MAR through glycosidic bonds to preceding ADP-ribose units covalently to form a poly-ADP-ribose (PAR) chain. A macrodomain is a conserved protein fold, existing either as a single protein or embedded within a larger protein, capable of binding to and, in some cases, reversing this modification [19,20,23]. The macrodomain structure consists of a three-layered α/β/α fold and a conserved ADP-ribose binding pocket (Figure 1a,b) [24].

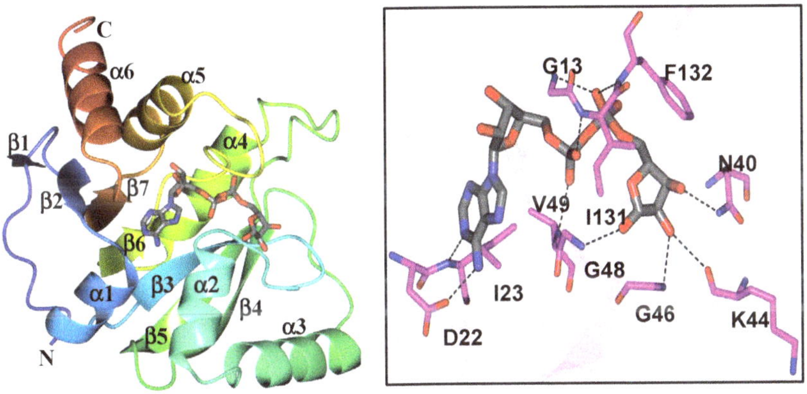

(a) (b)

Figure 1. (**a**) The structure of SARS-CoV-2 macrodomain complexed with ADP-ribose (6WOJ); (**b**) hydrogen bonds (dashed lines) between amino acids in the binding pocket and ADP-ribose. Obtained from Alhammad et al., 2020.

These macrodomains are identified in all kingdoms of life [22], including a subset of plus-strand RNA viruses: alphaviruses, coronaviruses, rubella virus, and hepatitis E virus (HEV) [19,20,23]. Viral macrodomains bind to ADP-ribose, its derivatives, and protein-conjugated ADP-ribose. Viral macrodomains are highly conserved in the nonstructural proteins of both alphaviruses and coronaviruses and belong to the MacroD subclass that has ADP-ribosylhydrolase activity [19,25–30]. All alphaviruses contain a single macrodomain in the N-terminal portion of nonstructural protein 3 (nsP3) while the highly pathogenic β-coronaviruses SARS-CoV, SARS-CoV-2, and MERS-CoV contain two to three tandem macrodomains in their nsP3, but only the first (Mac1) possesses ADP-ribosylhydrolase activity [31]. The coronavirus and alphavirus macrodomains are primarily MAR-hydrolases, although alphavirus and HEV macrodomains may also have PAR-hydrolase activity, especially when coupled with other proteins [27,32,33]. Therefore, these viral macrodomains, while conserved across different viral families, may have distinct functions.

4. Viral Macrodomains Are Critical for Viral Replication and Disease Pathogenesis

Some mutations in the ADP-ribose binding regions of coronavirus, alphavirus, and HEV macrodomains are not tolerated and viruses cannot replicate, indicating an essential function for this domain [28,34–36]. Alphavirus macrodomain mutants without binding or hydrolase activity are not viable due to defects in initiation of infection and viral RNA synthesis [34,35]. In addition, hydrolase activity is critical later in infection for translation of the sub-genomic RNA to produce the viral structural proteins and for disruption of stress granules, which are enriched with translation initiation factors [28,33,34].

Other viruses without macrodomain activity, such as SARS-CoV N1040A, can replicate normally in some tissue culture cells, but these mutants are attenuated in mice [37]. Studies in animals have shown that macrodomain ADP-ribosylhydrolase activity is critical for both coronavirus and alphavirus pathogenesis (e.g., Figure 2a,b) [20,38]. These data indicate that, while there may be virus- and cell-type specific differences in macrodomain function, the macrodomain is universally necessary for virus virulence.

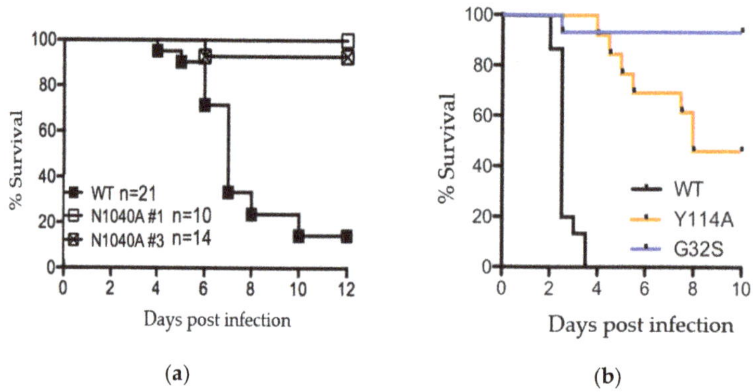

Figure 2. SARS-CoV and CHIKV macrodomain activity is required for viral pathogenesis. (**a**) Female Balb/C mice were infected with a lethal dose of SARS-CoV and equivalent amount of 2 separate clones of macrodomain mutant (N1040A) virus and monitored for survival over 12 days. Data from Fehr et al., 2016; (**b**) 2-day old CD-1 mice (N = 24–28/group) were infected with CHIKV or nsP3 macrodomain mutants Y114A and G32S and monitored for survival over 10 days. Data from McPherson et al., 2017.

The coronavirus macrodomain is also required for full repression of the interferon (IFN) response during infection. Mouse hepatitis virus (MHV) and SARS-CoV macrodomain mutant viruses increase innate immune responses following infection [37,39]. Furthermore, it was demonstrated in a co-infection model that this antiviral response helped protect mice from a lethal SARS-CoV infection [37]. Importantly, an early IFN response after infection is critical for its protective effects. Early administration of IFN-I or IFN-III is protective in mouse models of SARS-CoV and MERS-CoV infection, but administration of IFN in the later stages of infection is not [40,41]. Unlike the wild-type virus which suppresses IFN response, macrodomain mutant virus infection elicits IFN induction in the very early stages of infection [37,39]. Thus, macrodomain inhibitors have the potential to restore the host's protective early and robust IFN response in addition to blocking virus replication.

5. Viral Macrodomain Counteracts Host Antiviral Responses Mediated through PARPs

Several PARPs are induced by virus infection and by IFN, indicating a role for PARPs and ADP-ribosylation in the antiviral response [39]. Amongst the 17 PARPs in humans, 4 are capable of adding PAR (PARPs 1, 2, 5a, and 5b), 11 add MAR (PARPs 3, 4, 6, 7, 8, 10, 11, 12, 14, 15, and 16), and 2 are catalytically inactive by themselves (PARPs 9 and 13) [21]. PARPs 9, 12, 13, and 14 are amongst the 62 IFN-stimulated genes conserved across vertebrates as part of the innate response to infection [42]. Induction of PARPs is observed in the brains of alphavirus and MHV-infected mice and simian immunodeficiency virus (SIV)-infected macaques, in SARS-CoV-2-infected human lung and bronchial cells, ferrets, and COVID-19 autopsy samples [35,43,44], and likely many other viral infection models.

Several pieces of evidence indicate that some PARPs may be critical for the host antiviral response [20,45]. Overexpression of PARPs 7, 10, and 12 strongly represses the replication of several classes of RNA viruses [46,47]. More recently, PARP11 and PARP12 were found to cooperatively inhibit Zika virus replication by ADP-ribosylating non-structural proteins, which targeted them for degradation [48,49]. Furthermore, PARP12 was identified in a screen for proteins that interact with SARS-CoV-2 genomic RNA, and PARP12 knockdown enhanced replication of SARS-CoV-2 in Calu-3 cells [50]. In addition, PARP7 and PARP11 ADP-ribosyltransferase activity reduces IFN signaling, resulting in enhanced replication of influenza virus and HSV-1 [51,52].

Most virus-induced PARPs add MAR to proteins, and viral macrodomains primarily bind and remove MAR from protein. Therefore, one intriguing hypothesis is that viruses may circumvent host defenses or regulate replication by binding or removing specific classes of ADP-ribosylation (Figure 3). Indeed, inhibition of PARP activity with 3-AB enhances the replication of a MHV macrodomain mutant virus (N1347A), indicating that the macrodomain counters PARP activity [39]. Similarly, knockdown of PARP12 or PARP14 partially restored the mutant virus replication, demonstrating the ability of these MAR-adding PARPs to restrict MHV replication [39]. On the contrary, enhancing PARP activity by supplementing cells with NAD^+ precursors, such as nicotinamide riboside, further restricts the replication of these mutant viruses [44], supporting the critical importance of ADP-ribosylation removal by coronaviruses.

Figure 3. Working model: arms-race on ADP-ribosylation between host PARPs and virus macrodomains.

6. High Conservation of the Viral Macrodomain Indicates the Potential to Develop Broad Spectrum Antivirals

The macrodomain is highly conserved in all coronaviruses, including all seven human coronaviruses, with 100% conservation of key residues critical for ADP-ribosylhydrolase activity (Figure 4). These key residues are also conserved in both New World encephalitic and Old World arthritic alphaviruses, such as CHIKV, MAYV, and eastern, western, and Venezuelan equine encephalitis viruses (EEEV, WEEV, and VEEV). The conservation extends to animal reservoir coronaviruses that pose future threats of zoonotic disease.

Mutagenesis has revealed that structural and amino acid configurations of the macrodomain ADP-ribose binding site are critical for viral replication and virulence. Some mutant viruses are completely non-viable. For example, no mutant viruses can be recovered from cells transfected with CHIKV RNA with mutations of D10A, G32E, G112E, or R144A, and recovered viruses are all reverted to wild type [28]. Similarly, SINV with an N10A mutation is non-viable, and recovered viruses are reverted to the wild type, mutated to D or T, or have developed an A31G compensatory mutation [53]. A similar phenomenon is observed in coronaviruses: most recovered G1439V mutant MHVs are reverted to wild type or have developed a compensatory mutation, A1438T, and the D1329A/N1347A double mutant cannot be recovered [36]. Because these residues line ADP-ribose binding sites, these data indicate strong selective pressure to maintain the macrodomain structural and amino acid configurations. Consistent with this premise, our recent genomic analyses of 440,212 SARS-CoV-2 sequences, including those from all variants of concern, revealed that key residues for binding ADP-ribose remain constant [54]. The high conservation observed suggests the potential of developing antivirals targeting particular, or even across, virus families.

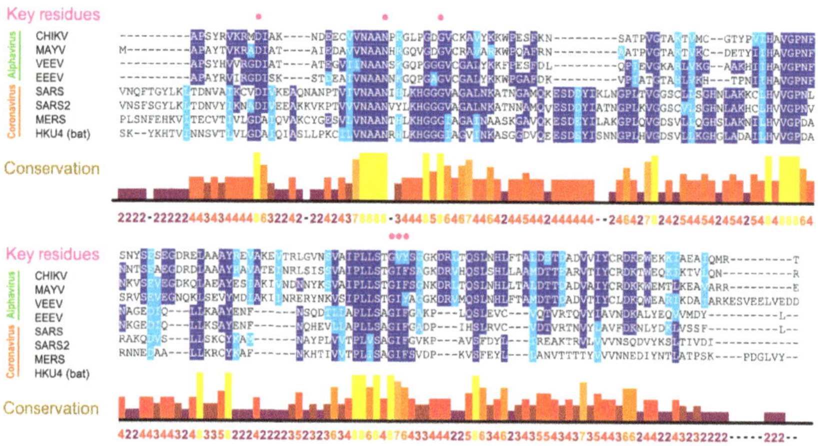

Figure 4. Amino acid analyses revealed high conservation of key residues responsible for macrodomain ADP-ribosylhydrolase activities (pink circles). Alphavirus: CHIKV; MAYV, VEEV, EEEV; Coronavirus: SARS-CoV (SARS), SARS-CoV-2 (SARS2), MERS-CoV (MERS), and a bat coronavirus (HKU4).

7. Structural and Biochemical Data Indicate Feasibility of Developing Macrodomain-Targeted Antivirals

One advantage of choosing the macrodomain as a drug target is that over 500 structures have been deposited in the Protein Data Bank, including 314 from viruses and 130 from humans. Cross-comparison of this rich structural dataset provides us with foundational information on the commonalities and differences across virus and human macrodomains. Structural analyses of SARS-CoV-2 and CHIKV macrodomains revealed three defined druggable "pockets" near the active site as potential targets for small molecule inhibitors (Figure 5a) [32,55–57]. The largest pocket P1 is where ADP-ribose binds, whereas adjacent pockets P2 and P3 could be explored for structure-based drug discovery. Several groups, using computational docking, fragment-based screens, thermal shift assays, and crystallographic screening, independently identified several fragments/compounds, including the metabolite of remdesivir GS-441524, that bind to the SARS-CoV-2 macrodomain [58–67]. Furthermore, using a displacement assay, some fragments identified in crystallographic screening inhibited ADP-ribose binding, albeit at very high concentrations [67]. In addition, a fragment-based screen showed binding of 2-oxo-1,2-dihydro-4-quinazoline-carboxylic acid (SRI-43750) to the CHIKV macrodomain [66].

Importantly, although macrodomains are also present in human proteins, structural and electrostatic differences should permit virus-specific targeting [54,68]. For example, although MacroD2 is the closest human homolog of SARS-CoV-2 with ~30% sequence identity and a similar 3D structure, the amino acids surrounding key residues for catalytic activity are different (Figure 5b), resulting in a less charged pocket [54]. As expected, the P1 pocket is well conserved for ADP-ribose binding, but P2 and P3 pockets are much less conserved and can be exploited for selective drug targeting against the viral macrodomain (Figure 5a). Biochemical data have confirmed that MacroD2 has activities that are significantly different from those of the viral macrodomains. MacroD2 binds much more efficiently to ADP-ribose than coronavirus macrodomains but ADP-ribosylhydrolase activity is reduced [32,54]. Consistent with these structural and biochemical findings, our recently developed ADP-ribosylhydrolase activity assay identified dasatinib as an inhibitor of SARS-CoV-2 Mac1, but not human MacroD2 [54]. Our data demonstrate the feasibility of identifying selective inhibitors based on ADP-ribosylhydrolase activity, paving the way for screening larger libraries to identify improved macrodomain inhibitors.

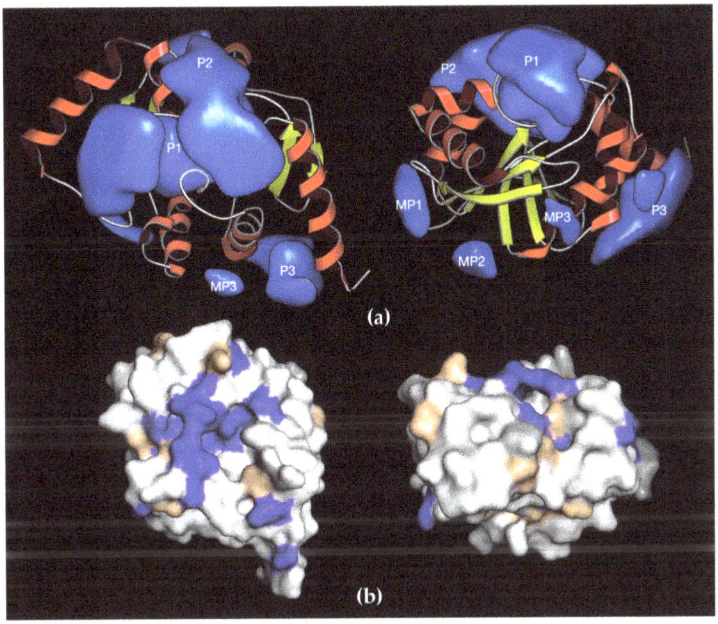

Figure 5. Druggable pockets of SARS-CoV-2 macrodomain and similarities with human MacroD2. (**a**) Ribbon representation of SARS-CoV-2 macrodomain with surface representation of the druggable pockets (P1, P2, P3). Additional minor pockets MP1, MP2, and MP3 are generally too small to be considered a good exploitable binding site as their volume is <200 Å^3 and would only allow small fragments to bind with typically low inhibitory potential. The right panel shows a 90°-rotated view along two axes; (**b**) identities between SARS-CoV-2 Mac1 macrodomain and the closest human homolog MacroD2 are shown in blue, conserved residues in wheat, and different residues in white. Views in b are identical to the orientation in a.

8. High-Throughput Assays for Compound Screening

Not only is the macrodomain a suitable target both virologically and biochemically, but high-throughput screening assays are now available to quickly identify hit compounds. These assays can efficiently screen for compounds that can inhibit macrodomain-ADP-ribose binding or macrodomain ADP-ribosylhydrolase activity. One of the first macrodomain high-throughput screening (HTS) assays, developed by Schuller et al., is an ADP-ribose displacement assay using AlphaScreen technology [69]. It was initially described in a screen used to identify inhibitors of the 2nd macrodomain of PARP14 and has since been used for the SARS-CoV-2 macrodomain [67]. AlphaScreen is a bead-based, non-radioactive **A**mplified **L**uminescent **P**roximity **H**omogenous **A**ssay, where a "donor" bead converts ambient oxygen to singlet oxygen, which interacts with an "acceptor" bead generating chemiluminescence at 370 nm and in turn activates additional fluorophores in the bead with emission at 520-620 nm. To give off a signal, the two beads must be in close proximity to each other or the singlet oxygen will go undetected. To make this assay suitable to measure macrodomain-ADP-ribose binding, a peptide was developed that has a biotin molecule attached to one lysine and a non-hydrolysable ADP-ribose on another lysine. Histidine-tagged macrodomains that are bound to this peptide will interact with a donor streptavidin bead and an Ni^{2+} acceptor bead, which will then give off the light signal (Figure 6a). Another ADP-ribose binding assay recently developed by Sowa et al. is a FRET-based assay [70]. In this assay, a YFP-tagged $G\alpha_i$ protein-based peptide is ADP-ribosylated by pertussis toxin at a cysteine residue, which cannot be hydrolyzed by viral macrodomains. This protein is then incubated with a CFP-labeled macrodomain, which will then give off a FRET signal upon binding. Using this assay, the authors identified suramin as a non-specific inhibitor of multiple macrodomain-containing proteins which, interestingly, was previously shown to inhibit alphavirus replication in cell culture [71].

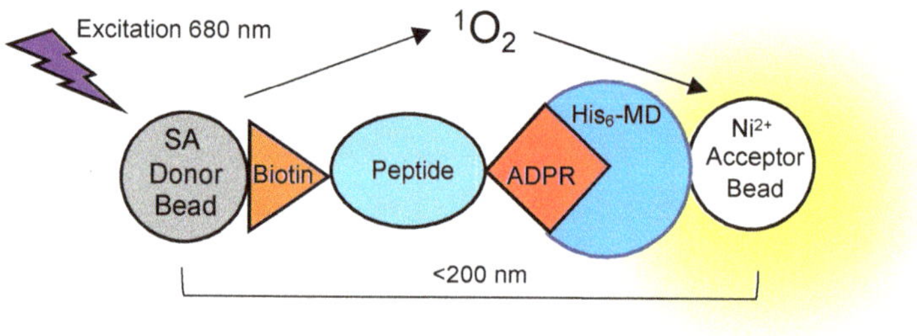

(**a**)

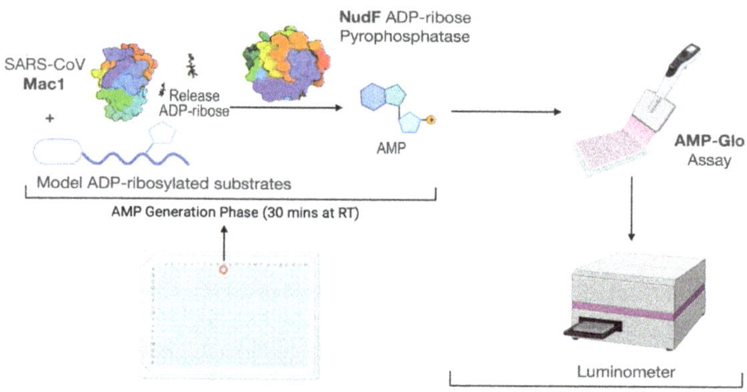

(**b**)

Figure 6. Assays developed to use high-throughput screening for macrodomain inhibitors. (**a**) Cartoon diagram depicting a bead-based AlphaScreen assay for measuring macrodomain interaction with an ADP-ribosylated peptide. SA—streptavidin; Ni^{2+}—Nickel; His_6-MD—Histidine-tagged macrodomain; ADPR—ADP-ribose; (**b**) Schematics of ADPr-Glo assay (see text for more details).

In addition, the binding of ADP-ribose or chemical compounds to the macrodomain can be indirectly assessed by differential scanning fluorimetry (DSF). DSF involves the use of a fluorescent dye (e.g., SYPRO Orange) with affinity for hydrophobic portions of proteins, which are exposed as proteins unfold. Binding of ADP-ribose to the macrodomain reduced the increase in fluorescence upon heat denaturation, indicating an increase in the melting temperature, and thus the stability of the SARS-CoV-2 macrodomain when bound to ADP-ribose [32,58]. Virdi et al. used this assay to screen ~2500 compounds for their ability to alter the melting temperature of the macrodomain. This screen identified nucleotides, steroids, antibiotics, and benzimidazoles as potential macrodomain binders. Molecular docking experiments suggested that some of these compounds may interact in the ADP-ribose binding pocket of the SARS-CoV-2 macrodomain [58]. However, many compounds influence the observed protein melting temperature by either quenching fluorescence, increasing fluorescence, or interacting with the reporter dye when mixed with SYPRO Orange, and may therefore be scored as a false positive or false negative in a DSF screen [58].

Measurements of macrodomain ADP-ribosylhydrolase activity have historically relied on gel-based autoradiography and Western blot assays that are not practical for screening large numbers of compounds [26,29]. More recently, our group developed a novel assay that measures ADP-ribosylhydrolase activity in a high-throughput, luminescence-based format called ADPr-Glo (Figure 6b) [54]: First, ADP-ribose is released from a defined MARylated substrate by the macrodomain ADP-ribosylhydrolase of interest. Second, the phosphodiesterase NudF cleaves the released ADP-ribose into phosphoribose and AMP. Finally, AMP is converted to luminescence with the commercially available AMP-Glo kit. This method takes advantage of the substrate selectivity of NudF, which cleaves free ADP-ribose but has no activity toward protein-conjugated ADP-ribose [72]. Therefore, the luminescence signal is controlled by the rate of the ADP-ribosylhydrolase. ADPr-Glo can be performed in 384-well plates with reaction volumes as low as 5 μL, greatly minimizing time and costs compared to traditional gel-based activity assays [28,32]. We established ADPr-Glo conditions for inhibitor screening and multiple macrodomains, including SARS-CoV-2 Mac1, MERS-CoV Mac1, and their closest human homolog, MacroD2. In a pilot screen of the 3233 pharmacologically active compounds, we identified dasatinib and dihydralazine as ADP-ribosylhydrolase inhibitors for both human and viral macrodomains tested. Importantly, dasatinib inhibits SARS-CoV-2 and MERS-CoV Mac1 but not the closest human homolog, MacroD2. The selectivity demonstrates it is possible to discover drugs that specifically inhibit viral macrodomains. Although cytotoxic when used at μM concentration, dasatinib has antiviral activities against SARS-CoV and MERS-CoV through an unknown mechanism [73]. These proof-of-concept data pave the way for screening large compound libraries to identify improved macrodomain inhibitors and explore their potential as antiviral therapies for SARS-CoV-2 and future viral threats [54].

Finally, Russo et al. also developed an immunofluorescence-based assay to assess the activity of SARS-CoV-2 Mac1 activity in cells [74]. Activation of IFN responses, by treating cells with IFN-γ or the double-stranded RNA mimic, poly(I:C), robustly induces ADP-ribosylation in cells, which can be removed by ectopic expression of wild type, but not catalytically inactive mutant, Mac1. Although the initial screen of a limited set of compounds did not yield any hits, this assay may potentially be used in the future for testing the target engagement of potential macrodomain inhibitors in cells.

9. Conclusions

Viral macrodomains offer an attractive antiviral target because they are highly conserved and have structurally well-defined (druggable) binding pockets. Because the viral macrodomain is mechanistically distinct from more common antiviral targets (e.g., viral polymerases or proteases), a macrodomain inhibitor would facilitate development of combination therapies for optimal treatment (as successfully developed for HIV). Besides inhibiting viral replication, macrodomain inhibitors may also boost immune responses that contribute to the recovery from disease [75].

Furthermore, coronaviruses and alphaviruses are prominent veterinary pathogens, so macrodomain inhibitors may also be useful for treatment of animals. While the current focus is on targeting macrodomains for antiviral therapy, drug development efforts may identify compounds that can inhibit human macrodomains as well. These inhibitors may have important uses in other human diseases such as cancer, metabolic disorders, and inflammatory diseases. In addition, first-generation macrodomain inhibitors may also serve as tools to probe pathways regulated by ADP-ribosylation that may be attractive novel targets for development of therapeutic interventions.

Funding: This research was funded by COVID-19 PreClinical Research Discovery Fund from Johns Hopkins University (A.K.L.L.) and discretionary funds from Johns Hopkins Bloomberg School of Public Health (A.K.L.L.). Macrodomain biology investigations have been funded by a Johns Hopkins Catalyst Award (A.K.L.L.), pilot grants from the Johns Hopkins University School of Medicine's Sherrilyn and Ken Fisher Center for Environmental Infectious Disease (D.E.G. and A.K.L.L.), and NIH grants R56AI137264 (D.E.G. and A.K.L.L.), R01GM104135 (A.K.L.L.), R35GM138029 (A.R.F.), K22AI134993-01 (A.R.F.), P20GM113117 (A.R.F.), and startup funds from the University of Kansas (A.R.F).

Acknowledgments: We thank Morgan Dasovich and Junlin Zhuo for making Figures 4 and 6b, respectively.

Conflicts of Interest: D.E.G. is on advisory boards for Takeda Pharmaceuticals, GlaxoSmithKline, and GreenLight Biosciences. The other authors declare no conflict of interest. The funders had no role in the design of the study; in the collection, analyses, or interpretation of data; in the writing of the manuscript; or in the decision to publish the results.

References

1. Wang, Y.; Grunewald, M.; Perlman, S. Coronaviruses: An Updated Overview of Their Replication and Pathogenesis. *Methods Mol. Biol.* **2020**, *2203*, 1–29. [CrossRef]
2. Weaver, S.C.; Winegar, R.; Manger, I.D.; Forrester, N.L. Alphaviruses: Population genetics and determinants of emergence. *Antivir. Res.* **2012**, *94*, 242–257. [CrossRef]
3. Zeller, H.; Van Bortel, W.; Sudre, B. Chikungunya: Its History in Africa and Asia and Its Spread to New Regions in 2013–2014. *J. Infect. Dis.* **2016**, *214*, S436–S440. [CrossRef] [PubMed]
4. Girard, M.; Nelson, C.B.; Picot, V.; Gubler, D.J. Arboviruses: A global public health threat. *Vaccine* **2020**, *38*, 3989–3994. [CrossRef] [PubMed]
5. Hoarau, J.-J.; Bandjee, M.-C.J.; Trotot, P.K.; Das, T.; Li-Pat-Yuen, G.; Dassa, B.; Denizot, M.; Guichard, E.; Ribera, A.; Henni, T.; et al. Persistent Chronic Inflammation and Infection by Chikungunya Arthritogenic Alphavirus in Spite of a Robust Host Immune Response. *J. Immunol.* **2010**, *184*, 5914–5927. [CrossRef]
6. Borgherini, G.; Poubeau, P.; Jossaume, A.; Gouix, A.; Cotte, L.; Michault, A.; Arvin-Berod, C.; Paganin, F. Persistent Arthralgia Associated with Chikungunya Virus: A Study of 88 Adult Patients on Reunion Island. *Clin. Infect. Dis.* **2008**, *47*, 469–475. [CrossRef]
7. Ceulemans, L.J.; Khan, M.; Yoo, S.-J.; Zapiec, B.; Van Gerven, L.; Van Slambrouck, J.; Vanstapel, A.; Van Raemdonck, D.; Vos, R.; Wauters, E.; et al. Persistence of SARS-CoV-2 RNA in lung tissue after mild COVID-19. *Lancet Respir. Med.* **2021**, *9*, e78–e79. [CrossRef]
8. Gaspar-Rodríguez, A.; Padilla-González, A.; Rivera-Toledo, E. Coronavirus persistence in human respiratory tract and cell culture: An overview. *Braz. J. Infect. Dis.* **2021**, *25*, 101632. [CrossRef]
9. Lin, W.-H.W.; Kouyos, R.; Adams, R.J.; Grenfell, B.T.; Griffin, D.E. Prolonged persistence of measles virus RNA is characteristic of primary infection dynamics. *Proc. Natl. Acad. Sci. USA* **2012**, *109*, 14989–14994. [CrossRef]
10. Metcalf, T.U.; Griffin, D.E. Alphavirus-Induced Encephalomyelitis: Antibody-Secreting Cells and Viral Clearance from the Nervous System. *J. Virol.* **2011**, *85*, 11490–11501. [CrossRef]
11. Nelson, A.N.; Lin, W.-H.W.; Shivakoti, R.; Putnam, N.E.; Mangus, L.M.; Adams, R.J.; Hauer, D.; Baxter, V.K.; Griffin, D.E. Association of persistent wild-type measles virus RNA with long-term humoral immunity in rhesus macaques. *JCI Insight* **2020**, *5*, e134992. [CrossRef]
12. Yang, B.; Fan, J.; Huang, J.; Guo, E.; Fu, Y.; Liu, S.; Xiao, R.; Liu, C.; Lu, F.; Qin, T.; et al. Clinical and molecular characteristics of COVID-19 patients with persistent SARS-CoV-2 infection. *Nat. Commun.* **2021**, *12*, 3501. [CrossRef]
13. Levine, B.; Hardwick, J.M.; Griffin, D.E. Persistence of alphaviruses in vertebrate hosts. *Trends Microbiol.* **1994**, *2*, 25–28. [CrossRef]
14. De Andrade, D.C.; Jean, S.; Clavelou, P.; Dallel, R.; Bouhassira, D. Chronic pain associated with the Chikungunya Fever: Long lasting burden of an acute illness. *BMC Infect. Dis.* **2010**, *10*, 31. [CrossRef]
15. Soumahoro, M.-K.; Gérardin, P.; Boelle, P.-Y.; Perrau, J.; Fianu, A.; Pouchot, J.; Malvy, D.; Flahault, A.; Favier, F.; Hanslik, T. Impact of Chikungunya Virus Infection on Health Status and Quality of Life: A Retrospective Cohort Study. *PLoS ONE* **2009**, *4*, e7800. [CrossRef] [PubMed]
16. Shyr, Z.A.; Gorshkov, K.; Chen, C.Z.; Zheng, W. Drug Discovery Strategies for SARS-CoV-2. *J. Pharmacol. Exp. Ther.* **2020**, *375*, 127–138. [CrossRef] [PubMed]
17. Battisti, V.; Urban, E.; Langer, T. Antivirals against the Chikungunya Virus. *Viruses* **2021**, *13*, 1307. [CrossRef]
18. Kaur, P.; Chu, J.J.H. Chikungunya virus: An update on antiviral development and challenges. *Drug Discov. Today* **2013**, *18*, 969–983. [CrossRef]
19. Fehr, A.R.; Jankevicius, G.; Ahel, I.; Perlman, S. Viral Macrodomains: Unique Mediators of Viral Replication and Pathogenesis. *Trends Microbiol.* **2018**, *26*, 598–610. [CrossRef]

20. Leung, A.K.L.; McPherson, R.L.; Griffin, D.E. Macrodomain ADP-ribosylhydrolase and the pathogenesis of infectious diseases. *PLoS Pathog.* **2018**, *14*, e1006864. [CrossRef]
21. Lüscher, B.; Ahel, I.; Altmeyer, M.; Ashworth, A.; Bai, P.; Chang, P.; Cohen, M.; Corda, D.; Dantzer, F.; Daugherty, M.D.; et al. ADP-ribosyltransferases, an update on function and nomenclature. *FEBS J.* **2021**. Available online: https://doi.org/10.1111/febs.16142 (accessed on 7 January 2022). [CrossRef] [PubMed]
22. Palazzo, L.; Mikolčević, P.; Mikoč, A.; Ahel, I. ADP-ribosylation signalling and human disease. *Open Biol.* **2019**, *9*, 190041. [CrossRef] [PubMed]
23. Rack, J.G.M.; Perina, D.; Ahel, I. Macrodomains: Structure, Function, Evolution, and Catalytic Activities. *Annu. Rev. Biochem.* **2016**, *85*, 431–454. [CrossRef]
24. Karras, G.; Kustatscher, G.; Buhecha, H.R.; Allen, M.D.; Pugieux, C.; Sait, F.; Bycroft, M.; Ladurner, A.G. The macro domain is an ADP-ribose binding module. *EMBO J.* **2005**, *24*, 1911–1920. [CrossRef] [PubMed]
25. Eckei, L.; Krieg, S.; Bütepage, M.; Lehmann, A.; Gross, A.; Lippok, B.; Grimm, A.R.; Kümmerer, B.M.; Rossetti, G.; Lüscher, B.; et al. The conserved macrodomains of the non-structural proteins of Chikungunya virus and other pathogenic positive strand RNA viruses function as mono-ADP-ribosylhydrolases. *Sci. Rep.* **2017**, *7*, 41746. [CrossRef]
26. Jankevicius, G.; Hassler, M.; Golia, B.; Rybin, V.; Zacharias, M.; Timinszky, G.; Ladurner, A.G. A family of macrodomain proteins reverses cellular mono-ADP-ribosylation. *Nat. Struct. Mol. Biol.* **2013**, *20*, 508–514. [CrossRef]
27. Li, C.; Debing, Y.; Jankevicius, G.; Neyts, J.; Ahel, I.; Coutard, B.; Canard, B. Viral Macro Domains Reverse Protein ADP-Ribosylation. *J. Virol.* **2016**, *90*, 8478–8486. [CrossRef]
28. McPherson, R.L.; Abraham, R.; Sreekumar, E.; Ong, S.-E.; Cheng, S.-J.; Baxter, V.K.; Kistemaker, H.A.V.; Filippov, D.V.; Griffin, D.E.; Leung, A.K.L. ADP-ribosylhydrolase activity of Chikungunya virus macrodomain is critical for virus replication and virulence. *Proc. Natl. Acad. Sci. USA* **2017**, *114*, 1666–1671. [CrossRef]
29. Rosenthal, F.; Feijs, K.L.; Frugier, E.; Bonalli, M.; Forst, A.H.; Imhof, R.; Winkler, H.C.; Fischer, D.; Caflisch, A.; Hassa, P.O.; et al. Macrodomain-containing proteins are new mono-ADP-ribosylhydrolases. *Nat. Struct. Mol. Biol.* **2013**, *20*, 502–507. [CrossRef]
30. Daniels, C.M.; Ong, S.-E.; Leung, A.K. Phosphoproteomic Approach to Characterize Protein Mono- and Poly(ADP-ribosyl)ation Sites from Cells. *J. Proteome Res.* **2014**, *13*, 3510–3522. [CrossRef]
31. Neuman, B.W. Bioinformatics and functional analyses of coronavirus nonstructural proteins involved in the formation of replicative organelles. *Antivir. Res.* **2016**, *135*, 97–107. [CrossRef]
32. Alhammad, Y.M.O.; Kashipathy, M.M.; Roy, A.; Gagné, J.-P.; McDonald, P.; Gao, P.; Nonfoux, L.; Battaile, K.P.; Johnson, D.K.; Holmstrom, E.D.; et al. The SARS-CoV-2 Conserved Macrodomain Is a Mono-ADP-Ribosylhydrolase. *J. Virol.* **2021**, *95*, e01969-20. [CrossRef]
33. Jayabalan, A.K.; Adivarahan, S.; Koppula, A.; Abraham, R.; Batish, M.; Zenklusen, D.; Griffin, D.E.; Leung, A.K.L. Stress granule formation, disassembly, and composition are regulated by alphavirus ADP-ribosylhydrolase activity. *Proc. Natl. Acad. Sci. USA* **2021**, *118*, e2021719118. [CrossRef]
34. Abraham, R.; Hauer, D.; McPherson, R.L.; Utt, A.; Kirby, I.T.; Cohen, M.S.; Merits, A.; Leung, A.K.L.; Griffin, D.E. ADP-ribosyl–binding and hydrolase activities of the alphavirus nsP3 macrodomain are critical for initiation of virus replication. *Proc. Natl. Acad. Sci. USA* **2018**, *115*, E10457–E10466. [CrossRef]
35. Abraham, R.; McPherson, R.L.; Dasovich, M.; Badiee, M.; Leung, A.K.L.; Griffin, D.E. Both ADP-Ribosyl-Binding and Hydrolase Activities of the Alphavirus nsP3 Macrodomain Affect Neurovirulence in Mice. *mBio* **2020**, *11*, e03253-19. [CrossRef] [PubMed]
36. Voth, L.S.; O'Connor, J.J.; Kerr, C.M.; Doerger, E.; Schwarting, N.; Sperstad, P.; Johnson, D.K.; Fehr, A.R. Unique Mutations in the Murine Hepatitis Virus Macrodomain Differentially Attenuate Virus Replication, Indicating Multiple Roles for the Macrodomain in Coronavirus Replication. *J. Virol.* **2021**, *95*, e0076621. [CrossRef]
37. Fehr, A.R.; Channappanavar, R.; Jankevicius, G.; Fett, C.; Zhao, J.; Athmer, J.; Meyerholz, D.K.; Ahel, I.; Perlman, S. The Conserved Coronavirus Macrodomain Promotes Virulence and Suppresses the Innate Immune Response during Severe Acute Respiratory Syndrome Coronavirus Infection. *mBio* **2016**, *7*, e01721-16. [CrossRef] [PubMed]
38. Alhammad, Y.M.O.; Fehr, A.R. The Viral Macrodomain Counters Host Antiviral ADP-Ribosylation. *Viruses* **2020**, *12*, 384. [CrossRef]
39. Grunewald, M.E.; Chen, Y.; Kuny, C.; Maejima, T.; Lease, R.; Ferraris, D.; Aikawa, M.; Sullivan, C.S.; Perlman, S.; Fehr, A.R. The coronavirus macrodomain is required to prevent PARP-mediated inhibition of virus replication and enhancement of IFN expression. *PLoS Pathog.* **2019**, *15*, e1007756. [CrossRef]
40. Channappanavar, R.; Fehr, A.; Vijay, R.; Mack, M.; Zhao, J.; Meyerholz, D.; Perlman, S. Dysregulated Type I Interferon and Inflammatory Monocyte-Macrophage Responses Cause Lethal Pneumonia in SARS-CoV-Infected Mice. *Cell Host Microbe* **2016**, *19*, 181–193. [CrossRef] [PubMed]
41. Channappanavar, R.; Fehr, A.R.; Zheng, J.; Wohlford-Lenane, C.; Abrahante, J.E.; Mack, M.; Sompallae, R.; McCray, P.B., Jr.; Meyerholz, D.K.; Perlman, S. IFN-I response timing relative to virus replication determines MERS coronavirus infection outcomes. *J. Clin. Investig.* **2019**, *129*, 3625–3639. [CrossRef] [PubMed]
42. Shaw, A.E.; Hughes, J.; Gu, Q.; Behdenna, A.; Singer, J.B.; Dennis, T.; Orton, R.J.; Varela, M.; Gifford, R.J.; Wilson, S.J.; et al. Fundamental properties of the mammalian innate immune system revealed by multispecies comparison of type I interferon responses. *PLoS Biol.* **2017**, *15*, e2004086. [CrossRef] [PubMed]

43. Mavian, C.; Ramirez-Mata, A.S.; Dollar, J.J.; Nolan, D.J.; Cash, M.; White, K.; Rich, S.N.; Magalis, B.R.; Marini, S.; Prosperi, M.C.F.; et al. Brain tissue transcriptomic analysis of SIV-infected macaques identifies several altered metabolic pathways linked to neuropathogenesis and poly (ADP-ribose) polymerases (PARPs) as potential therapeutic targets. *J. NeuroVirol.* **2021**, *27*, 101–115. [CrossRef] [PubMed]
44. Heer, C.D.; Sanderson, D.J.; Voth, L.S.; Alhammad, Y.M.O.; Schmidt, M.S.; Trammell, S.A.J.; Perlman, S.; Cohen, M.S.; Fehr, A.R.; Brenner, C. Coronavirus infection and PARP expression dysregulate the NAD metabolome: An actionable component of innate immunity. *J. Biol. Chem.* **2020**, *295*, 17986–17996. [CrossRef]
45. Fehr, A.R.; Singh, S.A.; Kerr, C.M.; Mukai, S.; Higashi, H.; Aikawa, M. The impact of PARPs and ADP-ribosylation on inflammation and host–pathogen interactions. *Genes Dev.* **2020**, *34*, 341–359. [CrossRef]
46. Atasheva, S.; Akhrymuk, M.; Frolova, E.I.; Frolov, I. New PARP Gene with an Anti-Alphavirus Function. *J. Virol.* **2012**, *86*, 8147–8160. [CrossRef]
47. Atasheva, S.; Frolova, E.I.; Frolov, I. Interferon-Stimulated Poly(ADP-Ribose) Polymerases Are Potent Inhibitors of Cellular Translation and Virus Replication. *J. Virol.* **2014**, *88*, 2116–2130. [CrossRef]
48. Li, L.; Zhao, H.; Liu, P.; Li, C.; Quanquin, N.; Ji, X.; Sun, N.; Du, P.; Qin, C.-F.; Lu, N.; et al. PARP12 suppresses Zika virus infection through PARP-dependent degradation of NS1 and NS3 viral proteins. *Sci. Signal.* **2018**, *11*, eaas9332. [CrossRef]
49. Li, L.; Shi, Y.; Li, S.; Liu, J.; Zu, S.; Xu, X.; Gao, M.; Sun, N.; Pan, C.; Peng, L.; et al. ADP-ribosyltransferase PARP11 suppresses Zika virus in synergy with PARP12. *Cell Biosci.* **2021**, *11*, 116. [CrossRef]
50. Lee, S.; Lee, Y.-S.; Choi, Y.; Son, A.; Park, Y.; Lee, K.-M.; Kim, J.; Kim, J.-S.; Kim, V.N. The SARS-CoV-2 RNA interactome. *Mol. Cell* **2021**, *81*, 2838–2850.e2836. [CrossRef]
51. Guo, T.; Zuo, Y.; Qian, L.; Liu, J.; Yuan, Y.; Xu, K.; Miao, Y.; Feng, Q.; Chen, X.; Jin, L.; et al. ADP-ribosyltransferase PARP11 modulates the interferon antiviral response by mono-ADP-ribosylating the ubiquitin E3 ligase β-TrCP. *Nat. Microbiol.* **2019**, *4*, 1872–1884. [CrossRef]
52. Yamada, T.; Horimoto, H.; Kameyama, T.; Hayakawa, S.; Yamato, H.; Dazai, M.; Takada, A.; Kida, H.; Bott, D.; Zhou, A.C.; et al. Constitutive aryl hydrocarbon receptor signaling constrains type I interferon–mediated antiviral innate defense. *Nat. Immunol.* **2016**, *17*, 687–694. [CrossRef] [PubMed]
53. Park, E.; Griffin, D.E. The nsP3 macro domain is important for Sindbis virus replication in neurons and neurovirulence in mice. *Virology* **2009**, *388*, 305–314. [CrossRef] [PubMed]
54. Dasovich, M.; Zhuo, J.; Goodman, J.A.; Thomas, A.G.; McPherson, R.L.; Jayabalan, A.K.; Busa, V.F.; Cheng, S.-J.; Murphy, B.A.; Redinger, K.R.; et al. High-Throughput Activity Assay for Screening Inhibitors of the SARS-CoV-2 Mac1 Macrodomain. *ACS Chem. Biol.* **2021**. [CrossRef] [PubMed]
55. Malet, H.; Coutard, B.; Jamal, S.; Dutartre, H.; Papageorgiou, N.; Neuvonen, M.; Ahola, T.; Forrester, N.; Gould, E.A.; Lafitte, D.; et al. The Crystal Structures of Chikungunya and Venezuelan Equine Encephalitis Virus nsP3 Macro Domains Define a Conserved Adenosine Binding Pocket. *J. Virol.* **2009**, *83*, 6534–6545. [CrossRef]
56. Rack, J.G.M.; Palazzo, L.; Ahel, I. (ADP-ribosyl)hydrolases: Structure, function, and biology. *Genes Dev.* **2020**, *34*, 263–284. [CrossRef]
57. Frick, D.N.; Virdi, R.S.; Vuksanovic, N.; Dahal, N.; Silvaggi, N.R. Molecular Basis for ADP-Ribose Binding to the Mac1 Domain of SARS-CoV-2 nsp3. *Biochemistry* **2020**, *59*, 2608–2615. [CrossRef]
58. Virdi, R.S.; Bavisotto, R.V.; Hopper, N.C.; Vuksanovic, N.; Melkonian, T.R.; Silvaggi, N.R.; Frick, D.N. Discovery of Drug-Like Ligands for the Mac1 Domain of SARS-CoV-2 Nsp3. *SLAS Discov.* **2020**, *25*, 1162–1170. [CrossRef]
59. Babar, Z.; Khan, M.; Zahra, M.; Anwar, M.; Noor, K.; Hashmi, H.F.; Suleman, M.; Waseem, M.; Shah, A.; Ali, S.; et al. Drug similarity and structure-based screening of medicinal compounds to target macrodomain-I from SARS-CoV-2 to rescue the host immune system: A molecular dynamics study. *J. Biomol. Struct. Dyn.* **2020**, *40*, 523–537. [CrossRef]
60. Bajusz, D.; Wade, W.S.; Satała, G.; Bojarski, A.J.; Ilaš, J.; Ebner, J.; Grebien, F.; Papp, H.; Jakab, F.; Douangamath, A.; et al. Exploring protein hotspots by optimized fragment pharmacophores. *Nat. Commun.* **2021**, *12*, 3201. [CrossRef]
61. Debnath, P.; Debnath, B.; Bhaumik, S.; Debnath, S. In Silico Identification of Potential Inhibitors of ADP-Ribose Phosphatase of SARS-CoV-2 nsP3 by Combining E-Pharmacophore- and Receptor-Based Virtual Screening of Database. *ChemistrySelect* **2020**, *5*, 9388–9398. [CrossRef]
62. Jung, L.S.; Gund, T.M.; Narayan, M. Comparison of Binding Site of Remdesivir and Its Metabolites with NSP12-NSP7-NSP8, and NSP3 of SARS-CoV-2 Virus and Alternative Potential Drugs for COVID-19 Treatment. *Protein J.* **2020**, *39*, 619–630. [CrossRef] [PubMed]
63. Ni, X.; Schröder, M.; Olieric, V.; Sharpe, M.E.; Hernandez-Olmos, V.; Proschak, E.; Merk, D.; Knapp, S.; Chaikuad, A. Structural Insights into Plasticity and Discovery of Remdesivir Metabolite GS-441524 Binding in SARS-CoV-2 Macrodomain. *ACS Med. Chem. Lett.* **2021**, *12*, 603–609. [CrossRef]
64. Selvaraj, C.; Dinesh, D.C.; Panwar, U.; Boura, E.; Singh, S.K. High-Throughput Screening and Quantum Mechanics for Identifying Potent Inhibitors Against Mac1 Domain of SARS-CoV-2 Nsp3. *IEEE/ACM Trans. Comput. Biol. Bioinform.* **2021**, *18*, 1262–1270. [CrossRef] [PubMed]
65. Singh, A.K.; Kushwaha, P.P.; Prajapati, K.S.; Shuaib, M.; Gupta, S.; Kumar, S. Identification of FDA approved drugs and nucleoside analogues as potential SARS-CoV-2 A1pp domain inhibitor: An in silico study. *Comput. Biol. Med.* **2021**, *130*, 104185. [CrossRef]

66. Zhang, S.; Garzan, A.; Haese, N.; Bostwick, R.; Martinez-Gzegozewska, Y.; Rasmussen, L.; Streblow, D.N.; Haise, M.T.; Pathak, A.K.; Augelli-Szafran, C.E.; et al. Pyrimidone inhibitors targeting Chikungunya Virus nsP3 macrodomain by fragment-based drug design. *PLoS ONE* **2021**, *16*, e0245013. [CrossRef]
67. Schuller, M.; Correy, G.J.; Gahbauer, S.; Fearon, D.; Wu, T.; Díaz, R.E.; Young, I.D.; Martins, L.C.; Smith, D.H.; Schulze-Gahmen, U.; et al. Fragment binding to the Nsp3 macrodomain of SARS-CoV-2 identified through crystallographic screening and computational docking. *Sci. Adv.* **2021**, *7*, eabf8711. [CrossRef]
68. Rack, J.G.M.; Zorzini, V.; Zhu, Z.; Schuller, M.; Ahel, D.; Ahel, I. Viral macrodomains: A structural and evolutionary assessment of the pharmacological potential. *Open Biol.* **2020**, *10*, 200237. [CrossRef]
69. Schuller, M.; Riedel, K.; Gibbs-Seymour, I.; Uth, K.; Sieg, C.; Gehring, A.P.; Ahel, I.; Bracher, F.; Kessler, B.M.; Elkins, J.M.; et al. Discovery of a Selective Allosteric Inhibitor Targeting Macrodomain 2 of Polyadenosine-Diphosphate-Ribose Polymerase 14. *ACS Chem. Biol.* **2017**, *12*, 2866–2874. [CrossRef] [PubMed]
70. Sowa, S.T.; Galera-Prat, A.; Wazir, S.; Alanen, H.I.; Maksimainen, M.M.; Lehtiö, L. A molecular toolbox for ADP-ribosyl binding proteins. *Cell Rep. Methods* **2021**, *1*, 100121. [CrossRef]
71. Albulescu, I.C.; White-Scholten, L.; Tas, A.; Hoornweg, T.E.; Ferla, S.; Kovacikova, K.; Smit, J.M.; Brancale, A.; Snijder, E.J.; van Hemert, M.J. Suramin Inhibits Chikungunya Virus Replication by Interacting with Virions and Blocking the Early Steps of Infection. *Viruses* **2020**, *12*, 314. [CrossRef] [PubMed]
72. Daniels, C.; Thirawatananond, P.; Ong, S.-E.; Gabelli, S.B.; Leung, A.K.L. Nudix hydrolases degrade protein-conjugated ADP-ribose. *Sci. Rep.* **2015**, *5*, 18271. [CrossRef] [PubMed]
73. Dyall, J.; Coleman, C.M.; Hart, B.J.; Venkataraman, T.; Holbrook, M.R.; Kindrachuk, J.; Johnson, R.F.; Olinger, G.G., Jr.; Jahrling, P.B.; Laidlaw, M.; et al. Repurposing of clinically developed drugs for treatment of Middle East respiratory syndrome coronavirus infection. *Antimicrob. Agents Chemother.* **2014**, *58*, 4885–4893. [CrossRef] [PubMed]
74. Russo, L.C.; Tomasin, R.; Araújo Matos, I.; Manucci, A.C.; Sowa, S.T.; Dale, K.; Caldecott, K.W.; Lehtiö, L.; Schechtman, D.; Meotti, F.C.; et al. The SARS-CoV-2 Nsp3 macrodomain reverses PARP9/DTX3L-dependent ADP-ribosylation induced by interferon signalling. *J. Biol. Chem.* **2021**, *297*, 101041. [CrossRef] [PubMed]
75. Lazear, H.M.; Schoggins, J.W.; Diamond, M.S. Shared and Distinct Functions of Type I and Type III Interferons. *Immunity* **2019**, *50*, 907–923. [CrossRef] [PubMed]

pathogens

Review

PARP1 as an Epigenetic Modulator: Implications for the Regulation of Host-Viral Dynamics

Asher A. Sobotka [1,2] and Italo Tempera [1,*]

1 Wistar Institute, Philadelphia, PA 19104, USA
2 Department of Biology, University of Pennsylvania, Philadelphia, PA 19104, USA
* Correspondence: itempera@wistar.org

Abstract: The principal understanding of the Poly(ADP-ribose) polymerase (PARP) regulation of genomes has been focused on its role in DNA repair; however, in the past few years, an additional role for PARPs and PARylation has emerged in regulating viral-host interactions. In particular, in the context of DNA virus infection, PARP1-mediated mechanisms of gene regulations, such as the involvement with cellular protein complexes responsible for the folding of the genome into the nucleus, the formation of chromatin loops connecting distant regulatory genomic regions, and other methods of transcriptional regulation, provide additional ways through which PARPs can modulate the function of both the host and the viral genomes during viral infection. In addition, potential viral amplification of the activity of PARPs on the host genome can contribute to the pathogenic effect of viral infection, such as viral-driven oncogenesis, opening the possibility that PARP inhibition may represent a potential therapeutic approach to target viral infection. This review will focus on the role of PARPs, particularly PARP1, in regulating the infection of DNA viruses.

Keywords: PARP1; DNA virus; epigenetic; viral gene regulation

check for updates

Citation: Sobotka, A.A.; Tempera, I. PARP1 as an Epigenetic Modulator: Implications for the Regulation of Host-Viral Dynamics. *Pathogens* **2024**, *13*, 131. https://doi.org/10.3390/pathogens13020131

Academic Editors: Anthony K L Leung, Anthony Fehr and Rachy Abraham

Received: 5 April 2023
Revised: 23 January 2024
Accepted: 24 January 2024
Published: 30 January 2024

1. PARP Overview

Poly (ADP-ribose) polymerase (PARP) proteins are a family of proteins responsible for the transfer of single or multiple ADP-ribose moieties to protein acceptors utilizing the NAD+ as substrate, a process labeled Mono-ADP-ribosylation (MARylation) or Poly-ADP-ribosylation (PARylation), respectively [1–3]. The addition of these long moieties to target proteins, especially by PARP1 and PARP-2, has been primarily associated with several regulatory mechanisms, including DNA repair and programmed cell death [4–7]. However, the discovery of an increasing number of novel proteins capable of PARylation has been accompanied by an increase in known targets of PAR modifications [7]. The ability of PARylation to directly modify the structure and activity of proteins and associated elements in addition to functioning as a recruiter for additional proteins indicates that PARylation is capable of multiple means of regulation [7].

Early work regarding PARPs (especially PARP1) has focused on the DNA repair mechanisms of PARPs and PARP1-induced cell death [1,4,8]. This mechanism relies on charged ADP-ribose moiety polymers recruiting proteins responsible for DNA repair [9]. PARP1-dependent cell death—parthanatos—is enabled by high levels of PARP1 activity resulting in DNA fragmentation along with depletion of cellular NAD+ and ATP [10,11]. In addition to DNA repair mechanisms, the presence of PAR can recruit proteins responsible for other regulatory pathways [12]. While seventeen unique PARPs have been discovered, only five (PARP1, PARP-2, PARP-3, tankyrase-1, and tankyrase-2) have been directly associated with PARylation, while the others have not been shown to build ADP-ribose polymers [13–15]. Additionally, in regard to DNA damage, PARP1 is responsible for the majority of PARylation activity, while the other PARPs have more limited—and potentially more specific—regulatory roles [14]. While less work has been conducted regarding Mono-ADP-Ribosylation, there is evidence of similar roles of MARylation in cellular processes [16].

2. The Role of PARP1 in the Regulation of the Epigenome

PARylation is commonly associated with DNA repair. However, several epigenetic regulatory pathways utilize PARP1 and PARylation to modify chromatin structure and regulate gene expression. It is beyond the scope of this review to provide an exhaustive and comprehensive overview of the role of PARP1 on transcription, for which we refer readers to the works of Kraus, Ko and Ren, and Huang and Kraus [17–19]. Nevertheless, essential epigenetic mechanisms regulated by PARP1 and are eventually relevant to virus infection are further discussed.

2.1. PARPs' Regulation of Nucleosome Structure

PARP1 and PARylation regulate transcription through multiple mechanisms that directly interact with and remodel chromatin structure, due to the ability of PARP1 to bind to DNA [17]. PARP1 can create compact chromatin structures comparable in function to H1 repressed chromatin structures [17,20]. Specifically, the DNA-binding domain (DBD) containing PARP1's Zinc regions is involved individually in transcriptional repression and works cooperatively with the catalytic domain (CAT) to form condensed chromatin structures [20]. This N-terminal DNA binding domain consists of two Zinc fingers responsible for DNA interaction. PARP1's zinc finger 2 has a strong affinity for DNA breaks (enabling PARP1's DNA repair mechanism), while zinc finger 1 is responsible for the activation of PARylation activity [9]. The DBD, in conjunction with the catalytic domain of PARP1, can both condense individual nucleosome regions and form condensed structures of adjacent nucleosomes [20]. The DBD is crucial for the recruitment of PARP1 to chromatin, enabling the CAT activity on chromatin by "tethering" PARP1 to the DNA, which in turn can condense individual nucleosomes [20]. Furthermore, the CAT was found to bring together multiple nucleosomes into a denser structure through the catalytic activity that promotes PARP1 dimerization, drawing together multiple PARP1-bound nucleosomes, which creates a further repressed chromatin structure [20]. PARP1 is also involved in inhibiting this process, as PARP1 loses its nucleosome affinity when it is PARylated at the BRCT region, located in the middle core domain of the protein, which is usually referred to as the auto-modification domain [21–25]. PARP1 auto-PARylation results in the loss of the chromatin-compacted structures [17,20]. Together, these functions provide a picture of the specific roles of the domains of PARP1 in condensing chromatin structure: the DBD recruits PARP to the chromatin structure, whereas the CAT provides the enzymatic activity required for creating the structure, and the BRCT provides a spot for the auto modification responsible for reversing this process. This role of BRCT implies that the control of PARP enzymatic function determines both the condensing and opening of chromatin through nucleosome PARylation and auto-PARylation, respectively.

In addition to the cooperation of the DBD and CAT domains to repress chromatin, the DBD of PARP1 is capable of inhibiting transcription through the repression of RNA Pol II activity when bound to chromatin, meaning that when attached to chromatin, PARP1 has multiple methods to regulate transcription [20]. This method of transcriptional regulation is independent of the previously described nucleosome compaction but functions simultaneously, providing a cooperative method of transcriptional repression [20].

2.2. The Regulation of CTCF by PARP1

The CCCTC-binding factor (CTCF)—a highly conserved zinc-finger protein—has been implicated in transcriptional regulation by remodeling the three-dimensional structure of the chromatin [26]. CTCF has been identified in the formation of chromatin loops associated with transcriptional activation, insulation, and silencing [26,27]. Additionally, CTCF can function as an activator of PARP1 and is regulated in part via PARylation [17,28–31] (Figure 1). CTCF recruits PARP1—in the absence of the usual stress signals such as DNA damage—both to stabilize its own binding to chromatin and to recruit PARP1 for chromatin modifications [17,31]. While PARP1 recruited by CTCF can have multiple targets (such as CTCF itself), a common target of PARylation by CTCF-recruited PARP1 is PARP1 itself, suggesting that CTCF can function as a regulator of auto-PARylation [17,31]. Additionally, PARG has been identified as reversing the modifications made by CTCF-recruited PARP1, suggesting that these modifications are transient in many cases [29]. The high levels of PARG and associated PAR depletion have even been shown to remove CTCF binding from key regulatory locations [32].

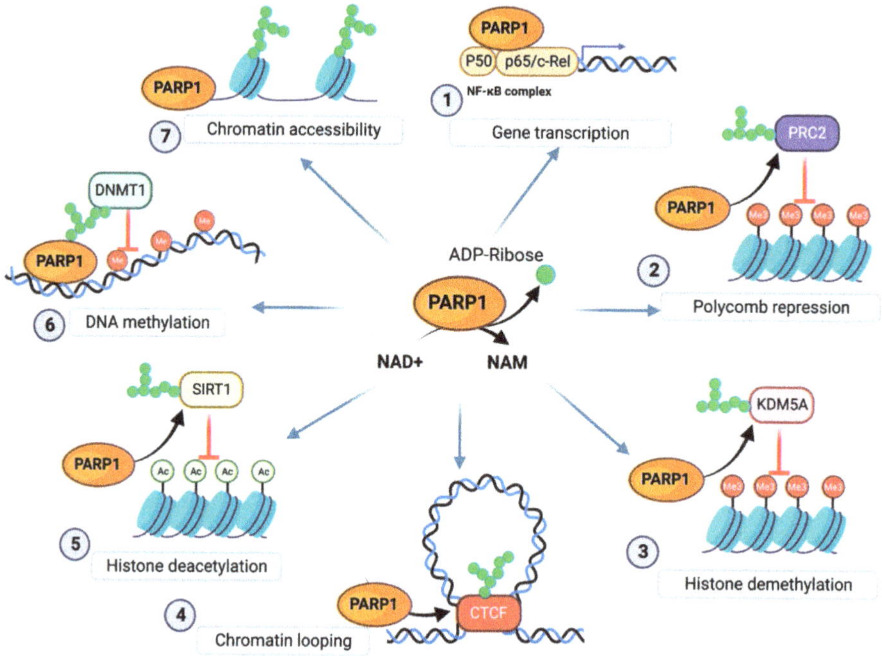

Figure 1. Overview of PARP1 regulation of epigenetic modification. (1) PARP1 binding to the NF-κB complex enables gene transcription of immune responses (2) PARP1 PARylation of the EZH2 component of the PRC2 complex downregulates H3K27 methylation. (3) PARylation of KDM5A inhibits histone demethylation. (4) PARylation of CTCF enables stable formation of chromatin loops (5) PARylation of SIRT1 decreases SIRT1 histone deacetylation activity. (6) PARylation of DNMT1 decreases DMNT1 methyltransferase activity. (7) Direct PARylation of histones prevents the formation of dense inaccessible chromatin structure. NAM, nicotinamide. The figure was created with Biorender.

2.3. *The Regulation of Histone Modifications by PARP1*

PARP1 is associated with the repression of transcription through direct interactions with chromatin structure; conversely, PARP1 has been associated with the upregulation of transcription through histone interactions and modifications (Figure 1). PARP1 promotes transcription through the PARylation of histone H1 and core histones, which results in increased transcription [33]. This PARylation resulted in the opening of condensed chromatin structures due to the negative charge of the PAR polymer, weakening the chromatin–histone interaction and allowing for increased transcription. Although the primary histone acceptor of PARylation is H1, additional core histones are essential targets of PARylation, especially in H1-depleted chromatin structures [33]. For example, PARP1 activity has been associated with chromatin opening due to the PARylation of chromatin-bound histone H2B, which inhibits normal histone interaction with the DNA, similarly to its effect on H1 [34,35]. These PARP1 modifications have specifically been suggested to target the amino-terminal of histone tails, indicating that potentially histone acetylation at this location would have an inhibitory effect on PARylation at this site and associated upregulation in transcription [36].

Although PARylated histones are still found to be associated with chromatin, they are incapable of forming either the H1-chromatin structures typically found in H1 condensed chromatin or other canonical H1 interactions found in condensed chromatin. PARylation addition to the H1 chromatin binding sites was found to competitively bind to linker histones, displacing these linker histones and leading to increased chromatin opening [36]. In addition to the regulation of nuclear histones, cytoplasm-localized H2B, H2A, H3, and H4 were all established as targets of PARylation at specific lysine residues localized near histone tails [36]. Furthermore, the exclusion of H1 from chromatin structures due to PARylation has been demonstrated to increase transcription due to the increasing accessibility of the transcription start promoter site [37].

In addition to directly modifying histones, PARP1 has been associated with the regulation of histone modifications, such as the downregulation of the repressive chromatin marker H3K27me3. PARP1 inhibition was associated with the increased activity of Polycomb Repressive Complex 2 (PRC2)—a protein complex responsible for the methyltransferase activity of lysine 27 on H3—resulting in decreased condensed chromatin [38]. Additionally, in BRCA2 proficient cancers, this PARP1-PRC2 interaction has been shown to inhibit NF-kB immune activity [39–41]. This data would suggest that PARP1 supports transcriptionally activated chromatin through the downregulation of PRC2 activity (Figure 1). This inhibition of PRC2 was determined to be a consequence of the PARylation of the EZH2 enzymatic component of PRC2, which prevents the methyltransferase activity of PRC2 due to decreasing the strength of EZH2's interaction with histone H3 [41]. This PARylation of EZH2 results in EZH2 disassociation from the PRC2 complex in addition to the downregulation of EZH2 [42]. An additional pathway through which PARP1 modifies H3 methylation is through the PARylation of the demethylase protein KDM5B (Figure 1) [37]. PARP1 was found to preserve transcriptionally activating H3K4 trimethylation through the inhibition of KDM5B, which is responsible for H3K4me3 demethylation [37]. PARP1 was established as a critical element in preventing the KDM5B repression of target genes through H3K4me3 demethylation, as PARylated KDM5B is incapable of binding target histones [37]. Corroborating data have shown that the PARP1 PARylation of EZH2 results in EZH2 disassociation from the PRC2 complex, and subsequent downregulation of H3K27me3 [42].

Beyond the modification of histone methylation, PARP1 plays an additional secondary role in histone modification through the regulation of proteins responsible for histone acetylation. Histone H3 and H4 were found to have higher levels of acetylation in the presence of PARP (despite, of course, PARP1 having no HAT activity) [43]. PARP1, independent of activation by DNA breaks, participates in the ERK/MAP kinase signaling cascade, heightening the ERK-mediated histone H3 and H4 acetylation activity [43]. This is enabled by a pERK2-PARP1 complex. When this complex is formed, PARylated ERK2 increases the phosphorylation of ELK1, which in turn is responsible for the activation of HATs such as CBP/p300 [44]. Additionally, the ERK2-PARP1 complex is responsible for the increased PARylation of histone H1 [44]. The PARG-mediated downregulation of this activity is associated with the removal of H3 and H4 acetylation, further suggesting that the presence of PAR moieties serves to regulate H3 and H4 acetylation [45]. This allows for PARP1 to act as a promotor of H3 and H4 acetylation, where in turn PARG acts to regulate this PARP1-mediated acetylation [45].

2.4. The Regulation of DNA Methylation by PARP1

PARP-mediated PARylation has been associated with the regulation of DNA methylation (Figure 1). It has been identified that active PARP1 is responsible for regulating DNA methylation by interacting with the DNA-methyltransferase DNMT1 [31,45,46]. This process is regulated through auto-PARylation, which determines the extent of PARP1 activity on such protein [9]. While PARP1 itself has no direct involvement with DNA methylation, it has been identified as a regulator of DNMT1 activity [31,45,46]. Based on the proposed model, it is suggested that when PARP1 is auto-modified (PARylated) or when PARs are present, they cause DNMT1 to become catalytically inactive and, therefore, less effective at carrying out DNA methylation. [31,45,46]. These changes have been implicated in the formation of less-dense chromatin structure, presumably a cooperative process with PARP1-mediated histone H1 modifications as DNA methylation's transcriptional inhibition has relied upon the presence of linker histones, potentially suggesting a synergistic effect [47,48]. A proposed mechanism for the PARP1-mediated regulation of methylation suggests also that PARP1, activated by the CTCF binding factor, could deactivate DNMT1 activity through the PARylation of two DNA binding domains [46,49,50]. Furthermore, CTCF binding is strengthened in a complex with DNMT1 and auto-PARylated PARP1 [2]. In addition to PARylation directly inhibiting DNMT1 activity, EZH2 (a component of the PRC2 complex) has been implicated as a "recruitment platform" for DNMT1 [28]. These data, taken with PARP1's role in inhibiting EZH2 activity, could suggest that PARP1 is able to further regulate DNA methylation through EZH2.

2.5. The Regulation of Protein Acetylation by PARP

In addition to the role of PARP1 in modifying histones through acetylation pathways, there is evidence of PARP1 having a role in protein acetylation beyond histones (Figure 1) [51–53]. In mouse models, PARP inhibition was associated with increased Sirtuin activity, suggesting that PARP1 and PARP-2 are responsible for maintaining protein acetylation [53]. The acetylation of High-Mobility Group Box-1 (HMGB1), a mediator in the inflammatory response, was found to be upregulated through multiple PARP1-mediated mechanisms, which is critical for the transport of HMGB1 to the cytoplasm and eventual release [52]. PARP1 was found to play a role in both the activation of acetyltransferases and the deactivation of deacetyltransferases [53]. Additionally, the PARylation of HMGB1 was found to facilitate acetylation [53].

3. Viral Utilization of PARylation

3.1. Immune Response

Studies over the past two decades have provided important insights into the role that PARPs and PARylation play in regulating the immune response of host cells to pathogens. PARP1 has been identified as a mediator in the activation of host antiviral immune responses. Nuclear factor-kappa B (NF-κB) is a crucial regulator of several antiviral immune responses, and PARP1 has been shown to have an interaction with both NF-κB subunits p50 and p65, in addition to interacting with p300—a coactivator of NF-κB [53] (Figure 1). In the absence of PARP1, NF-κB-dependent immune response proteins were found not to be expressed [54,55]. NF-κB is a central player in immune and inflammatory signaling responses and is responsible for various immune responses. While most studies have focused on the role of NF-κB as a transcription factor necessary for the development of B cells, NF-κB is also critical for the development and function of T cell thymocytes, dendritic cells, macrophages, and fibroblasts [56–58]. Additionally, NF-kB is responsible for antiviral inflammatory activation [59,60]. NF-kB can regulate these responses partly through its activity as a regulator of transcription; NF-kB has been shown to selectively activate and deactivate transcription through its direct interaction with HAT and HDAC enzymes which regulate H4 acetylation on histones near regions responsible for immune response genes [58,61]. PARP1—through the regulation of NF-kB—could regulate several parts of both the adaptive and innate immune systems by altering chromatin structure. These data suggest that PARP1, among other proteins, acts as an activator of the cellular immune response as a stress response to viral infection. NF-kB dysregulation has also been implicated in numerous pathological processes such as tumorigenesis and progression, multiple sclerosis, and inflammatory diseases such as arthritis [62,63]. In addition to the NF-kB-related inflammatory response, PARP has been implicated in regulating inflammation through HMGB1 [64–67]. Similarly to NF-kB, HMGB1 has been identified as a central protein in activating several innate immune responses related to dendritic cells, macrophages, and programmed cell death [53]. Similarly to NF-kB, HMGB1 dysfunction has also been implicated in malignancies [68–70]. Several PARPs have been further implicated in signaling pathways, such as the pathways related to IFN-1 production and JAK-STAT signaling [71].

Even beyond the activation of the immune response through the PARylation of host proteins related to immune pathways, PARP has antiviral activity through its PARylation of viral proteins responsible for viral maintenance, such as the Epstein–Barr nuclear antigen 1 (EBNA1) in Epstein–Barr virus (EBV), LANA1 in KSHV, nonstructural proteins in Zika and Chikungunya (CHIKV) virus, and the nucleocapsid protein in coronaviruses [72]. Some of these modifications have been identified as inhibiting critical viral functions, suggesting that PARylation may function as an immune response [72]. However, it is beyond the scope of this review to provide a comprehensive overview of the role of PARylation as an antiviral part of the innate immune system. For a more comprehensive overview of PARylation as an antiviral function, we suggest the work of Du et al. [72].

3.2. *DNA Virus's Utilization of PARP*

Despite PARP1 having an established antiviral role, several viruses have been implicated in utilizing PARP1 to evade immune detection and assist the virus in host-pathogen conflicts. The viral utilization of PARP1 to modify viral episomes or host genes has been implicated in helping long-term viral infection in several DNA viruses as viruses utilize PARylation to affect pro-viral changes to themselves or the host.

3.2.1. Herpes Simplex Virus Type 1

Herpes simplex virus type 1, or HSV-1, is characterized by chronic long-term infection of the peripheral nervous system, enabled by HSV-1 transportation through axons to neuronal ganglia, where the long-term infection is established through latent HSV-1 infection characterized by the expression of a limited set of viral genes to evade immune detection [72–74]. However, despite latent infection accounting for such a large portion of HSV-1 infection, HSV-1 lytic reactivation can occur due to various stress stimuli [75].

HSV-1 infection is associated with a substantial decrease in cellular NAD+ content and an increase in PARylation [76] (Figure 2). This increase in PARylation can be reversed with PARP1 and PARP-2 inhibition, suggesting that PARP activity is responsible for these changes [77]. Furthermore, in addition to changes in PARP activity, PARG is degraded during HSV-1 infection, further implying the role of PARylation in promoting viral infection [77]. These modifications may induce the auto-PARylation of PARP1 to prevent the PARP1-mediated activation of apoptosis or parthanatos (as a stress immune response), suggesting a viral hijacking of this regulatory mechanism [77]. In addition to boosting PARP1 activity, HSV-1 has been observed to use the E3 ubiquitin ligase ICP0 to break down PARG. This process helps elevate PAR levels [78,79]. In order to prevent PARP1/2-induced cell death as part of the immune response, HSV-1 increases NAD+ levels available to PARP1/2 while degrading PARG in order to increase and maintain auto-PARylation, inhibiting PARP1/2's stimulation of cell death. In addition to boosting PARP1 activity, HSV-1 has been observed to use the E3 ubiquitin ligase ICP0 to break down PARG. This process helps elevate PAR levels [78,79]. The aim is to safeguard against PARP1/2-triggered cell death in the immune response.

3.2.2. Kaposi's Sarcoma-Associated Herpesvirus

Kaposi's sarcoma-associated herpesvirus (KSHV) is a γ-herpesvirus associated with long-term infection and linked to several related malignancies [28,46,80]. Long-term KSHV infection is enabled by the latent infection expression of only a limited number of viral genes [80,81].

PARP1 has been associated with the repression of KSHV replication and viral expression [82]. PARP1, in conjunction with the Ste-20-like kinase hKFC, PARylate, and phosphorylate, is the KSHV replication and transcription activator (RTA), inhibiting its activity and preventing KSHV replication and transcription [83] (Figure 3). The PARP-hKFC complex enables the latent expression of KSHV by preventing the expression of genes activated through RTA [83]. The formation of the PARP1-hKFC complex is regulated through the KSHV viral processivity factor PF-8, which is responsible for the degradation of PARP1, resulting in lytic reactivation [84].

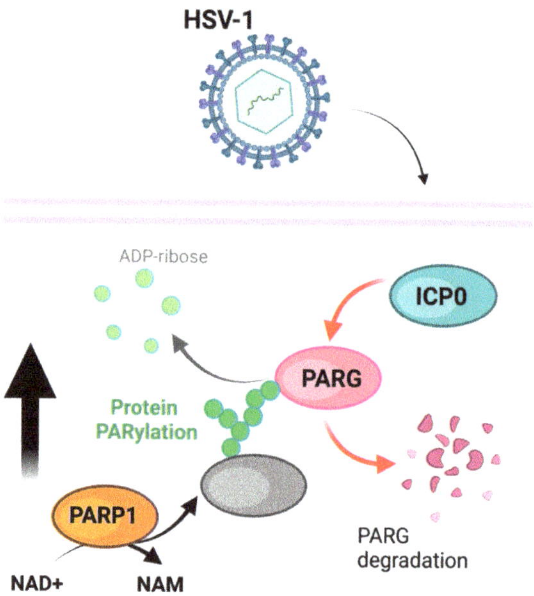

Figure 2. Model of PARylation during HSV-1 infection. PARP1 activity is increased, while concurrentlythe HSV-1 ICP0 degrades PARG, thereby preventing the removal of PARylation. The figure was created using Biorender.

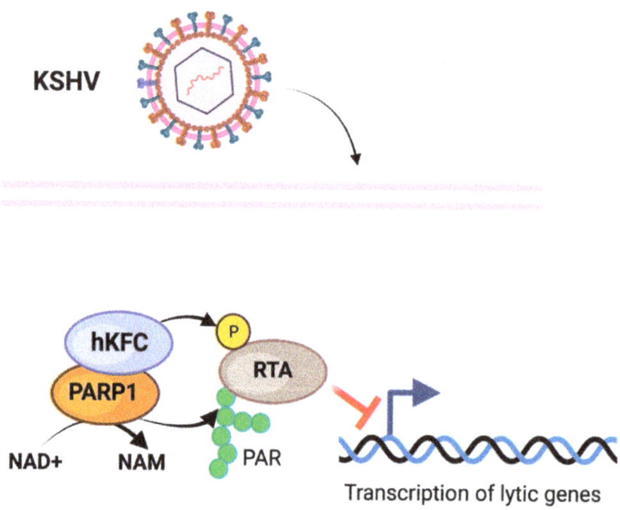

Figure 3. Model of PARylation during KSHV infection. Modification of KSHV transcriptional and replicative activator RTA through PARP1 PARylation along with SLK phosphorylation inhibits KSHV lytic activity, enabling latent infection. The figure was created using Biorender.

In addition to regulating KSHV latent infection, PARP1 is responsible for enabling the replication of KSHV during latent infection [84]. PARP1 has been shown to bind to

KSHV's terminal repeat sequence (which, during latent infection, functions as an origin of replication) and PARylate the latency-associated nuclear antigen associated with the terminal repeat sequence during latent infection [85].

3.2.3. Epstein–Barr Virus

Epstein–Barr Virus (EBV) is a γ-herpesvirus that establishes a persistent life-long latent infection in the host cell [80,85]. However, unlike KSHV, EBV exhibits three types of latency defined by the expression of a limited number of viral genes (latent genes) in different patterns [86].

Multiple PARPs have been ascribed a regulatory role related to EBV replication. PARP1 has been shown to directly bind to the dyad symmetry element of the EBV origin of plasmid replication (OriP), downregulating the expression of several EBV-associated genes [86–88] Moreover, telomere-associated PARPs (Tankerases) were also related to OriP regulation. Tankerases have been shown to bind to the dyad-symmetry elements and family of repeats region of the EBV OriP locus, a region that serves as the origin of replication of the viral genome during latent infection [89]. In addition to binding the OriP, PARP1 can modify EBV latency by binding with an additional host protein—CTCF—to the viral promoter region for the *BZLF1* viral gene, which codes the Zta protein that activates the expression of viral proteins responsible for the viral lytic replication [90,91]. This binding enables PARP1 to regulate the 3D remodeling of the EBV genome through its colocalization with other proteins, including CTCF [30] (Figure 4). The PARylation of CTCF enables loop formation to occur across the EBV genome, permitting the expression of specific viral gene programs [88,92]. In addition to maintaining latent infection, PARP1 activity has been implicated in regulating the lytic EBV cycle through similar methods [32]. This is enabled by the mechanism elucidated in Section 3.2.

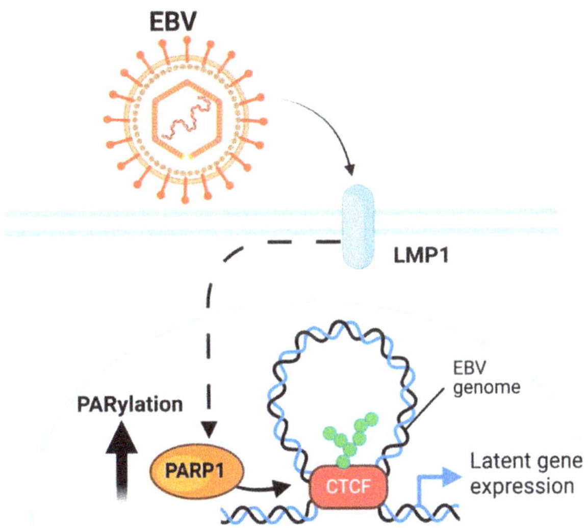

Figure 4. Model of PARylation during EBV infection. EBV, through the EBV LMP1 protein, increases PARP1 activity and PARylation of CTCF, enabling loop formation and expression of latent gene programs. The figure was created using Biorender.

In addition to PARPs modifying the EBV viral genome, PARP1 can be recruited by EBV to modify elements of the host genome. The EBV latent membrane protein 1 (LMP1) can activate hypoxia-inducible factor 1-alpha (HIF-1α) through the LMP1 activation of PARP1 mediated HIF-1α PARylation [93]. This LMP1-mediated activation of genes through PARP1 is not restricted to HIF-1α; as in the absence of competitive PARylation, increased levels of repressive histone marker H3k27me3 were found on several host genes regulated by LMP1 [94]. In addition to LMP1, EZH2, an enzymatic component of the Polycomb Repressive Complex 2 (PRC2) responsible for methyltransferase activity on H3 lysine 27, is downregulated by PARP1 PARylation (PARP1 upregulates host gene expression by decreasing global H3K27me3 through the mechanism explained above) [95]. EZH2 has also been identified as a regulator of EBV latency [38].

Besides PARP1 role in regulating the viral epigenome, EBV-transformed B-lymphocytes depend on PARP1 DNA repair functions to survive. The immune transcriptional regulator STAT3—activated during the transition from lytic EBV infection to latent infection—compromises homologous DNA repair through the inhibition of ART phosphorylation of Chk1 [96]. This results in DNA repair relying upon PARP1 contingent microhomology-mediated end-joining to the maintenance of genome integrity [96]. It is important to note that other proteins implicated in DNA mismatch repair such as PCNA are also implicated in the maintenance of EBV latency through modulating the transition from replication to transcription [97]; however, these concurrent mechanisms are beyond the scope of this review.

These data suggest that PARylation—especially PARP1-catalyzed PARylation—is responsible for both sides of the crosstalk between the EBV and the host genome. Not only does PARP1 regulate the expression of the host genome, but it is also responsible for regulating the modification of host genes relevant to EBV survival.

3.2.4. Cytomegalovirus

PARP1 and PARylation also play a role in an additional herpesvirus: the human cytomegalovirus (HCMV), a β-herpesvirus [98]. Similarly to EBV's establishment of latent infection, HCMV has been found to recruit host factors to regulate its own gene expression and enable chronic silent infection [99]. In addition to utilizing host proteins to modify viral expression, HCMV has been shown to modify cells to create a more pro-viral environment, such as by inhibiting cell death [99]. PARP1 has been linked to this regulation of HCMV infection, with HCMV infection enabling increased PARylation by PARP1, possibly due to increased DNA damage during viral infection or as a promotor of viral replication [100,101]. Further supporting the pro-viral role of PARylation in the case of HCMV, PARG has been shown to have an antiviral effect on HCMV replication [102].

3.2.5. Polyomavirus

PARP1 is also implicated in the life cycle of several Polyomaviruses, including simian vacuolating virus 40 (SV40) and human polyomavirus 2 (JC virus). PARP1 is activated by and directly binds to SV40 capsid proteins [102]. PARP1 has been shown to be activated by SV40 protein VP3, inducing apoptosis which in turn enables SV40 to escape from the cell membrane [103]. However, while PARP plays a pro-viral role in SV40 replication, it has also been linked with antiviral mechanisms in response to other polyomaviruses. PARP has been implicated in several mechanisms that inhibit the replication of SV40 DNA [103]. PARP binds to the end of viral DNA Strands, competitively inhibiting the binding and activity of exonuclease III, DNA ligase, and Polα during leading strand synthesis [104]. Consequently, the presence of PARP during SV40 replication acts to limit lytic reactivation of SV40 by preventing SV40 replication of the viral genome.

3.2.6. PARylation as a Tool for RNA Viruses

Even though PARPs are primarily associated with either protein or chromatin modification, there is evidence that in addition to regulating the infection of DNA viruses, PARPs also have regulatory mechanisms relevant to RNA virus function. PARP12—while only having MARylation activity—has been linked to antiviral functions regarding several alphaviruses, such as the Venezuelan Equine Encephalitis Virus (VEEV) [105].

There is emerging evidence that viral macrodomains in RNA virus families Coronaviradae, Togoviridae, and Hepeviridae themselves have enzymatic de-ADP-ribosylation activity that can offset antiviral activity by PARPs [103]. The SARS-CoV-2 MacroD-like macrodomain (Mac1)—in addition to highly similar macrodomains found in SARS-CoV and MERS-CoV—has been identified as having mono-ADP-ribose glycohydrolase activity [106,107].

There is evidence that some RNA viruses—in order to counteract PARP1's antiviral activity—code for viral proteins with glycohydrolase activity. The SARS-CoV-2 virus encodes the nonstructural protein 3 (Nsp3) which contains a macrodomain. This macrodomain has been found to remove ADP-ribose from proteins, counteracting the increase in PARylation that occurs in response to the interferon signaling cascade. Due to the highly conserved nature of these macrodomains (26% of the amino acids differ from the SARS-CoV-2 macrodomain compared to other coronaviruses), it has been suggested that this de-PARylating activity is common among viral macrodomains [106,108]. However, despite several macrodomains being similarly capable of PAR removal, the effects of these macrodomains on viral infection vary. The macrodomain of CHIKV—a Togavirus—was identified as promoting viral replication through the de-PARylation activity [109]. Conversely, in the Mac1 macrodomain found in several coronaviruses (the macrodomain responsible for activity related to PAR), there is no evidence that it has a similar role to the CHIKV macrodomain in viral replication [106]. The coronavirus domain instead was associated with the modification of pathogenesis [106,110].

Human immunodeficiency virus (HIV-1) has been found to be able to inhibit the PARP1 activation of the NF-κB immune pathway, a pathway that, as discussed above, is highly regulated by PARP1. The HIV-1 Vpr and glucocorticoid receptor cooperatively function to form a complex with PARP1, preventing PARP1 activation of the host immune system and PARP1 antiviral activity [111]. However, there is some evidence of a pro-viral role of PARP1 in HIV-1 infection. While there is disagreement regarding to what extent PARP1 is dispensable for HIV-1, it is clear that PARP1 plays a prominent role in enabling quick integration of HIV-1 due to its ability to repair DNA breaks left by HIV-1 enzyme integrase's insertion into the host genome of HIV-1 DNA from HIV-1 reverse transcriptase, utilizing the previously elucidated DNA repair mechanism [112,113].

4. Conclusions

Summarizing the role of PARP1 and PARylation in regulating host-viral conflicts is difficult as PARylation has the aforementioned mechanisms of epigenetic regulation that can both increase and decrease gene expression. Furthermore, its interaction with even genetically similar viruses can vary greatly. While PARylation has multiple antiviral mechanisms, several viruses can inhibit these antiviral functions and recruit and utilize PARPs to make pro-viral modifications to the host or the viruses themselves. These increasing number of observations have revealed the importance of PARylation and PARPs' activity in regulating virus–host interaction. However, the increasing number of findings has brought up critical theoretical issues related to the role of PARylation and PARP1. For example, since PARylation requires adequate levels of NAD+, either antiviral or pro-viral functions of PARP necessitate a rewire of cellular metabolism to allow proper PARP function. While the modifications made by PARPs to promote viral infection are well established, how cellular metabolism is modified to enable this activity is less understood. While extensive data demonstrate the role PARylation plays in promoting viral infection, how exactly the cellular environment is changed to allow for this modification needs to be further established.

While PARPs (especially PARP1) have been identified both as vital parts of several regulatory pathways and as directors of host-viral interactions, the role glycohydrolase activity and PARG play beyond its response to PARPs is much less examined. Most work regarding PARG relates to its role in reversing the activity of PARP1; however, recent data regarding glycohydrolase activity in the macrodomains of several viruses such as SARS-CoV-2 would suggest that further investigation into glycohydrolase activity—independent of PARP1—is warranted. Even though the importance of PAR for several pro and antiviral mechanisms is well established, very little is known about PARG's role in mediating viral activity relative to PARP1.

A better understanding of the role of PARPs and PARylation during viral infection is fundamental; while several FDA-approved PARP inhibitors exist and are used as targeted cancer drugs, their potential efficacy in treating viral infection and wider viral-associated pathogenesis has not been tested yet.

Author Contributions: Conceptualization, A.A.S. and I.T.; writing—original draft preparation, A.A.S. and I.T.; writing—review and editing, A.A.S. and I.T.; visualization, I.T.; supervision, I.T.; project administration, I.T.; funding acquisition, I.T. All authors have read and agreed to the published version of the manuscript.

Funding: This research received no external funding.

Institutional Review Board Statement: Not applicable.

Informed Consent Statement: Not applicable.

Data Availability Statement: Not applicable.

Acknowledgments: We thank the entire Tempera laboratory team for their invaluable input and feedback, which has greatly enriched the quality of our review. While we have taken great care to ensure that we have cited the most relevant sources, we recognize that there may be other important contributions that we have unintentionally overlooked. We apologize to any authors whose work may have been inadvertently omitted. IT was supported by R01AI130209, R01GM124449, and P01CA269043.

Conflicts of Interest: The authors declare no conflict of interest.

References

1. Amé, J.C.; Spenlehauer, C.; de Murcia, G. The PARP superfamily. *BioEssays News Rev. Mol. Cell. Dev. Biol.* **2004**, *26*, 882–893. [CrossRef]
2. de Murcia, G.; de Murcia, J.M. Poly(ADP-ribose) polymerase: A molecular nick-sensor. *Trends Biochem. Sci.* **1994**, *19*, 172–176. [CrossRef]
3. Hayaishi, O.; Ueda, K. Poly(ADP-ribose) and ADP-ribosylation of proteins. *Annu. Rev. Biochem.* **1977**, *46*, 95–116. [CrossRef] [PubMed]
4. Bürkle, A. Physiology and pathophysiology of poly(ADP-ribosyl)ation. *BioEssays* **2001**, *23*, 795–806. [CrossRef]
5. Woodhouse, B.C.; Dianov, G.L. Poly ADP-ribose polymerase-1: An international molecule of mystery. *DNA Repair* **2008**, *7*, 1077–1086. [CrossRef] [PubMed]
6. Herceg, Z.; Wang, Z.Q. Functions of poly(ADP-ribose) polymerase (PARP) in DNA repair, genomic integrity and cell death. *Mutat. Res.* **2001**, *477*, 97–110. [CrossRef] [PubMed]
7. Morales, J.; Li, L.; Fattah, F.J.; Dong, Y.; Bey, E.A.; Patel, M.; Gao, J.; Boothman, D.A. Review of poly (ADP-ribose) polymerase (PARP) mechanisms of action and rationale for targeting in cancer and other diseases. *Crit. Rev. Eukaryot. Gene Expr.* **2014**, *24*, 15–28. [CrossRef] [PubMed]
8. Alvarez-Gonzalez, R.; Althaus, F.R. Poly(ADP-ribose) catabolism in mammalian cells exposed to DNA-damaging agents. *Mutat. Res.* **1989**, *218*, 67–74. [CrossRef]
9. Javle, M.; Curtin, N.J. The role of PARP in DNA repair and its therapeutic exploitation. *Br. J. Cancer* **2011**, *105*, 1114–1122. [CrossRef]
10. Wang, Y.; Luo, W.; Wang, Y. PARP-1 and its associated nucleases in DNA damage response. *DNA Repair* **2019**, *81*, 102651. [CrossRef]
11. Huang, P.; Chen, G.; Jin, W.; Mao, K.; Wan, H.; He, Y. Molecular Mechanisms of Parthanatos and Its Role in Diverse Diseases. *Int. J. Mol. Sci.* **2022**, *23*, 7292. [CrossRef]
12. Gagné, J.P.; Isabelle, M.; Lo, K.S.; Bourassa, S.; Hendzel, M.J.; Dawson, V.L.; Dawson, T.M.; Poirier, G.G. G. Proteome-wide identification of poly(ADP-ribose) binding proteins and poly(ADP-ribose)-associated protein complexes. *Nucleic Acids Res.* **2008**, *36*, 6959–6976. [CrossRef] [PubMed]
13. Carter-O'Connell, I.; Cohen, M.S. Identifying Direct Protein Targets of Poly-ADP-Ribose Polymerases (PARPs) Using Engineered PARP Variants-Orthogonal Nicotinamide Adenine Dinucleotide (NAD+) Analog Pairs. *Curr. Protoc. Chem. Biol.* **2015**, *7*, 121–139. [CrossRef] [PubMed]
14. Kamaletdinova, T.; Fanaei-Kahrani, Z.; Wang, Z.Q. The Enigmatic Function of PARP1: From PARylation Activity to PAR Readers. *Cells* **2019**, *8*, 1625. [CrossRef]

15. Sanderson, D.J.; Cohen, M.S. Mechanisms governing PARP expression, localization, and activity in cells. *Crit. Rev. Biochem. Mol. Biol.* **2020**, *55*, 541–554. [CrossRef] [PubMed]
16. Hopp, A.K.; Hottiger, M.O. Uncovering the Invisible: Mono-ADP-ribosylation Moved into the Spotlight. *Cells* **2021**, *10*, 680. [CrossRef]
17. Kraus, W.L. Transcriptional control by PARP-1: Chromatin modulation, enhancer-binding, coregulation, and insulation. *Curr. Opin. Cell Biol.* **2008**, *20*, 294–302. [CrossRef]
18. Ko, H.L.; Ren, E.C. Functional Aspects of PARP1 in DNA Repair and Transcription. *Biomolecules* **2012**, *2*, 524–548. [CrossRef] [PubMed]
19. Huang, D.; Kraus, W.L. The expanding universe of PARP1-mediated molecular and therapeutic mechanisms. *Mol. Cell* **2022**, *82*, 2315–2334. [CrossRef] [PubMed]
20. Wacker, D.A.; Ruhl, D.D.; Balagamwala, E.H.; Hope, K.M.; Zhang, T.; Kraus, W.L. The DNA binding and catalytic domains of poly(ADP-ribose) polymerase 1 cooperate in the regulation of chromatin structure and transcription. *Mol. Cell. Biol.* **2007**, *27*, 7475–7485. [CrossRef]
21. Langelier, M.F.; Pascal, J.M. PARP-1 mechanism for coupling DNA damage detection to poly(ADP-ribose) synthesis. *Curr. Opin. Struct. Biol.* **2013**, *23*, 134–143. [CrossRef]
22. Yelamos, J.; Farres, J.; Llacuna, L.; Ampurdanes, C.; Martin-Caballero, J. PARP-1 and PARP-2: New players in tumour development. *Am. J. Cancer Res.* **2011**, *1*, 328–346. [PubMed]
23. Morrone, S.; Cheng, Z.; Moon, R.T.; Cong, F.; Xu, W. Crystal structure of a Tankyrase-Axin complex and its implications for Axin turnover and Tankyrase substrate recruitment. *Proc. Natl. Acad. Sci. USA* **2012**, *109*, 1500–1505. [CrossRef] [PubMed]
24. Smith, S. The world according to PARP. *Trends Biochem. Sci.* **2001**, *26*, 174–179. [CrossRef]
25. Pascal, J.M. The comings and goings of PARP-1 in response to DNA damage. *DNA Repair* **2018**, *71*, 177–182. [CrossRef]
26. Phillips, J.E.; Corces, V.G. G. CTCF: Master weaver of the genome. *Cell* **2009**, *137*, 1194–1211. [CrossRef] [PubMed]
27. Nichols, M.H.; Corces, V.G. A CTCF Code for 3D Genome Architecture. *Cell* **2015**, *162*, 703–705. [CrossRef] [PubMed]
28. Guastafierro, T.; Cecchinelli, B.; Zampieri, M.; Reale, A.; Riggio, G.; Sthandier, O.; Zupi, G.; Calabrese, L.; Caiafa, P. CCCTC-binding factor activates PARP-1 affecting DNA methylation machinery. *J. Biol. Chem.* **2008**, *283*, 21873–21880. [CrossRef]
29. Farrar, D.; Rai, S.; Chernukhin, I.; Jagodic, M.; Ito, Y.; Yammine, S.; Ohlsson, R.; Murrell, A.; Klenova, E. Mutational analysis of the poly(ADP-ribosyl)ation sites of the transcription factor CTCF provides an insight into the mechanism of its regulation by poly(ADP-ribosyl)ation. *Mol. Cell. Biol.* **2010**, *30*, 1199–1216. [CrossRef]
30. Lupey-Green, L.N.; Moquin, S.A.; Martin, K.A.; McDevitt, S.M.; Hulse, M.; Caruso, L.B.; Pomerantz, R.T.; Miranda, J.L.; Tempera, I. PARP1 restricts Epstein Barr Virus lytic reactivation by binding the BZLF1 promoter. *Virology* **2017**, *507*, 220–230. [CrossRef]
31. Lupey-Green, L.N.; Caruso, L.B.; Madzo, J.; Martin, K.A.; Tan, Y.; Hulse, M.; Tempera, I. PARP1 Stabilizes CTCF Binding and Chromatin Structure To Maintain Epstein-Barr Virus Latency Type. *J. Virol.* **2018**, *92*, e00755-18. [CrossRef] [PubMed]
32. Guastafierro, T.; Catizone, A.; Calabrese, R.; Zampieri, M.; Martella, O.; Bacalini, M.G.; Reale, A.; Di Girolamo, M.; Miccheli, M.; Farrar, D.; et al. ADP-ribose polymer depletion leads to nuclear Ctcf re-localization and chromatin rearrangement(1). *Biochem. J.* **2013**, *449*, 623–630. [CrossRef] [PubMed]
33. Huletsky, A.; de Murcia, G.; Muller, S.; Hengartner, M.; Ménard, L.; Lamarre, D.; Poirier, G.G. The effect of poly(ADP-ribosyl)ation on native and H1-depleted chromatin. A role of poly(ADP-ribosyl)ation on core nucleosome structure. *J. Biol. Chem.* **1989**, *264*, 8878–8886. [CrossRef] [PubMed]
34. Kim, M.Y.; Mauro, S.; Gévry, N.; Lis, J.T.; Kraus, W.L. NAD+-dependent modulation of chromatin structure and transcription by nucleosome binding properties of PARP-1. *Cell* **2004**, *119*, 803–814. [CrossRef]
35. D'Amours, D.; Desnoyers, S.; D'Silva, I.; Poirier, G.G. Poly(ADP-ribosyl)ation reactions in the regulation of nuclear functions. *Biochem. J.* **1999**, *342 Pt 2*, 249–268. [CrossRef]
36. Messner, S.; Altmeyer, M.; Zhao, H.; Pozivil, A.; Roschitzki, B.; Gehrig, P.; Rutishauser, D.; Huang, D.; Caflisch, A.; Hottiger, M.O. PARP1 ADP-ribosylates lysine residues of the core histone tails. *Nucleic Acids Res.* **2010**, *38*, 6350–6362. [CrossRef]
37. Krishnakumar, R.; Kraus, W.L. PARP-1 regulates chromatin structure and transcription through a KDM5B-dependent pathway. *Mol. Cell* **2010**, *39*, 736–749. [CrossRef]
38. Martin, K.A.; Cesaroni, M.; Denny, M.F.; Lupey, L.N.; Tempera, I. Global Transcriptome Analysis Reveals That Poly(ADP-Ribose) Polymerase 1 Regulates Gene Expression through EZH2. *Mol. Cell. Biol.* **2015**, *35*, 3934–3944. [CrossRef]
39. Yang, A.Y.; Choi, E.B.; So Park, M.; Kim, S.K.; Park, M.S.; Kim, M.Y. PARP1 and PRC2 double deficiency promotes BRCA-proficient breast cancer growth by modification of the tumor microenvironment. *FEBS J.* **2021**, *288*, 2888–2910. [CrossRef] [PubMed]
40. Chen, M.K. Efficacy of PARP inhibition combined with EZH2 inhibition depends on BRCA mutation status and microenvironment in breast cancer. *FEBS J.* **2021**, *288*, 2884–2887. [CrossRef] [PubMed]
41. Caruso, L.B.; Martin, K.A.; Lauretti, E.; Hulse, M.; Siciliano, M.; Lupey-Green, L.N.; Abraham, A.; Skorski, T.; Tempera, I. Poly(ADP-ribose) Polymerase 1, PARP1, modifies EZH2 and inhibits EZH2 histone methyltransferase activity after DNA damage. *Oncotarget* **2018**, *9*, 10585–10605. [CrossRef]
42. Yamaguchi, H.; Du, Y.; Nakai, K.; Ding, M.; Chang, S.S.; Hsu, J.L.; Yao, J.; Wei, Y.; Nie, L.; Jiao, S.; et al. EZH2 contributes to the response to PARP inhibitors through its PARP-mediated poly-ADP ribosylation in breast cancer. *Oncogene* **2018**, *37*, 208–217. [CrossRef]

43. Cohen-Armon, M.; Visochek, L.; Rozensal, D.; Kalal, A.; Geistrikh, I.; Klein, R.; Bendetz-Nezer, S.; Yao, Z.; Seger, R. DNA-independent PARP-1 activation by phosphorylated ERK2 increases Elk1 activity: A link to histone acetylation. *Mol. Cell* **2007**, *25*, 297–308. [CrossRef]
44. Camps, M.; Nichols, A.; Gillieron, C.; Antonsson, B.; Muda, M.; Chabert, C.; Boschert, U.; Arkinstall, S. Catalytic activation of the phosphatase MKP-3 by ERK2 mitogen-activated protein kinase. *Science* **1998**, *280*, 1262–1265. [CrossRef]
45. Cohen-Armon, M.; Yeheskel, A.; Pascal, J.M. Signal-induced PARP1-Erk synergism mediates IEG expression. *Sig. Transduct. Target. Ther.* **2019**, *4*, 8. [CrossRef]
46. Caiafa, P.; Guastafierro, T.; Zampieri, M. Epigenetics: Poly(ADP-ribosyl)ation of PARP-1 regulates genomic methylation patterns. *FASEB J.* **2009**, *23*, 672–678. [CrossRef]
47. Zardo, G.; D'Erme, M.; Reale, A.; Strom, R.; Perilli, M.; Caiafa, P. Does poly(ADP-ribosyl)ation regulate the DNA methylation pattern? *Biochemistry* **1997**, *36*, 7937–7943. [CrossRef] [PubMed]
48. de Capoa, A.; Febbo, F.R.; Giovannelli, F.; Niveleau, A.; Zardo, G.; Marenzi, S.; Caiafa, P. Reduced levels of poly(ADP-ribosyl)ation result in chromatin compaction and hypermethylation as shown by cell-by-cell computer-assisted quantitative analysis. *FASEB J.* **1999**, *13*, 89–93. [CrossRef] [PubMed]
49. Karymov, M.A.; Tomschik, M.; Leuba, S.H.; Caiafa, P.; Zlatanova, J. DNA methylation-dependent chromatin fiber compaction in vivo and in vitro: Requirement for linker histone. *FASEB J.* **2001**, *15*, 2631–2641. [CrossRef] [PubMed]
50. Zampieri, M.; Guastafierro, T.; Calabrese, R.; Ciccarone, F.; Bacalini, M.G.; Reale, A.; Perilli, M.; Passananti, C.; Caiafa, P. ADP-ribose polymers localized on Ctcf-Parp1-Dnmt1 complex prevent methylation of Ctcf target sites. *Biochem. J.* **2012**, *441*, 645–652. [CrossRef]
51. Viré, E.; Brenner, C.; Deplus, R.; Blanchon, L.; Fraga, M.; Didelot, C.; Morey, L.; Van Eynde, A.; Bernard, D.; Vanderwinden, J.M.; et al. The Polycomb group protein EZH2 directly controls DNA methylation. *Nature* **2006**, *439*, 871–874. [CrossRef]
52. Chacon-Cabrera, A.; Fermoselle, C.; Salmela, I.; Yelamos, J.; Barreiro, E. MicroRNA expression and protein acetylation pattern in respiratory and limb muscles of Parp-1(−/−) and Parp-2(−/−) mice with lung cancer cachexia. *Biochim. Biophys. Acta* **2015**, *1850*, 2530–2543. [CrossRef]
53. Yang, Z.; Li, L.; Chen, L.; Yuan, W.; Dong, L.; Zhang, Y.; Wu, H.; Wang, C. PARP-1 mediates LPS-induced HMGB1 release by macrophages through regulation of HMGB1 acetylation. *J. Immunol.* **2014**, *193*, 6114–6123. [CrossRef]
54. Hassa, P.; Burki, C.; Lombardi, C.; Imhof, R.; Hottiger, M. Transcriptional coactivation of NF-kB by p300 is mediated by PARP-1. *Med. Sci. Monit.* **2003**, *9*, 27.
55. Vuong, B.; Hogan-Cann, A.D.; Alano, C.C.; Stevenson, M.; Chan, W.Y.; Anderson, C.M.; Swanson, R.A.; Kauppinen, T.M. NF-κB transcriptional activation by TNFα requires phospholipase C, extracellular signal-regulated kinase 2 and poly(ADP-ribose) polymerase-1. *J. Neuroinflamm.* **2015**, *12*, 229. [CrossRef]
56. Castri, P.; Lee, Y.J.; Ponzio, T.; Maric, D.; Spatz, M.; Bembry, J.; Hallenbeck, J. Poly(ADP-ribose) polymerase-1 and its cleavage products differentially modulate cellular protection through NF-kappaB-dependent signaling. *Biochim. Biophys. Acta* **2014**, *1843*, 640–651. [CrossRef] [PubMed]
57. Tripathi, P.; Aggarwal, A. NF-kB transcription factor: A key player in the generation of immune response. *Curr. Sci.* **2006**, *90*, 519–531.
58. Li, Q.; Verma, I.M. NF-kappaB regulation in the immune system. *Nat. Rev. Immunol.* **2002**, *2*, 725–734. [CrossRef]
59. Caamaño, J.; Hunter, C.A. NF-kappaB family of transcription factors: Central regulators of innate and adaptive immune functions. *Clin. Microbiol. Rev.* **2002**, *15*, 414–429. [CrossRef] [PubMed]
60. Lai, J.L.; Liu, Y.H.; Liu, C.; Qi, M.P.; Liu, R.N.; Zhu, X.F.; Zhou, Q.G.; Chen, Y.Y.; Guo, A.Z.; Hu, C.M. Indirubin Inhibits LPS-Induced Inflammation via TLR4 Abrogation Mediated by the NF-kB and MAPK Signaling Pathways. *Inflammation* **2017**, *40*, 1–12. [CrossRef] [PubMed]
61. Hunter, C.J.; De Plaen, I.G. Inflammatory signaling in NEC: Role of NF-κB, cytokines and other inflammatory mediators. *Pathophysiology* **2014**, *21*, 55–65. [CrossRef] [PubMed]
62. Ito, K.; Barnes, P.J.; Adcock, I.M. Glucocorticoid receptor recruitment of histone deacetylase 2 inhibits interleukin-1beta-induced histone H4 acetylation on lysines 8 and 12. *Mol. Cell. Biol.* **2000**, *20*, 6891–6903. [CrossRef]
63. Ashburner, B.P.; Westerheide, S.D.; Baldwin, A.S., Jr. The p65 (RelA) subunit of NF-kappaB interacts with the histone deacetylase (HDAC) corepressors HDAC1 and HDAC2 to negatively regulate gene expression. *Mol. Cell. Biol.* **2001**, *21*, 7065–7077. [CrossRef]
64. Demchenko, Y.N.; Kuehl, W.M. A critical role for the NFkB pathway in multiple myeloma. *Oncotarget* **2010**, *1*, 59–68. [CrossRef] [PubMed]
65. Zhou, Y.; Eppenberger-Castori, S.; Eppenberger, U.; Benz, C.C. The NFkappaB pathway and endocrine-resistant breast cancer. *Endocr.-Relat. Cancer* **2005**, *12* (Suppl. 1), S37–S46. [CrossRef]
66. Ilchovska, D.D.; Barrow, D.M. An Overview of the NF-kB mechanism of pathophysiology in rheumatoid arthritis, investigation of the NF-kB ligand RANKL and related nutritional interventions. *Autoimmun. Rev.* **2021**, *20*, 102741. [CrossRef]
67. Dolcet, X.; Llobet, D.; Pallares, J.; Matias-Guiu, X. NF-kB in development and progression of human cancer. *Virchows Arch. Int. J. Pathol.* **2005**, *446*, 475–482. [CrossRef]
68. Dumitriu, I.E.; Baruah, P.; Manfredi, A.A.; Bianchi, M.E.; Rovere-Querini, P. HMGB1: Guiding immunity from within. *Trends Immunol.* **2005**, *26*, 381–387. [CrossRef] [PubMed]

69. Park, J.S.; Arcaroli, J.; Yum, H.K.; Yang, H.; Wang, H.; Yang, K.Y.; Choe, K.H.; Strassheim, D.; Pitts, T.M.; Tracey, K.J.; et al. Activation of gene expression in human neutrophils by high mobility group box 1 protein. *Am. J. Physiol. Cell Physiol.* **2003**, *284*, C870–C879. [CrossRef]
70. Messmer, D.; Yang, H.; Telusma, G.; Knoll, F.; Li, J.; Messmer, B.; Tracey, K.J.; Chiorazzi, N. High mobility group box protein 1: An endogenous signal for dendritic cell maturation and Th1 polarization. *J. Immunol.* **2004**, *173*, 307–313. [CrossRef]
71. Pullerits, R.; Jonsson, I.M.; Verdrengh, M.; Bokarewa, M.; Andersson, U.; Erlandsson-Harris, H.; Tarkowski, A. High mobility group box chromosomal protein 1, a DNA binding cytokine, induces arthritis. *Arthritis Rheum.* **2003**, *48*, 1693–1700. [CrossRef]
72. Du, Q.; Miao, Y.; He, W.; Zheng, H. ADP-Ribosylation in Antiviral Innate Immune Response. *Pathogens* **2023**, *12*, 303. [CrossRef]
73. Decman, V.; Freeman, M.L.; Kinchington, P.R.; Hendricks, R.L. Immune control of HSV-1 latency. *Viral Immunol.* **2005**, *18*, 466–473. [CrossRef]
74. Divito, S.; Cherpes, T.L.; Hendricks, R.L. A triple entente: Virus, neurons, and CD8+ T cells maintain HSV-1 latency. *Immunol. Res.* **2006**, *36*, 119–126. [CrossRef]
75. Margolis, T.P.; Sedarati, F.; Dobson, A.T.; Feldman, L.T.; Stevens, J.G. Pathways of viral gene expression during acute neuronal infection with HSV-1. *Virology* **1992**, *189*, 150–160. [CrossRef]
76. Bloom, D.C.; Giordani, N.V.; Kwiatkowski, D.L. Epigenetic regulation of latent HSV-1 gene expression. *Biochim. Biophys. Acta* **2010**, *1799*, 246–256. [CrossRef] [PubMed]
77. Grady, S.L.; Hwang, J.; Vastag, L.; Rabinowitz, J.D.; Shenk, T. Herpes simplex virus 1 infection activates poly(ADP-ribose) polymerase and triggers the degradation of poly(ADP-ribose) glycohydrolase. *J. Virol.* **2012**, *86*, 8259–8268. [CrossRef] [PubMed]
78. Smith, S.; Weller, S.K. K. HSV-I and the cellular DNA damage response. *Future Virol.* **2015**, *10*, 383–397. [CrossRef] [PubMed]
79. Lanfranca, M.P.; Mostafa, H.H.; Davido, D.J. HSV-1 ICP0: An E3 Ubiquitin Ligase That Counteracts Host Intrinsic and Innate Immunity. *Cells* **2014**, *3*, 438–454. [CrossRef] [PubMed]
80. Giffin, L.; Damania, B. KSHV: Pathways to tumorigenesis and persistent infection. *Adv. Virus Res.* **2014**, *88*, 111–159. [CrossRef] [PubMed]
81. Ganem, D. KSHV infection and the pathogenesis of Kaposi's sarcoma. *Annu. Rev. Pathol.* **2006**, *1*, 273–296. [CrossRef]
82. Gregory, S.M.; West, J.A.; Dillon, P.J.; Hilscher, C.; Dittmer, D.P.; Damania, B. Toll-like receptor signaling controls reactivation of KSHV from latency. *Proc. Natl. Acad. Sci. USA* **2009**, *106*, 11725–11730. [CrossRef] [PubMed]
83. Gwack, Y.; Nakamura, H.; Lee, S.H.; Souvlis, J.; Yustein, J.T.; Gygi, S.; Kung, H.J.; Jung, J.U. Poly(ADP-ribose) polymerase 1 and Ste20-like kinase hKFC act as transcriptional repressors for gamma-2 herpesvirus lytic replication. *Mol. Cell. Biol.* **2003**, *23*, 8282–8294. [CrossRef]
84. Chung, W.C.; Park, J.H.; Kang, H.R.; Song, M.J. Downregulation of Poly(ADP-Ribose) Polymerase 1 by a Viral Processivity Factor Facilitates Lytic Replication of Gammaherpesvirus. *J. Virol.* **2015**, *89*, 9676–9682. [CrossRef]
85. Ohsaki, E.; Ueda, K.; Sakakibara, S.; Do, E.; Yada, K.; Yamanishi, K. Poly(ADP-ribose) polymerase 1 binds to Kaposi's sarcoma-associated herpesvirus (KSHV) terminal repeat sequence and modulates KSHV replication in latency. *J. Virol.* **2004**, *78*, 9936–9946. [CrossRef]
86. Tempera, I.; Lieberman, P.M. Epigenetic regulation of EBV persistence and oncogenesis. *Semin. Cancer Biol.* **2014**, *26*, 22–29. [CrossRef]
87. Young, L.S.; Murray, P.G. Epstein-Barr virus and oncogenesis: From latent genes to tumours. *Oncogene* **2003**, *22*, 5108–5121. [CrossRef] [PubMed]
88. Morgan, S.M.; Tanizawa, H.; Caruso, L.B.; Hulse, M.; Kossenkov, A.; Madzo, J.; Keith, K.; Tan, Y.; Boyle, S.; Lieberman, P.M.; et al. The three-dimensional structure of Epstein-Barr virus genome varies by latency type and is regulated by PARP1 enzymatic activity. *Nat. Commun.* **2022**, *13*, 187. [CrossRef] [PubMed]
89. Tempera, I.; Deng, Z.; Atanasiu, C.; Chen, C.J.; D'Erme, M.; Lieberman, P.M. Regulation of Epstein-Barr virus OriP replication by poly(ADP-ribose) polymerase 1. *J. Virol.* **2010**, *84*, 4988–4997. [CrossRef]
90. Deng, Z.; Lezina, L.; Chen, C.J.; Shtivelband, S.; So, W.; Lieberman, P.M. Telomeric proteins regulate episomal maintenance of Epstein-Barr virus origin of plasmid replication. *Mol. Cell* **2002**, *9*, 493–503. [CrossRef]
91. Deng, Z.; Atanasiu, C.; Zhao, K.; Marmorstein, R.; Sbodio, J.I.; Chi, N.W.; Lieberman, P.M. Inhibition of Epstein-Barr virus OriP function by tankyrase, a telomere-associated poly-ADP ribose polymerase that binds and modifies EBNA1. *J. Virol.* **2005**, *79*, 4640–4650. [CrossRef]
92. Tempera, I.; Wiedmer, A.; Dheekollu, J.; Lieberman, P.M. CTCF prevents the epigenetic drift of EBV latency promoter Qp. *PLoS Pathog.* **2010**, *6*, e1001048. [CrossRef]
93. Mattiussi, S.; Tempera, I.; Matusali, G.; Mearini, G.; Lenti, L.; Fratarcangeli, S.; Mosca, L.; D'Erme, M.; Mattia, E. Inhibition of Poly(ADP-ribose)polymerase impairs Epstein Barr Virus lytic cycle progression. *Infect. Agents Cancer* **2007**, *2*, 18. [CrossRef]
94. Hulse, M.; Caruso, L.B.; Madzo, J.; Tan, Y.; Johnson, S.; Tempera, I. Poly(ADP-ribose) polymerase 1 is necessary for coactivating hypoxia-inducible factor-1-dependent gene expression by Epstein-Barr virus latent membrane protein 1. *PLoS Pathog.* **2018**, *14*, e1007394. [CrossRef]
95. Martin, K.A.; Lupey, L.N.; Tempera, I. Epstein-Barr Virus Oncoprotein LMP1 Mediates Epigenetic Changes in Host Gene Expression through PARP1. *J. Virol.* **2016**, *90*, 8520–8530. [CrossRef]

96. McIntosh, M.T.; Koganti, S.; Boatwright, J.L.; Li, X.; Spadaro, S.V.; Brantly, A.C.; Ayers, J.B.; Perez, R.D.; Burton, E.M.; Burgula, S.; et al. STAT3 imparts BRCAness by impairing homologous recombination repair in Epstein-Barr virus-transformed B lymphocytes. *PLoS Pathog.* **2020**, *16*, e1008849. [CrossRef] [PubMed]
97. Xu, H.; Akinyemi, I.A.; Haley, J.; McIntosh, M.T.; Bhaduri-McIntosh, S. ATM, KAP1 and the Epstein-Barr virus polymerase processivity factor direct traffic at the intersection of transcription and replication. *Nucleic Acids Res.* **2023**, *51*, 11104–11122. [CrossRef] [PubMed]
98. Ichikawa, T.; Okuno, Y.; Sato, Y.; Goshima, F.; Yoshiyama, H.; Kanda, T.; Kimura, H.; Murata, T. Regulation of Epstein-Barr Virus Life Cycle and Cell Proliferation by Histone H3K27 Methyltransferase EZH2 in Akata Cells. *mSphere* **2018**, *3*, e00478-18. [CrossRef] [PubMed]
99. Goodrum, F. Human Cytomegalovirus Latency: Approaching the Gordian Knot. *Annu. Rev. Virol.* **2016**, *3*, 333–357. [CrossRef] [PubMed]
100. Brune, W.; Andoniou, C.E. Die Another Day: Inhibition of Cell Death Pathways by Cytomegalovirus. *Viruses* **2017**, *9*, 249. [CrossRef] [PubMed]
101. Fliss, P.M.; Brune, W. Prevention of cellular suicide by cytomegaloviruses. *Viruses* **2012**, *4*, 1928–1949. [CrossRef]
102. Zhang, W.; Guo, J.; Chen, Q. Role of PARP-1 in Human Cytomegalovirus Infection and Functional Partners Encoded by This Virus. *Viruses* **2022**, *14*, 2049. [CrossRef] [PubMed]
103. Gordon-Shaag, A.; Yosef, Y.; Abd El-Latif, M.; Oppenheim, A. The abundant nuclear enzyme PARP participates in the life cycle of simian virus 40 and is stimulated by minor capsid protein VP3. *J. Virol.* **2003**, *77*, 4273–4282. [CrossRef] [PubMed]
104. Eki, T.; Hurwitz, J. Influence of poly(ADP-ribose) polymerase on the enzymatic synthesis of SV40 DNA. *J. Biol. Chem.* **1991**, *266*, 3087–3100. [CrossRef] [PubMed]
105. Atasheva, S.; Akhrymuk, M.; Frolova, E.I.; Frolov, I. New PARP gene with an anti-alphavirus function. *J. Virol.* **2012**, *86*, 8147–8160. [CrossRef]
106. Fehr, A.R.; Jankevicius, G.; Ahel, I.; Perlman, S. Viral Macrodomains: Unique Mediators of Viral Replication and Pathogenesis. *Trends Microbiol.* **2018**, *26*, 598–610. [CrossRef]
107. Brosey, C.A.; Houl, J.H.; Katsonis, P.; Balapiti-Modarage LP, F.; Bommagani, S.; Arvai, A.; Moiani, D.; Bacolla, A.; Link, T.; Warden, L.S.; et al. Targeting SARS-CoV-2 Nsp3 macrodomain structure with insights from human poly(ADP-ribose) glycohydrolase (PARG) structures with inhibitors. *Prog. Biophys. Mol. Biol.* **2021**, *163*, 171–186. [CrossRef]
108. Alhammad, Y.M.O.; Kashipathy, M.M.; Roy, A.; Gagné, J.P.; McDonald, P.; Gao, P.; Nonfoux, L.; Battaile, K.P.; Johnson, D.K.; Holmstrom, E.D.; et al. The SARS-CoV-2 Conserved Macrodomain Is a Mono-ADP-Ribosylhydrolase. *J. Virol.* **2021**, *95*, e01969-20. [CrossRef]
109. Frick, D.N.; Virdi, R.S.; Vuksanovic, N.; Dahal, N.; Silvaggi, N.R. Molecular Basis for ADP-Ribose Binding to the Mac1 Domain of SARS-CoV-2 nsp3. *Biochemistry* **2020**, *59*, 2608–2615. [CrossRef]
110. Alhammad, Y.M.; Parthasarathy, S.; Ghimire, R.; O'Connor, J.J.; Kerr, C.M.; Pfannenstiel, J.J.; Chanda, D.; Miller, C.A.; Unckless, R.L.; Zuniga, S.; et al. SARS-CoV-2 Mac1 is required for IFN antagonism and efficient virus replication in mice. *Proc. Natl. Acad. Sci. USA* **2023**, *120*, e2302083120. [CrossRef]
111. Muthumani, K.; Choo, A.Y.; Zong, W.X.; Madesh, M.; Hwang, D.S.; Premkumar, A.; Thieu, K.P.; Emmanuel, J.; Kumar, S.; Thompson, C.B.; et al. The HIV-1 Vpr and glucocorticoid receptor complex is a gain-of-function interaction that prevents the nuclear localization of PARP-1. *Nat. Cell Biol.* **2006**, *8*, 170–179. [CrossRef] [PubMed]
112. Ha, H.C.; Juluri, K.; Zhou, Y.; Leung, S.; Hermankova, M.; Snyder, S.H. Poly(ADP-ribose) polymerase-1 is required for efficient HIV-1 integration. *Proc. Natl. Acad. Sci. USA* **2001**, *98*, 3364–3368. [CrossRef] [PubMed]
113. Ariumi, Y.; Turelli, P.; Masutani, M.; Trono, D. DNA damage sensors ATM, ATR, DNA-PKcs, and PARP-1 are dispensable for human immunodeficiency virus type 1 integration. *J. Virol.* **2005**, *79*, 2973–2978. [CrossRef] [PubMed]

pathogens

Review

PARPs and ADP-Ribosylation in Chronic Inflammation: A Focus on Macrophages

Diego V. Santinelli-Pestana [1], Elena Aikawa [1,2], Sasha A. Singh [1] and Masanori Aikawa [1,2,3,*]

1 Center for Interdisciplinary Cardiovascular Sciences, Division of Cardiovascular Medicine, Department of Medicine, Brigham and Women's Hospital, Harvard Medical School, Boston, MA 02115, USA; dsantinellipestana@partners.org (D.V.S.-P.); eaikawa@bwh.harvard.edu (E.A.); sasingh@bwh.harvard.edu (S.A.S.)

2 Center for Excellence in Vascular Biology, Division of Cardiovascular Medicine, Department of Medicine, Brigham and Women's Hospital, Harvard Medical School, Boston, MA 02115, USA

3 Channing Division of Network Medicine, Department of Medicine, Brigham and Women's Hospital, Harvard Medical School, Boston, MA 02115, USA

* Correspondence: maikawa@bwh.harvard.edu; Tel.: +617-730-7777

Abstract: Aberrant adenosine diphosphate-ribose (ADP)-ribosylation of proteins and nucleic acids is associated with multiple disease processes such as infections and chronic inflammatory diseases. The poly(ADP-ribose) polymerase (PARP)/ADP-ribosyltransferase (ART) family members promote mono- or poly-ADP-ribosylation. Although evidence has linked PARPs/ARTs and macrophages in the context of chronic inflammation, the underlying mechanisms remain incompletely understood. This review provides an overview of literature focusing on the roles of PARP1/ARTD1, PARP7/ARTD14, PARP9/ARTD9, and PARP14/ARTD8 in macrophages. PARPs/ARTs regulate changes in macrophages during chronic inflammatory processes not only via catalytic modifications but also via non-catalytic mechanisms. Untangling complex mechanisms, by which PARPs/ARTs modulate macrophage phenotype, and providing molecular bases for the development of new therapeutics require the development and implementation of innovative technologies.

Keywords: immunity; mass spectrometry; proteomics; ADP-ribosylation; poly(ADP-ribose) glycohydrolase; Diphtheria toxin-like ADP-ribosyltransferases; chronic infection; arboviruses; cardiovascular disease; emphysema; alcoholic liver disease; SARS-CoV-2; host–pathogen interactions

Citation: Santinelli-Pestana, D.V.; Aikawa, E.; Singh, S.A.; Aikawa, M. PARPs and ADP-Ribosylation in Chronic Inflammation: A Focus on Macrophages. *Pathogens* **2023**, *12*, 964. https://doi.org/10.3390/pathogens12070964

Academic Editor: Lawrence S. Young

Received: 1 April 2023
Revised: 25 June 2023
Accepted: 15 July 2023
Published: 23 July 2023

1. Introduction

Poly(ADP-ribose) polymerases (PARPs), or ADP-ribosyltransferases (ARTs), catalyze the covalent transfer of ADP-ribose (ADPr) groups from NAD+ onto target biological macromolecules including nucleic acids (DNA, mRNA), transcription factors (e.g., NF-κB), or enzymes (e.g., PARP1/ARTD1 auto-ADP-ribosylation) [1]. The process of adding a single ADPr moiety is known as mono-ADP-ribosylation (MARylation), whereas adding multiple ADPr moieties is known as poly-ADP-ribosylation (PARylation); the latter occurs in a sequential way, starting with the transfer of one ADPr unit followed by the transfer of additional ADPr units onto a growing chain. The PARP enzyme family comprises 17 members, with variable functionality. PARP1/ARTD1, PARP2/ARTD2, PARP5a/TNKS1/ARTD5, and PARP5b/TNKS2/ARTD6 are poly-ARTs and have PARylation activity, while PARP3/ARTD3, PARP4/ARTD4, PARP6/ARTD17, PARP7/ARTD14, PARP8/ARTD16, PARP9/ARTD9, PARP10/ARTD10, PARP11/ARTD11, PARP12/ARTD12, PARP14/ARTD8, PARP15/ARTD7, and PARP16/ARTD15 are mono-ARTs and have MARylation activity [2]. PARP13/ARTD13 is catalytically inactive [3]. The poly-ARTs' catalytic activity is counterbalanced by poly(adenosine diphosphate-ribose)-glycohydrolase (PARG) that hydrolyzes PARylation to MARylation. PARG is completely unable to hydrolyze the MAR covalently attached to proteins [4], however, the biological significance of this limitation remains unclear [5]. Nonetheless, this PARG enzymatic property is convenient for mass spectrometry-enabled ribosylome profiling (more below), since only the MARylated form of the modification is conducive to mass spectrometric analysis [6]. The mono-ARTs' catalytic activity can be counterbalanced by enzymes other than PARG, such as ADP-ribosylhydrolase 3 (ARH3) [7,8], terminal ADP-ribose protein glycohydrolase (TARG)/C6orf130 [9,10], and MacroD1 [11] and MacroD2 [12], which are able to hydrolyze the MAR attached to proteins, functioning as mono-ADP-ribosylhydrolases. PARPs/ARTs also orchestrate biological processes via non-catalytic activities, such as directly binding to nuclear DNA or binding to transcription factors, but these roles remain to be further explored [13].

The subcellular locations of the PARPs/ARTs also dictate their biological functions. In a comprehensive analysis of human somatic cell lineages, Vyas et al. used N-terminal green fluorescent protein (GFP) and affinity-purified peptide antibodies to study the cellular localization of PARPs/ARTs and the occurrence of PARylation during the cell cycle [14]. PARP1/ARTD1 localized to the nucleus; PARP5a/TNKS1/ARTD5, PARP5b/TNKS2/ARTD6, PARP12/ARTD12, PARP13/ARTD13, PARP6/ARTD17, PARP8/ARTD16, PARP10/ARTD10, and PARP16/ARTD15 localized to the cytoplasm; and PARP2/ARTD2, PARP3/ARTD3, PARP7/ARTD14, PARP9/ARTD9, PARP14/ARTD8, PARP4/ARTD4, and PARP11/ARTD11 localized to the nucleus and cytoplasm. Their findings also suggested that: firstly, the expression of most PARPs/ARTs was pervasive across human tissues; secondly, while PARPs/ARTs could be found in the nucleus and in the cytoplasm, they were predominantly found in the cytoplasm; thirdly, PAR levels were influenced by the cell cycle, and the proportion of PAR identified in the nucleus versus cytoplasm changed during the cell cycle. Leung et al. demonstrated that PARP5a/TNKS1/ARTD5, PARP12/ARTD12, two isoforms of PARP13/ARTD13, PARP14/ARTD8, and PARP15/ARTD7 coordinate the assembly of stress granules in the cytoplasm, modifying each other within this cellular compartment [15]. Ryu et al. demonstrated that low concentrations of NAD+ can limit PARP1/ARTD1 activity in the nucleus [16]. The influx or efflux of NAD+ thus interferes with PARP1/ARTD1's activity, leading to alternative gene expression signatures in the early process of adipogenesis [17]. Additionally, the predominant type of ADP-ribosylation in distinct cellular compartments seems to vary: PARylation appears to occur primarily in the nucleus [18], whereas MARylation in the cytoplasm [19]. These examples illustrate that PARPs/ARTs' functions vary based on ADP-ribosylation catalytic activity, cellular compartment location, and the physiological and/or pathological microenvironment.

PARP/ART biology has been studied in the context of the innate immune system, with a particular focus on macrophages [20]. PARPs/ARTs were associated with biological responses mediated by IFN-γ, TNF-α, IL-1β, IL-6, and NF-κB, such as host–pathogen interactions in viral infections, vascular inflammation, and others [20]. In 1985, Singh et al. generated DNA double-strand breaks to induce PARylation in human monocytes [21]; and in 1991, Berton et al. reported that PARylation levels increased after IFN-γ stimulation in human macrophages [22]. More recently, Heer et al. demonstrated that the catalytic activities of PARP7/ARTD14, PARP10/ARTD10, PARP12/ARTD12, and PARP14/ARTD8 were closely connected to nicotinamide and derivates in the establishment of cellular innate immune response during COVID-19 infection [23]. In another pathological setting, Wang et al. reported that PARP1/ARTD1 and PARP2/ARTD2 inhibition with olaparib [24–27] (a PARP1/ARTD1/2 inhibitor approved by the Food and Drug Administration for the treatment of ovarian, breast, pancreatic, and prostate cancer) induced macrophage reprogramming towards an anti-tumor, pro-inflammatory phenotype [28]. Macrophages are found within the microenvironments of solid tumors [29] and chronic inflammatory conditions such as diabetes [30], neurodegenerative diseases [31], prolonged bacterial infections [32] but exhibit distinct functions. In tumors, they are associated with an anti-tumor innate immune response [33] but are hypothesized to promote and sustain a pro-inflammatory tissue milieu [20]. To date, research has focused primarily on the role of PARPs/ARTs in cancer biology, whereas their roles in other macrophage-driven pathologies are only just beginning to be explored.

As we will highlight further below, recent studies have begun to investigate more the role(s) of PARPs/ARTs and ADP-ribosylation in macrophage activation, with the aim to identify therapeutic avenues for acute and chronic inflammation. Olaparib is already available for the treatment of cancer and is also being investigated as a potential therapy for pulmonary arterial hypertension (Clinical Trial No.: NCT03782818), although these applications focus on DNA damage and repair features of PARP1/ARTD1 inhibition. Nevertheless, a phase-I trial has been initiated to test the PARP14/ARTD8 inhibitor, RBN-3143, as a potential therapy for atopic dermatitis (Clinical Trial No.: NCT05215808), focusing on inflammation control. Macrophages are key players in the sustained inflammation occurring in atopic dermatitis [34,35], thus representing the initiative to translate the interplay between ADP-ribosylation and macrophage biology to the clinic.

2. PARylation, Macrophages, and Chronic Inflammation

Among the PARPs/ARTs mediating PARylation during inflammation, PARP1/ARTD1 is the most studied [36,37]. Various stimuli promote PARP1/ARTD1 activity in macrophages, often leading to the expression of pro-inflammatory genes and downstream inflammatory responses. In the current section, we will review the most recent articles exploring the distinct mechanisms of action of PARP1/ARTD1 in macrophage activity in the setting of chronic or prolonged inflammation.

2.1. ADP-Ribosylation and DNA Damage

The catalytic activity of PARP1/ARTD1 increases with DNA damage following genotoxic stimuli. Dawicki-McKenna et al. used hydrogen/deuterium exchange–mass spectrometry to demonstrate that breaks in the DNA strand led to structural changes in PARP1/ARTD1′s helical subdomain (HD), which is part of the catalytic domain [38]. The helical subdomain functions as an autoinhibitory portion of the catalytic domain, unfolding in the presence of DNA strand breaks and thus promoting PARP1/ARTD1′s catalytic activity. Eustermann et al. [39] demonstrated that a sequential multidomain unfolding occurs in PARP1/ARTD1 in response to DNA single-strand breaks (SSBs). Firstly, the F2 domain recognizes and detects SSBs; secondly, the F1 domain binds to the complex, exposing the 5′ cryptic site and orienting the assembly of remaining PARP1/ARTD1 domains; thirdly, the F3, WGR, and CAT domains also bind the exposed strand, culminating in the unfolding of the autoinhibitory helical subdomain. This cooperative process generates a specific recognition of sites of SSBs by PARP1/ARTD1, promoting PAR-mediated signaling and modulation of chromatin structure upon DNA damage. Figure 1 provides a graphical representation [40] of PARP1/ARTD1 domains and their structure, as well as a flowchart indicating the dual action of PARP1/ARTD1 during inflammation.

These works also aided in the paradox involving PARP1/ARTD1 *cis* versus *trans* (another PARP1/ARTD1 molecule) modification during response to DNA damage. While PARP1/ARTD1 dimers have been reported [41–43], suggesting that the *trans* modification occurs, results from Dawicki-McKenna et al. [38] and Eustermann et al. [39] indicated that PARP1/ARTD1 automodifies itself, unless two DNA binding sites are closely adjacent, leading to *trans* modification activity.

More recently, other reports have described the dynamic nature of the interactions between PARP1/ARTD1 and DNA [44], either using its DNA-binding domain (DBD) along with zinc finger domains I and II (ZI and ZII, respectively) for short-term interactions [45], or using histone H4, which leads to a prolonged interaction with the DNA strand [46]. Short-term interactions between the DBD of PARP1/ARTD1 and DNA were associated with activation of DNA repair pathways at specific stages of DNA damage, while long-term interactions between the C-terminal domain of PARP1/ARTD1 and histone H4 were associated with promotion of gene expression [44]. This dual action of PARP1/ARTD1 on DNA illustrates the complexity of this enzyme and provides indications that PARP1/ARTD1 may be associated with chronic inflammation not only as a repair mechanism secondary to inflammation-driven DNA damage [47] but also promoting the expression of pro-inflammatory and/or anti-inflammatory genes.

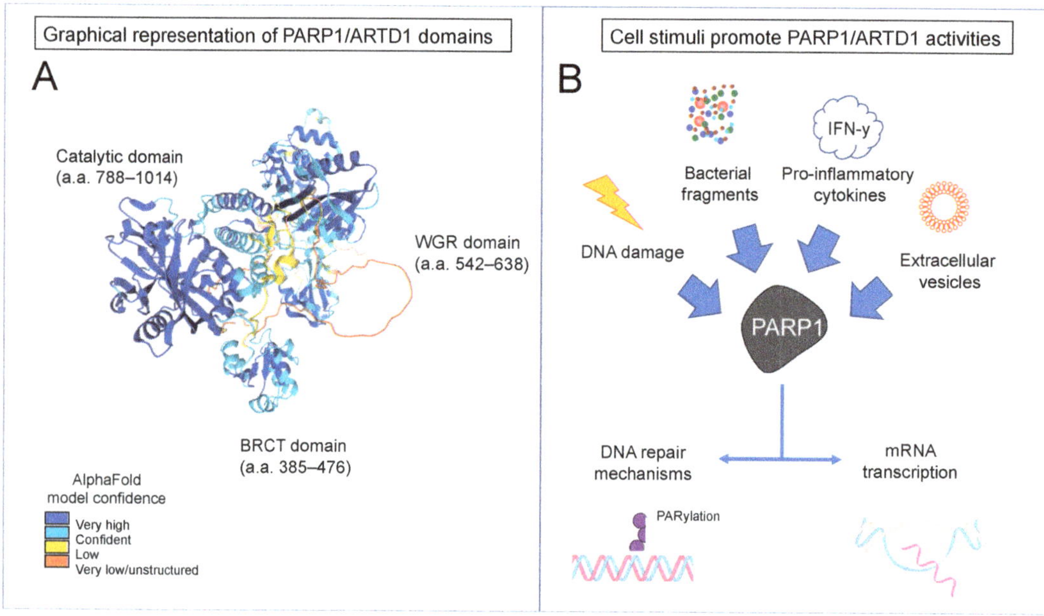

Figure 1. Graphical representation of PARP1/ARTD1 and its role during inflammation. (**A**): Tridimensional representation of PARP1/ARTD1 structure, indicating the catalytic, the WGR, and the BRCT domains. (**B**): Flowchart depicting the cell stimuli able to promote PARP1/ARTD1 activity (catalytic and non-catalytic) during prolonged inflammatory processes, leading to PARylation and transcription of mRNA.

Reactive oxygen species (ROS) generated during pro-inflammatory responses lead to DNA damage [48], triggering short-term PARP1/ARTD1-DNA interactions [44,45]. However, in a study using a model of elastase-induced emphysema and chronic lung inflammation in mice, prolonged inhibition of PARP1/ARTD1 with olaparib reduced the number of macrophages in the bronco-alveolar lavage after 21 days of treatment when compared with the control group [49]. Levels of ROS and malondialdehyde (MDA, a marker of lipid peroxidation) increased in lung tissues of the control group (four-fold and seven-fold, respectively) due to the inflammation and macrophage activity induced by elastase, but daily treatment with olaparib restored ROS and MDA to normal levels, indicating an improvement in the inflammatory and redox balances [49,50]. These results exemplify how PARP1/ARTD1 may have a dual and contrasting role in chronic inflammation, repairing DNA following ROS while promoting the production of ROS in macrophages.

2.2. PARP1/ARTD1 Promotes Transcription of Pro-Inflammatory and Apoptosis-Related Genes

Inhibition of PARP1/ARTD1 ameliorates inflammation in chronic conditions and innate immune responses, and this effect was found in multiple pathologies driven by long-term inflammatory processes. Kunze et al. [51] demonstrated that stimulation of bone marrow-derived monocytes (BMDMs) from genetically modified mice expressing catalytically inactive PARP1/ARTD1 induced the expression of a pro-inflammatory signature of almost 2500 genes, including genes regulating IL-12, IFN-γ, and TNF-α production. In the same study, they reported that mice transplanted with catalytically inactive PARP1/ARTD1 myeloid progenitors were colonized by H. pylori at higher levels when compared to their control littermates [51], suggesting that PARP1/ARTD1 contributed to controlling gastric bacterial colonization. In another disease model, inhibition of PARP1/ARTD1 with 3-aminobenzamide, an anti-inflammatory compound classically used for PARP1/ARTD1 inhibition [52], improved rectal hemorrhage, blood sugar levels, blood IL-1β levels, weight loss, and the histological score of colonic sections in mice with colitis-associated diabetes [53]. Similar findings were reported by Kovács et al. [54] after using olaparib to inhibit PARP1/ARTD1 activity in a mouse model of Crohn's disease (a type of inflammatory bowel disease). They found that olaparib increased the levels of IL-10, while it suppressed the concentration of IL-1β and IL-6 [54]. Also, olaparib generated a reduction in the number of monocytes in the blood of treated mice when compared with controls [54]. Gupte et al. [55] stimulated BMDMs from wild-type and PARP1/ARTD1-deficient mice, demonstrating that PARP1/ARTD1-mediated STAT1-α PARylation influenced the transcriptional program upon IFN-γ stimulation [55].

The regulation of PARPs/ARTs' catalytic activities in chronic inflammation also relates to NAD+ metabolism. Gerner et al. [56] inhibited nicotinamide phosphoribosyltransferase (NAMPT), a rate-limiting enzyme in the NAD+ salvage pathway, to reduce NAD+ levels in human cells and mice with intestinal colitis. They found that depletion of NAD+ reduced PARP1/ARTD1 catalytic activity, suppressed the expression of IL-6, IL-1β, and TNF-α, and skewed monocytes/macrophages from pro-inflammatory towards anti-inflammatory phenotypes [56]. In the same line, reduction of NAMPT-derived NAD+ via pharmacological inhibition of NAMPT reduced the pathological changes in psoriasis [57] and atopic dermatitis [58] and diminished the expression of pro-inflammatory biomarkers.

In addition to promoting cytokine/chemokine gene expression, PARP1/ARTD1 also influences the cellular fate in apoptosis [59], a fundamental element of inflammation [60]. PARP1/ARTD1 has been extensively associated with caspases in a mechanism known as parthanatos [61,62] (not reviewed in this manuscript). For instance, Zhang et al. [63] analyzed cleaved caspase 3 in liver samples from mice with chronic alcoholic liver injury [63], and demonstrated that pharmacological inhibition with PJ-34 [64] or genetic depletion of PARP1/ARTD1 decreased the number of cleaved caspase 3-positive cells in diseased livers when compared to controls. They found that long-term ethanol consumption promoted PARP/ART activation, hepatic steatosis, and intense cytokine expression in liver samples, while in vivo pharmacologic inhibition of PARP1/ARTD1 with PJ-34 attenuated triglyceride content and serum alanine transaminase levels in liver, suggesting a milder injury phenotype [63]. Erener et al. [65] also identified an association between caspase 1, caspase 7, and PARP1/ARTD1. They found that stimulation with LPS promoted the translocation of caspase 7 to the nucleus (mediated by caspase 1 and NLRP3 inflammasome activation), where it cleaves PARP1/ARTD1 at the caspase cleavage site D214, generating free PARP1/ARTD1 fragments, decondensation of chromatin, and expression of NF-κB dependent-genes. They generated human THP-1 cells expressing non-cleavable PARP1/ARTD1, stimulated them with LPS, and compared them with genetically unmodified controls, confirming that caspase 7 cleaved PARP1/ARTD1 mostly at the D214 site. Martínez-Morcillo et al. [66] found that PARP1/ARTD1 activation leads to skin inflammation and cell death via parthanatos-mediated apoptosis in psoriasis, and pharmacological inhibition of NAMPT decreased the expression of genes associated with psoriasis.

Together, those findings suggest that PARP1/ARTD1 can influence gene expression during chronic inflammation via ADP-ribosylation of macromolecules and can initiate apoptosis upon interaction with caspases. Controlling NAD+ levels via NAMPT regulation in such immune responses may be a potential source of new targets to suppress pathogenesis derived from ADP-ribosylation, although a deeper understanding of these mechanisms is still needed.

2.3. PARP1/ARTD1 Mediates Host–Pathogen Interactions in Chagas Heart Disease

Chagas heart disease is caused by the protozoan parasite *Trypanosoma cruzi* (*T. cruzi*). The classical phenotype seen in this condition is the result of chronic (years to decades) of sustained myocyte inflammation, oxidative stress, and macrophage infiltration into cardiac muscle [67,68]. Ba et al. [69] demonstrated that *T. cruzi* infection of cardiomyocytes leads to mitochondrial production of ROS that diffuse to the cytosol and nucleus, leading to DNA damage and PARP1/ARTD1 activation. As a result, the expression of genes related to pro-inflammatory cytokines increased either due to the interaction between ROS and cytosolic NF-κB or due to PARP1/ARTD1-mediated PARylation of proteins that interact with RelA(p65) (an NF-κB subunit). Further evidence indicated that depletion of PARP1/ARTD1 (with genetic deletion or PJ-34 administration) in infected mice prevented cardiac hypertrophy and left ventricle dysfunction and restored the mitochondrial antioxidant/oxidant balance [70]. PARP1/ARTD1 is associated with chromatin during *T. cruzi* infection but its mRNA levels did not change when compared to non-infected states, indicating that a translocation of PARP1/ARTD1 to chromatin-dense regions occurred [71,72]. These results suggest that PARP1/ARTD1 influences the response to mitochondrial stress during *T. cruzi* infection. Evidence also connects PARP1/ARTD1 to macrophages in the host–pathogen interaction. Macrophage-like RAW264.7 cells treated with extracellular vesicles (EVs) derived from infected mouse plasma released higher levels of TNF-α, IL-1β, and IL-6 than did control cells [73]. EVs derived from *T. cruzi*-infected RAW264.7 cells induced lower expression levels of TNF-α, IL-1β, and IL-6 in BMDMs harvested from PARP1/ARTD1-deficient mice compared to wild-type control [73]. Thus, it is possible that the previously described role of macrophages in Chagas heart disease [74] may be mediated by PARP1/ARTD1, but more studies are needed.

2.4. *PARP1/ARTD1 in Cardiovascular Inflammation*

Von Lukowicz et al. proposed that PARP1/ARTD1 mediates macrophage adhesion to endothelial cells in the process of atherogenesis [75]. Both PARP1/ARTD1 and PARP2/ARTD2 inhibition with PJ-34 and PARP1/ARTD1 genetic deletion without PJ-34 reduced plaque formation and the expression of adhesion molecules such as E-selectin, P-selectin, VCAM1, and iNOS. Another study linked high glucose and PARP1/ARTD1 levels in streptozocin-induced diabetes mellitus in apolipoprotein E-deficient mice [76]. In a rat model of cerebral aneurysms, treatment with 3-aminobenzamide, an anti-inflammatory compound classically used for PARP1/ARTD1 inhibition [52], decreased macrophage accumulation and PARP1/ARTD1 expression [77]. These studies indicate that different forms of inflammatory arterial injury (i.e., atherosclerosis, aneurysm formation, and hyperglycemia-induced inflammation) share PARP1/ARTD1 as a common mediator of the inflammatory process.

3. MARylation, Macrophages, and Chronic Inflammation

Although most studies to date have focused on PARylation and PARP1/ARTD1, evidence suggests that MARylation also regulates macrophage activation, inflammation, and host–pathogen interactions. For instance, in an evolutionary analysis of the PARP/ART genes, Daugherty et al. [78] demonstrated that PARP9/ARTD9, PARP14/ARTD8, and PARP15/ARTD7 had signs of genetic adaptation in primates, notably in their macrodomains, and evolved under positive selection. In another example, our own research demonstrated that PARP9/ARTD9 and PARP14/ARTD8 regulate pro-inflammatory activation of macrophages upon stimulation [79,80]. Therefore, considering the accumulated evidence that MARylation and mono-PARPs/ARTs are involved with the innate immune system, in this section we will review recent articles investigating the interplay between PARPs/ARTs, MARylation, and macrophage activation and explore how these findings provide insight into mechanisms that drive chronic inflammation and host–pathogen interactions.

3.1. *PARP7/ARTD14 Mediates Epithelial Inflammation*

In a mouse model of a dextran sodium sulfate-induced ulcerative colitis study, PARP7/ARTD14 deletion increased mRNA levels of IL-1β, IL-6, IL-17, and Lcn2 and decreased survival rate [81]. Aryl hydrocarbon receptor (AHR), which induces the expression of PARP7/ARTD14, mediates pro-inflammatory responses in this model. The PARP7/ARTD14 catalytic domain MARylates AHR, which represses AHR signaling in a negative feedback loop. AHR responsiveness was enhanced by short-chain fatty acids in mouse colonocytes [81], supporting the hypothesis that chronic inflammation related to toxic lipid particles in cells of epithelial origin involves PARP7/ARTD14 [82].

3.2. PARP9/ARTD9 Mediates Viral and Bacterial Host–Pathogen Interactions

In a cohort with patients infected with pulmonary tuberculosis (TB) and healthy controls, Chen et al. [83] identified an inversely proportional association between TB infection severity and methylation status of PARP9/ARTD9 DNA in PBMCs extracted from participants. Severe TB clinical phenotypes were associated with hypomethylation of the PARP9/ARTD9 gene, suggesting that lower expression of PARP9/ARTD9 may lead to impaired innate response to TB infection in individuals with that epigenotype. Novel data from Thirunavukkarasu et al. [84] further support this hypothesis. They reported that PARP9/ARTD9 mRNA was increased in humans and mice infected with TB, and *Parp9-/-* mice were more susceptible to TB infection and developed more severe phenotypes compared to controls.

Similarly, PARP9/ARTD9 appears to be involved in innate immune responses against RNA viruses. Xing et al. [85] demonstrated that PARP9/ARTD9 is able recognize and bind RNA virus in human and mouse dendritic cells and macrophages, deploying an IFN-mediated response independent of the mitochondrial anti-viral signaling (a major mechanism for recognizing RNA viruses during infection). Furthermore, *Parp9-/-* deletion made mice more susceptible to RNA virus infection [85], reinforcing that PARP9/ARTD9 participates in the host–pathogen interactions. Curiously, PARP9/ARTD9 was associated with persistent hepatitis B virus (HBV) infection in a transcriptome-wide association study, in which chronic HBV carriers had increased expression of PARP9/ARTD9 when compared to non-infected individuals [86]. HBV is a DNA virus with unique features that approximate it to RNA viruses [87], which may relate to the results above (PARP9/ARTD9 acting as a recognizer of viral RNA).

3.3. PARP14/ARTD8 Mediates Chronic Inflammation and Response to Arboviruses

Recent data indicate that PARP14/ARTD8 participates in the establishment of an immune response to arboviruses. Eckei et al. [88] reported that the macrodomains of Chikungunya virus (a positive single-strand RNA virus) have strong hydrolase activity on proteins that were ADP-ribosylated by PARP10/ARTD10, PARP14/ARTD8, and PARP15/ARTD7. Fernandez et al. [89] reported that Zika virus infection in human PBMCs induced the expression of PARP14, IL-6, CCL8, CXCL1, and CXCL5, suggesting that the infection promoted changes in the transcriptional and post-transcriptional levels. These results indicate that PARP14/ARTD8 influences the host–pathogen dynamic in arbovirus infections.

3.4. PARP9/ARTD9 and PARP14/ARTD8 Mediate Macrophage Activation in Atherosclerosis

PARP14/ARTD8 is also important in other chronic inflammatory responses. Using a systems approach based on unbiased network analysis and artificial intelligence, our previous studies discovered PARP14/ARTD8 and PARP9/ARTD9 as potential molecular switches of macrophage activation [79,80]. Proteome analyses from stimulated and non-stimulated human and mouse macrophage-like cells detected an increase in the ADP-ribosylated PARP14/ARTD8 and PARP9/ARTD9 peptide levels upon stimulation with IFN-γ, and network analysis identified a close link between those PARPs/ARTs and the human coronary artery disease gene module [79,80]. Additional in vitro experiments indicated that PARP9/ARTD9 and PARP14/ARTD8 may function upstream of pro-inflammatory STAT1 and anti-inflammatory STAT6 signaling pathways, respectively [79,80]. Iqbal et al. [90] reported that macrophages from PARP14/ARTD8-deficient mice express higher levels of tissue factor mRNA and protein than do wild-type mice [90]. Mehrotra et al. [91] reported that PARP14/ARTD8 specifically binds to STAT6, regulating its promoter activity upon stimulation with IL-4, and demonstrated that this interaction is dependent on PARP14/ARTD8 catalytic domain [91]. Figure 2 provides a summary of the different disease models mediated by PARPs/ARTs and macrophages in chronic inflammation.

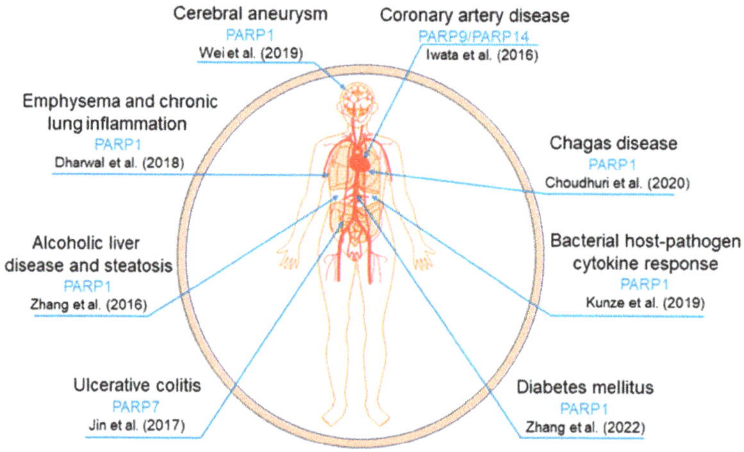

Figure 2. Representation of the disease models influenced by the interplay between PARPs/ARTs and macrophages in chronic inflammation [49,51,63,73,76,77,79,81].

4. SARS-CoV-2, ADP-Ribosylation, and Innate Immune Response

ADP-ribosylation and PARPs/ARTs are important in viral host–pathogen interactions and in the organization of the host's innate immune response [92]. PARP/ART genes are interferon-stimulated genes [93,94]. Stress granule formation is a major effect of interferon stimulation during innate immune responses, and it is closely related to PARPs/ARTs and ADP-ribosylation [95–97]. PARPs/ARTs and PARG probably mediate the assembly and maintenance of stress granules in a dynamic way: PARylation of stress granule proteins increases in stress conditions or with PARG silencing [15,98]. Together, these results suggest that hydrolysis of PAR/MAR could limit the effectiveness of the host's innate immune response against viruses.

Even before the severe acute respiratory syndrome coronavirus (SARS-CoV)-2 pandemic, different types of coronaviruses' macrodomains belonging to non-structural protein 3 (nsp3) were identified as ADP-ribose-binding modules [99]. In 2006, Egloff et al. reported that the crystal structure of the SARS-CoV macrodomain associates with ADP-ribose, being able to bind PAR and to function as an ADP-ribose 1″-phosphatase [100]. Evidence suggests that nsp3 and its macrodomains were part of coronaviruses' virulence mechanisms [101–103], promoting virus replication and suppressing interferon-mediated host responses (e.g., stress granule formation) [104–106].

With the onset of the pandemic, data connecting the new SARS-CoV-2 macrodomains and ADP-ribosylation quickly became available [107]. The crystal structures of SARS-CoV-2 nsp3 and its macrodomains were the initial focus of many research groups, often associating structural studies [108–111] and computational methods [112,113] to identify potential treatments for the infection. Alhammad et al. [114] reported that SARS-CoV-2 nsp3 macrodomain 1 (Mac1) hydrolyzes MARylated proteins, functioning as a mono-ADP-ribosylhydrolase. This macrodomain's function is preserved across the three coronaviruses that caused pandemics in the recent past: SARS-CoV, SARS-CoV-2, and Middle East respiratory syndrome (MERS) coronavirus. Brosey et al. [115] compared human PARG with Mac1 crystal structures and identified homology between their active sites, revealing that PARG inhibitor fragments PARG-345 and PARG-329 can fully interact with Mac1, appearing as potential inhibitors for Mac1. Chea et al. [116] proposed that Mac1 has specific targets and functionality when compared to Mac2 and Mac3. Their results indicate that Mac1 may act specifically in the ADP-ribose moieties on O- and N-linked groups, being able to cleave ADP-ribosylated substrates via a-NAD+, ADPr-1″phosphate, and O-acetyl-ADP-ribose, but not via b-NAD+, a-ADP-ribose-(arginine), and ADP-ribose-(serine)-histone H3.

Other studies also investigated the link between Mac1 and innate immunity responses against SARS-CoV-2. Russo et al. [117] demonstrated that ectopic nsp3 (macrodomain not specified) is able to hydrolyze downstream ADP-ribosylation mediated by PARP9-DTX3L dimers following IFN-γ stimulation. Preliminary data indicated [118] that deletion of Mac1 in SARS-CoV-2 (ΔMac1) led to a faster clearance of the virus in a mouse model of severe infection when compared to wild-type SARS-CoV-2. ΔMac1 also promoted the expression of ISGs and interferons and sharply reduced the number of inflammatory neutrophils and macrophages.

There is another mechanism by which SARS-CoV-2 may intervene in the host–pathogen interaction via ADP-ribosylation. PARP/ART catalytic activity depends on NAD+ for the covalent transference of ADP-ribose to biological macromolecules. It is also well established that increased PARP/ART catalytic activity leads to depletion of NAD+ [119,120]. Reports before SARS-CoV-2 already suggested that restoration of NAD+ would enhance host immune responses against viruses, aiding macrophage function and the interferon cascade [121]. Based on this background, authors hypothesized that NAD+ may also be a key element of pathogenesis in acute and chronic (post-acute sequelae of COVID-19) SARS-CoV-2 infection [120,122,123]. Heer et al. [23] demonstrated that varied human lung cell lineages infected with SARS-CoV-2 have increased expression of PARP7/ARTD14, PARP10/ARTD10, PARP12/ARTD12, and PARP14/ARTD8 (among other PARPs/ARTs) and that NAD+ concentration was the limiting factor for these enzymes. In addition, the authors of the same study demonstrated that infection of human cells with murine hepatitis virus (a model of coronaviruses) leads to NAD+ and NADP+ depletion and that SARS-CoV-2 changes the expression of genes related to NAD biosynthesis [23]. Using SARS-CoV-2-infected mice, Jiang et al. [124] confirmed that SARS-CoV-2 infection alters the expression of genes related to NAD and NADPH biosynthesis. They also demonstrated that NAD+ supplementation alleviated the pathological phenotypes of pneumonia in infected mice and partially rescued the imbalance in NAD+ genes. Table 1 provides an overview of the enzymes discussed in this review, with their respective activities and disease processes.

Table 1. PARPs/ARTs and ADP-ribose hydrolases and their action in chronic inflammation.

Enzyme	Activity	Disease/Biological Process(s)
PARP1/ARTD1	PARylation, MARylation, or non-catalytic activity	Emphysema/Chronic lung inflammation [49,50] H. pylori infection [51] Colitis/Inflammatory bowel diseases [53,54,56] Psoriasis [57,66] Atopic dermatitis [58] Alcoholic liver injury [63] Chagas heart disease [69–72]
PARP7/ARTD14	MARylation	SARS-CoV-2 infection [23] Colitis/Inflammatory bowel diseases [53,54,56]
PARP9/ARTD9	MARylation	Pulmonary tuberculosis [83,84] RNA-viruses infections [85,86] Atherosclerosis/arterial inflammation [79,80] SARS-CoV-2 infection [117]
PARP10/ARTD10	MARylation	SARS-CoV-2 infection [23] Arboviruses infections [88]
PARP12/ARTD12	MARylation	Assembly and maintenance of stress granules [15] SARS-CoV-2 infection [23]
PARP13/ARTD13	Non-catalytic activity	Assembly and maintenance of stress granules [15]

Table 1. *Cont.*

Enzyme	Activity	Disease/Biological Process(s)
PARP14/ARTD8	MARylation	Atopic dermatitis (Clinical Trial No.: NCT05215808) SARS-CoV-2 infection [23] Atherosclerosis/arterial inflammation [79,80,90,91] Arboviruses infections [88,89]
PARP15/ARTD7	MARylation	Assembly and maintenance of stress granules [15] Arboviruses infections [88]
PARG	Hydrolysis (PAR)	Assembly and maintenance of stress granules [15,98]
Macrodomain 1 (nsp3)	Hydrolysis (MAR)	SARS-CoV-2 infection [114,117,118]

5. Mass Spectrometry and ADP-Ribosylation

Enzyme-catalyzed covalent modifications of amino acids such as phosphorylation, ubiquitination, glycosylation, and ADP-ribosylation are post-translational modifications (PTMs). PTMs regulate various processes related to cellular homeostasis [125]. The biological complexity and the potentially ephemeral nature of PTMs make them challenging to decipher; but innovative mass spectrometry technologies have enabled their widespread investigations. Ribosylomics is the study of proteome-wide ADP-ribosylation, using mass spectrometry. The last ten years have witnessed technological advances that have overcome initial obstacles for ribosylome, including difficulties associated with precise identification of amino acid acceptor sites, the unstable nature of its covalent binding to the amino acid chain, and its complex pattern of fragmentation [126–132]. Several mass spectrometry-based workflows are currently available to study ADP-ribosylation, but PARylated peptides are not amenable to typical mass spectrometric acquisition methods. The conversion of PAR to MAR peptides using poly-PARG [80,131,132] or the complete reduction to a phosphoribose using a phosphodiesterase [133] provides the means to detect ADP-ribosylated proteins/peptides using mass spectrometry; however, these methods do not provide the nature of the original modification, MAR vs. PAR. In parallel to the mass spectrometry-based innovations to characterize ADP-ribosylated proteins are the ongoing efforts to develop computational resources to confidently characterize and report ribosylome data.

5.1. Enrichment Strategies and Activation Methods Influence the Identification of ADP-Ribosylated Proteins in Macrophages

Multiple research groups have tested the activation methods for identifying ADP-ribosylated peptides, their ADP-ribose acceptor sites, and unique enrichment strategies. Electron transfer dissociation (ETD) proved to be efficient in the identification of unambiguous ADP-ribosylated peptides and their acceptor sites [127], with the combination of ETD with higher-energy collisional dissociation (EThcD) being superior to ETD alone for the same purpose [128,132]. Also, the enrichment protocol using an Af1521-Sepharose bead workflow [134] can be combined with ETD for mass spectrometry analysis [131]. Different activation methods may provide the identification of specific ADP-ribosylated peptide groups, depending on their acceptor sites. Ion ETD seems to be superior to EthcD in the occurrence of non-dissociative electron transfer for ADP-ribosylated precursor peptides, and residues modified on arginine and lysine were more stable during HCD fragmentation, whereas the annotation of residues modified on serine, glutamate, tyrosine, and aspartate were more challenging [126]; this is interesting, as modifications on arginine were more frequent during physiological conditions, while modifications on serine were scarce in similar conditions [131], indicating that adjusting the activation method based on the biological condition may provide more reliable results. It is worth noting that the studies mentioned in this paragraph mainly used cancer cell lines and/or healthy mouse tissues, indicating that technical optimization would potentially be needed for the study of inflammatory biosystems related to macrophages and macrophage-like cells.

In 2019, our group demonstrated that different ADP-ribosylation enrichment strategies and activation methods influence the identification of ADP-ribosylated peptides in IFN-γ-stimulated human THP-1 macrophage-like cells [80]. We compared the mass spectrometry results of two enrichment strategies: the Af1521-based workflow [134], in which the macrodomain of the Af1521 compound is used to affinity-purify ADP-ribosylated peptides; and the 10H anti-ADP-ribose antibody-based workflow, in which the antibody is used in immunoprecipitation of ADP-ribosylated proteins. While the Af1521-based workflow provided spectra rich in MARylated peptides and amino acid acceptor sites, the antibody-based workflow only provides peptides suggesting candidate ADP-ribosylated proteins since ADP-ribosylated peptides themselves are not detected. The 10H strategy enriched 1,389 candidate ADP-ribosylated proteins, whereas the Af1521 strategy enriched 145 ADP-ribosylated proteins, resulting in 39 proteins commonly identified which included PARP14/ARTD8 and PARP9/ARTD9 [80].

We also compared distinct activation (peptide sequencing) methods and demonstrated that, while HCD provides a larger number of identified ADP-ribosylated peptides, ETD dissociation provides a more reliable identification of the ADP-ribosylation acceptor site in ADP-ribosylated peptides [80]. With these results, we were able to confirm that stimulation of human THP-1 macrophage-like cells with IFN-γ increased PARP9/ARTD9 and PARP14/ARTD8 ADP-ribosylation levels.

5.2. An Innovative Spectral Annotation Strategy Facilitates the Report of ADP-Ribosylated Peptides in IFN-γ-Stimulated Mice

The investigation of ADP-ribosylated proteins using mass spectrometry methods requires optimal annotation strategies to accurately identify such proteins after enrichment protocols. The ADPriboDB (initially published by Vivelo et al. [135] and updated by Ayyappan et al. [136]) was the first report of a publicly available database of ADP-ribosylated proteins, in which users can find information about proteins reported in the mass spectrometry literature, from as early as 1975. Each individual entry was revised manually by two independent reviewers before inclusion in the database. Likewise, manual annotation of peptide spectra is still commonly used in mass spectrometry studies in the ADP-ribosylation field.

In 2022, our group capitalized on our optimized ADP-ribosylation enrichment and activation methods to develop a new strategy for annotating ADP-ribosylated peptide spectra (named "RiboMap") from liver and spleen samples of IFN-γ-stimulated mice [129,137]. In this strategy, once a candidate ADP-ribosylated peptide fragment spectrum is assigned and scored by the standard spectral search engine, RiboMaP then annotates and scores the spectra for ADP-ribosylation-unique features [129].

With this unique spectral annotation tool, we could increase the confidence of the reported ADP-ribosylated peptide spectra associated with pro-inflammatory responses in liver and spleen. With that combination of mass spectrometry and computational techniques, even ADP-ribosylated peptides with overall low biological abundances, such as PARP9/ARTD9 and PARP14/ARTD8, could be identified. We further applied the RiboMap strategy to publicly available data sets and even to our own previously published human macrophage cell and mouse samples, and we found that, regardless of study and sample type, RiboMap increased the number of ADP-ribosylated peptide spectral annotations in all tests [129].

6. Future Perspectives

Since the seminal articles in the 1960s describing ADP-ribosylation as a post-translational modification occurring in stimulated human monocytes/macrophages [21,22,138], the field has expanded enormously. Although ADP-ribosylation is posited to regulate various biological or pathological processes, the mechanisms remain barely understood. Mass spectrometry and computational biology techniques appear to be some of the fundamental tools for overcoming knowledge gaps in the study of ADP-ribosylation. Mass spectrometry technologies are continually developing, with the aim to sequence deeper into a proteome. Ion mobility technology, such as field asymmetric waveform ion mobility spectrometry (FAIMS), is one such development. We envision FAIMS to be a promising technology to increase the sequencing depth of a given ribosylome, similar to what has been demonstrated for the phosphorylation field [139].

There are also promising perspectives for the field of ADP-ribosylation and macrophage-mediated chronic inflammation. A clinical trial is currently investigating the efficacy of a PARP1/ARTD14 inhibitor compound in the treatment of atopic dermatitis, a chronic inflammatory disease deeply associated with macrophage activation. Likewise, with the help of the strategies mentioned above, other novel drug targets may emerge from bench research. The cumulative evidence suggests that ADP-ribosylation is an important piece of the sustained inflammatory response of macrophages in cardiovascular, gastrointestinal, pulmonary, and hepatic diseases, as well as in prolonged infections. Therefore, we expect that other potential candidate drugs may appear in a near future, translating bench results into clinical tools for patient care.

Author Contributions: Conceptualization, D.V.S.-P., S.A.S., E.A. and M.A.; Methodology, D.V.S.-P. and M.A.; Investigation, D.V.S.-P.; Writing, Reviewing, and Editing, D.V.S.-P., S.A.S., E.A. and M.A. All authors have read and agreed to the published version of the manuscript.

Funding: This study was supported by research grants from Kowa Company Ltd., Nagoya, Japan (to MA), and the National Institutes of Health (R01HL126901 and R01HL149302 to MA).

Institutional Review Board Statement: Not applicable.

Informed Consent Statement: Not applicable.

Data Availability Statement: No new data were created or analyzed in this study. Data sharing is not applicable to this article.

Conflicts of Interest: The authors declare no conflict of interest.

References

1. Kim, D.S.; Challa, S.; Jones, A.; Kraus, W.L. PARPs and ADP-ribosylation in RNA biology: From RNA expression and processing to protein translation and proteostasis. *Genes. Dev.* **2020**, *34*, 302–320. [CrossRef] [PubMed]
2. Luscher, B.; Ahel, I.; Altmeyer, M.; Ashworth, A.; Bai, P.; Chang, P.; Cohen, M.; Corda, D.; Dantzer, F.; Daugherty, M.D.; et al. ADP-ribosyltransferases, an update on function and nomenclature. *FEBS J.* **2022**, *289*, 7399–7410. [CrossRef] [PubMed]
3. Xue, G.; Braczyk, K.; Goncalves-Carneiro, D.; Dawidziak, D.M.; Sanchez, K.; Ong, H.; Wan, Y.; Zadrozny, K.K.; Ganser-Pornillos, B.K.; Bieniasz, P.D.; et al. Poly(ADP-ribose) potentiates ZAP antiviral activity. *PLoS Pathog.* **2022**, *18*, e1009202. [CrossRef] [PubMed]
4. Slade, D.; Dunstan, M.S.; Barkauskaite, E.; Weston, R.; Lafite, P.; Dixon, N.; Ahel, M.; Leys, D.; Ahel, I. The structure and catalytic mechanism of a poly(ADP-ribose) glycohydrolase. *Nature* **2011**, *477*, 616–620. [CrossRef]
5. Poltronieri, P.; Miwa, M.; Masutani, M. ADP-Ribosylation as Post-Translational Modification of Proteins: Use of Inhibitors in Cancer Control. *Int. J. Mol. Sci.* **2021**, *22*, 10829. [CrossRef] [PubMed]
6. Hengel, S.M.; Shaffer, S.A.; Nunn, B.L.; Goodlett, D.R. Tandem mass spectrometry investigation of ADP-ribosylated kemptide. *J. Am. Soc. Mass. Spectrom.* **2009**, *20*, 477–483. [CrossRef]
7. Fontana, P.; Bonfiglio, J.J.; Palazzo, L.; Bartlett, E.; Matic, I.; Ahel, I. Serine ADP-ribosylation reversal by the hydrolase ARH3. *Elife* **2017**, *6*, e28533. [CrossRef]
8. Abplanalp, J.; Leutert, M.; Frugier, E.; Nowak, K.; Feurer, R.; Kato, J.; Kistemaker, H.V.A.; Filippov, D.V.; Moss, J.; Caflisch, A.; et al. Proteomic analyses identify ARH3 as a serine mono-ADP-ribosylhydrolase. *Nat. Commun.* **2017**, *8*, 2055. [CrossRef]
9. Abplanalp, J.; Hopp, A.K.; Hottiger, M.O. Mono-ADP-Ribosylhydrolase Assays. *Methods Mol. Biol.* **2018**, *1813*, 205–213. [CrossRef]
10. Stevens, L.A.; Kato, J.; Kasamatsu, A.; Oda, H.; Lee, D.Y.; Moss, J. The ARH and Macrodomain Families of alpha-ADP-ribose-acceptor Hydrolases Catalyze alpha-NAD(+) Hydrolysis. *ACS Chem. Biol.* **2019**, *14*, 2576–2584. [CrossRef]
11. Yang, X.; Ma, Y.; Li, Y.; Dong, Y.; Yu, L.L.; Wang, H.; Guo, L.; Wu, C.; Yu, X.; Liu, X. Molecular basis for the MacroD1-mediated hydrolysis of ADP-ribosylation. *DNA Repair.* **2020**, *94*, 102899. [CrossRef]
12. Chen, D.; Vollmar, M.; Rossi, M.N.; Phillips, C.; Kraehenbuehl, R.; Slade, D.; Mehrotra, P.V.; von Delft, F.; Crosthwaite, S.K.; Gileadi, O.; et al. Identification of macrodomain proteins as novel O-acetyl-ADP-ribose deacetylases. *J. Biol. Chem.* **2011**, *286*, 13261–13271. [CrossRef]
13. Zong, W.; Gong, Y.; Sun, W.; Li, T.; Wang, Z.Q. PARP1: Liaison of Chromatin Remodeling and Transcription. *Cancers* **2022**, *14*, 4162. [CrossRef] [PubMed]

14. Vyas, S.; Chesarone-Cataldo, M.; Todorova, T.; Huang, Y.H.; Chang, P. A systematic analysis of the PARP protein family identifies new functions critical for cell physiology. *Nat. Commun.* **2013**, *4*, 2240. [CrossRef] [PubMed]
15. Leung, A.K.; Vyas, S.; Rood, J.E.; Bhutkar, A.; Sharp, P.A.; Chang, P. Poly(ADP-ribose) regulates stress responses and microRNA activity in the cytoplasm. *Mol. Cell* **2011**, *42*, 489–499. [CrossRef]
16. Ryu, K.W.; Nandu, T.; Kim, J.; Challa, S.; DeBerardinis, R.J.; Kraus, W.L. Metabolic regulation of transcription through compartmentalized NAD(+) biosynthesis. *Science* **2018**, *360*, eaan5780. [CrossRef]
17. Erener, S.; Hesse, M.; Kostadinova, R.; Hottiger, M.O. Poly(ADP-ribose)polymerase-1 (PARP1) controls adipogenic gene expression and adipocyte function. *Mol. Endocrinol.* **2012**, *26*, 79–86. [CrossRef] [PubMed]
18. van Beek, L.; McClay, E.; Patel, S.; Schimpl, M.; Spagnolo, L.; Maia de Oliveira, T. PARP Power: A Structural Perspective on PARP1, PARP2, and PARP3 in DNA Damage Repair and Nucleosome Remodelling. *Int. J. Mol. Sci.* **2021**, *22*, 5112. [CrossRef] [PubMed]
19. Challa, S.; Stokes, M.S.; Kraus, W.L. MARTs and MARylation in the Cytosol: Biological Functions, Mechanisms of Action, and Therapeutic Potential. *Cells* **2021**, *10*, 313. [CrossRef]
20. Fehr, A.R.; Singh, S.A.; Kerr, C.M.; Mukai, S.; Higashi, H.; Aikawa, M. The impact of PARPs and ADP-ribosylation on inflammation and host-pathogen interactions. *Genes. Dev.* **2020**, *34*, 341–359. [CrossRef]
21. Singh, N.; Poirier, G.; Cerutti, P. Tumor promoter phorbol-12-myristate-13-acetate induces poly ADP-ribosylation in human monocytes. *Biochem. Biophys. Res. Commun.* **1985**, *126*, 1208–1214. [CrossRef] [PubMed]
22. Berton, G.; Sorio, C.; Laudanna, C.; Menegazzi, M.; Carcereri De Prati, A.; Suzuki, H. Activation of human monocyte-derived macrophages by interferon gamma is accompanied by increase of poly(ADP-ribose) polymerase activity. *Biochim. Biophys. Acta* **1991**, *1091*, 101–109. [CrossRef] [PubMed]
23. Heer, C.D.; Sanderson, D.J.; Voth, L.S.; Alhammad, Y.M.O.; Schmidt, M.S.; Trammell, S.A.J.; Perlman, S.; Cohen, M.S.; Fehr, A.R.; Brenner, C. Coronavirus infection and PARP expression dysregulate the NAD metabolome: An actionable component of innate immunity. *J. Biol. Chem.* **2020**, *295*, 17986–17996. [CrossRef]
24. Tutt, A.N.J.; Garber, J.E.; Kaufman, B.; Viale, G.; Fumagalli, D.; Rastogi, P.; Gelber, R.D.; de Azambuja, E.; Fielding, A.; Balmana, J.; et al. Adjuvant Olaparib for Patients with BRCA1- or BRCA2-Mutated Breast Cancer. *N. Engl. J. Med.* **2021**, *384*, 2394–2405. [CrossRef]
25. Tattersall, A.; Ryan, N.; Wiggans, A.J.; Rogozinska, E.; Morrison, J. Poly(ADP-ribose) polymerase (PARP) inhibitors for the treatment of ovarian cancer. *Cochrane Database Syst. Rev.* **2022**, *2*, CD007929. [CrossRef]
26. Yarchoan, M.; Myzak, M.C.; Johnson, B.A., 3rd; De Jesus-Acosta, A.; Le, D.T.; Jaffee, E.M.; Azad, N.S.; Donehower, R.C.; Zheng, L.; Oberstein, P.E.; et al. Olaparib in combination with irinotecan, cisplatin, and mitomycin C in patients with advanced pancreatic cancer. *Oncotarget* **2017**, *8*, 44073–44081. [CrossRef]
27. Hussain, M.; Mateo, J.; Fizazi, K.; Saad, F.; Shore, N.; Sandhu, S.; Chi, K.N.; Sartor, O.; Agarwal, N.; Olmos, D.; et al. Survival with Olaparib in Metastatic Castration-Resistant Prostate Cancer. *N. Engl. J. Med.* **2020**, *383*, 2345–2357. [CrossRef]
28. Wang, L.; Wang, D.; Sonzogni, O.; Ke, S.; Wang, Q.; Thavamani, A.; Batalini, F.; Stopka, S.A.; Regan, M.S.; Vandal, S.; et al. PARP-inhibition reprograms macrophages toward an anti-tumor phenotype. *Cell Rep.* **2022**, *41*, 111462. [CrossRef] [PubMed]
29. Demeny, M.A.; Virag, L. The PARP Enzyme Family and the Hallmarks of Cancer Part 2: Hallmarks Related to Cancer Host Interactions. *Cancers* **2021**, *13*, 2057. [CrossRef]
30. Xue, T.; Zhang, X.; Xing, Y.; Liu, S.; Zhang, L.; Wang, X.; Yu, M. Advances About Immunoinflammatory Pathogenesis and Treatment in Diabetic Peripheral Neuropathy. *Front. Pharmacol.* **2021**, *12*, 748193. [CrossRef]
31. Gupta, S.; You, P.; SenGupta, T.; Nilsen, H.; Sharma, K. Crosstalk between Different DNA Repair Pathways Contributes to Neurodegenerative Diseases. *Biology* **2021**, *10*, 163. [CrossRef]
32. Krug, S.; Parveen, S.; Bishai, W.R. Host-Directed Therapies: Modulating Inflammation to Treat Tuberculosis. *Front. Immunol.* **2021**, *12*, 660916. [CrossRef] [PubMed]
33. Siewe, N.; Friedman, A. Cancer therapy with immune checkpoint inhibitor and CSF-1 blockade: A mathematical model. *J. Theor. Biol.* **2023**, *556*, 111297. [CrossRef] [PubMed]
34. Kasraie, S.; Werfel, T. Role of macrophages in the pathogenesis of atopic dermatitis. *Mediat. Inflamm.* **2013**, *2013*, 942375. [CrossRef] [PubMed]
35. Fan, Y.; Hao, Y.; Gao, D.; Li, G.; Zhang, Z. Phenotype and function of macrophage polarization in monocrotaline-induced pulmonary arterial hypertension rat model. *Physiol. Res.* **2021**, *70*, 213–226. [CrossRef]
36. Kumar, V.; Kumar, A.; Mir, K.U.I.; Yadav, V.; Chauhan, S.S. Pleiotropic role of PARP1: An overview. *3 Biotech.* **2022**, *12*, 3. [CrossRef]
37. Luscher, B.; Butepage, M.; Eckei, L.; Krieg, S.; Verheugd, P.; Shilton, B.H. ADP-Ribosylation, a Multifaceted Posttranslational Modification Involved in the Control of Cell Physiology in Health and Disease. *Chem. Rev.* **2018**, *118*, 1092–1136. [CrossRef]
38. Dawicki-McKenna, J.M.; Langelier, M.F.; DeNizio, J.E.; Riccio, A.A.; Cao, C.D.; Karch, K.R.; McCauley, M.; Steffen, J.D.; Black, B.E.; Pascal, J.M. PARP-1 Activation Requires Local Unfolding of an Autoinhibitory Domain. *Mol. Cell* **2015**, *60*, 755–768. [CrossRef]
39. Eustermann, S.; Wu, W.F.; Langelier, M.F.; Yang, J.C.; Easton, L.E.; Riccio, A.A.; Pascal, J.M.; Neuhaus, D. Structural Basis of Detection and Signaling of DNA Single-Strand Breaks by Human PARP-1. *Mol. Cell* **2015**, *60*, 742–754. [CrossRef] [PubMed]

40. Jumper, J.; Evans, R.; Pritzel, A.; Green, T.; Figurnov, M.; Ronneberger, O.; Tunyasuvunakool, K.; Bates, R.; Zidek, A.; Potapenko, A.; et al. Highly accurate protein structure prediction with AlphaFold. *Nature* **2021**, *596*, 583–589. [CrossRef] [PubMed]
41. Mendoza-Alvarez, H.; Alvarez-Gonzalez, R. Poly(ADP-ribose) polymerase is a catalytic dimer and the automodification reaction is intermolecular. *J. Biol. Chem.* **1993**, *268*, 22575–22580. [CrossRef] [PubMed]
42. Pion, E.; Ullmann, G.M.; Ame, J.C.; Gerard, D.; de Murcia, G.; Bombarda, E. DNA-induced dimerization of poly(ADP-ribose) polymerase-1 triggers its activation. *Biochemistry* **2005**, *44*, 14670–14681. [CrossRef]
43. Ali, A.A.E.; Timinszky, G.; Arribas-Bosacoma, R.; Kozlowski, M.; Hassa, P.O.; Hassler, M.; Ladurner, A.G.; Pearl, L.H.; Oliver, A.W. The zinc-finger domains of PARP1 cooperate to recognize DNA strand breaks. *Nat. Struct. Mol. Biol.* **2012**, *19*, 685–692. [CrossRef] [PubMed]
44. Thomas, C.; Ji, Y.; Wu, C.; Datz, H.; Boyle, C.; MacLeod, B.; Patel, S.; Ampofo, M.; Currie, M.; Harbin, J.; et al. Hit and run versus long-term activation of PARP-1 by its different domains fine-tunes nuclear processes. *Proc. Natl. Acad. Sci. USA* **2019**, *116*, 9941–9946. [CrossRef] [PubMed]
45. Langelier, M.F.; Planck, J.L.; Roy, S.; Pascal, J.M. Crystal structures of poly(ADP-ribose) polymerase-1 (PARP-1) zinc fingers bound to DNA: Structural and functional insights into DNA-dependent PARP-1 activity. *J. Biol. Chem.* **2011**, *286*, 10690–10701. [CrossRef]
46. Thomas, C.J.; Kotova, E.; Andrake, M.; Adolf-Bryfogle, J.; Glaser, R.; Regnard, C.; Tulin, A.V. Kinase-mediated changes in nucleosome conformation trigger chromatin decondensation via poly(ADP-ribosyl)ation. *Mol. Cell* **2014**, *53*, 831–842. [CrossRef]
47. Ioannidou, A.; Goulielmaki, E.; Garinis, G.A. DNA Damage: From Chronic Inflammation to Age-Related Deterioration. *Front. Genet.* **2016**, *7*, 187. [CrossRef]
48. Bauer, M.; Goldstein, M.; Christmann, M.; Becker, H.; Heylmann, D.; Kaina, B. Human monocytes are severely impaired in base and DNA double-strand break repair that renders them vulnerable to oxidative stress. *Proc. Natl. Acad. Sci. USA* **2011**, *108*, 21105–21110. [CrossRef]
49. Dharwal, V.; Naura, A.S. PARP-1 inhibition ameliorates elastase induced lung inflammation and emphysema in mice. *Biochem. Pharmacol.* **2018**, *150*, 24–34. [CrossRef]
50. Dharwal, V.; Sandhir, R.; Naura, A.S. PARP-1 inhibition provides protection against elastase-induced emphysema by mitigating the expression of matrix metalloproteinases. *Mol. Cell Biochem.* **2019**, *457*, 41–49. [CrossRef]
51. Kunze, F.A.; Bauer, M.; Komuczki, J.; Lanzinger, M.; Gunasekera, K.; Hopp, A.K.; Lehmann, M.; Becher, B.; Muller, A.; Hottiger, M.O. ARTD1 in Myeloid Cells Controls the IL-12/18-IFN-gamma Axis in a Model of Sterile Sepsis, Chronic Bacterial Infection, and Cancer. *J. Immunol.* **2019**, *202*, 1406–1416. [CrossRef] [PubMed]
52. Durkacz, B.W.; Omidiji, O.; Gray, D.A.; Shall, S. (ADP-ribose)n participates in DNA excision repair. *Nature* **1980**, *283*, 593–596. [CrossRef]
53. Singla, S.; Kumar, V.; Jena, G. 3-aminobenzamide protects against colitis associated diabetes mellitus in male BALB/c mice: Role of PARP-1, NLRP3, SIRT-1, AMPK. *Biochimie* **2023**, *211*, 96–109. [CrossRef] [PubMed]
54. Kovacs, D.; Vantus, V.B.; Vamos, E.; Kalman, N.; Schicho, R.; Gallyas, F.; Radnai, B. Olaparib: A Clinically Applied PARP Inhibitor Protects from Experimental Crohn's Disease and Maintains Barrier Integrity by Improving Bioenergetics through Rescuing Glycolysis in Colonic Epithelial Cells. *Oxid. Med. Cell. Longev.* **2021**, *2021*, 7308897. [CrossRef] [PubMed]
55. Gupte, R.; Nandu, T.; Kraus, W.L. Nuclear ADP-ribosylation drives IFNgamma-dependent STAT1alpha enhancer formation in macrophages. *Nat. Commun.* **2021**, *12*, 3931. [CrossRef] [PubMed]
56. Gerner, R.R.; Klepsch, V.; Macheiner, S.; Arnhard, K.; Adolph, T.E.; Grander, C.; Wieser, V.; Pfister, A.; Moser, P.; Hermann-Kleiter, N.; et al. NAD metabolism fuels human and mouse intestinal inflammation. *Gut* **2018**, *67*, 1813–1823. [CrossRef]
57. Mercurio, L.; Morelli, M.; Scarponi, C.; Scaglione, G.L.; Pallotta, S.; Avitabile, D.; Albanesi, C.; Madonna, S. Enhanced NAMPT-Mediated NAD Salvage Pathway Contributes to Psoriasis Pathogenesis by Amplifying Epithelial Auto-Inflammatory Circuits. *Int. J. Mol. Sci.* **2021**, *22*, 6860. [CrossRef]
58. Arroyo, A.B.; Bernal-Carrion, M.; Canton-Sandoval, J.; Cabas, I.; Corbalan-Velez, R.; Martinez-Menchon, T.; Ferri, B.; Cayuela, M.L.; Garcia-Moreno, D.; Mulero, V. NAMPT and PARylation Are Involved in the Pathogenesis of Atopic Dermatitis. *Int. J. Mol. Sci.* **2023**, *24*, 7992. [CrossRef]
59. Soldani, C.; Scovassi, A.I. Poly(ADP-ribose) polymerase-1 cleavage during apoptosis: An update. *Apoptosis* **2002**, *7*, 321–328. [CrossRef]
60. Yang, Y.; Jiang, G.; Zhang, P.; Fan, J. Programmed cell death and its role in inflammation. *Mil. Med. Res.* **2015**, *2*, 12. [CrossRef]
61. Huang, P.; Chen, G.; Jin, W.; Mao, K.; Wan, H.; He, Y. Molecular Mechanisms of Parthanatos and Its Role in Diverse Diseases. *Int. J. Mol. Sci.* **2022**, *23*, 7292. [CrossRef]
62. Andrabi, S.A.; Dawson, T.M.; Dawson, V.L. Mitochondrial and nuclear cross talk in cell death: Parthanatos. *Ann. N. Y Acad. Sci.* **2008**, *1147*, 233–241. [CrossRef] [PubMed]
63. Zhang, Y.; Wang, C.; Tian, Y.; Zhang, F.; Xu, W.; Li, X.; Shu, Z.; Wang, Y.; Huang, K.; Huang, D. Inhibition of Poly(ADP-Ribose) Polymerase-1 Protects Chronic Alcoholic Liver Injury. *Am. J. Pathol.* **2016**, *186*, 3117–3130. [CrossRef]
64. Cohen-Armon, M. The Modified Phenanthridine PJ34 Unveils an Exclusive Cell-Death Mechanism in Human Cancer Cells. *Cancers* **2020**, *12*, 1628. [CrossRef] [PubMed]

65. Erener, S.; Petrilli, V.; Kassner, I.; Minotti, R.; Castillo, R.; Santoro, R.; Hassa, P.O.; Tschopp, J.; Hottiger, M.O. Inflammasome-activated caspase 7 cleaves PARP1 to enhance the expression of a subset of NF-kappaB target genes. *Mol. Cell* **2012**, *46*, 200–211. [CrossRef]
66. Martinez-Morcillo, F.J.; Canton-Sandoval, J.; Martinez-Navarro, F.J.; Cabas, I.; Martinez-Vicente, I.; Armistead, J.; Hatzold, J.; Lopez-Munoz, A.; Martinez-Menchon, T.; Corbalan-Velez, R.; et al. NAMPT-derived NAD+ fuels PARP1 to promote skin inflammation through parthanatos cell death. *PLoS Biol.* **2021**, *19*, e3001455. [CrossRef] [PubMed]
67. Lopez, M.; Tanowitz, H.B.; Garg, N.J. Pathogenesis of Chronic Chagas Disease: Macrophages, Mitochondria, and Oxidative Stress. *Curr. Clin. Microbiol. Rep.* **2018**, *5*, 45–54. [CrossRef]
68. Bonney, K.M.; Luthringer, D.J.; Kim, S.A.; Garg, N.J.; Engman, D.M. Pathology and Pathogenesis of Chagas Heart Disease. *Annu. Rev. Pathol.* **2019**, *14*, 421–447. [CrossRef]
69. Ba, X.; Gupta, S.; Davidson, M.; Garg, N.J. Trypanosoma cruzi induces the reactive oxygen species-PARP-1-RelA pathway for up-regulation of cytokine expression in cardiomyocytes. *J. Biol. Chem.* **2010**, *285*, 11596–11606. [CrossRef]
70. Wen, J.J.; Yin, Y.W.; Garg, N.J. PARP1 depletion improves mitochondrial and heart function in Chagas disease: Effects on POLG dependent mtDNA maintenance. *PLoS Pathog.* **2018**, *14*, e1007065. [CrossRef]
71. Florentino, P.T.V.; Vitorino, F.N.L.; Mendes, D.; da Cunha, J.P.C.; Menck, C.F.M. Trypanosoma cruzi infection changes the chromatin proteome profile of infected human cells. *J. Proteomics* **2023**, *272*, 104773. [CrossRef] [PubMed]
72. Florentino, P.T.V.; Mendes, D.; Vitorino, F.N.L.; Martins, D.J.; Cunha, J.P.C.; Mortara, R.A.; Menck, C.F.M. DNA damage and oxidative stress in human cells infected by Trypanosoma cruzi. *PLoS Pathog.* **2021**, *17*, e1009502. [CrossRef]
73. Choudhuri, S.; Garg, N.J. PARP1-cGAS-NF-kappaB pathway of proinflammatory macrophage activation by extracellular vesicles released during Trypanosoma cruzi infection and Chagas disease. *PLoS Pathog.* **2020**, *16*, e1008474. [CrossRef] [PubMed]
74. Macaluso, G.; Grippi, F.; Di Bella, S.; Blanda, V.; Gucciardi, F.; Torina, A.; Guercio, A.; Cannella, V. A Review on the Immunological Response against Trypanosoma cruzi. *Pathogens* **2023**, *12*, 282. [CrossRef] [PubMed]
75. von Lukowicz, T.; Hassa, P.O.; Lohmann, C.; Boren, J.; Braunersreuther, V.; Mach, F.; Odermatt, B.; Gersbach, M.; Camici, G.G.; Stahli, B.E.; et al. PARP1 is required for adhesion molecule expression in atherogenesis. *Cardiovasc. Res.* **2008**, *78*, 158–166. [CrossRef]
76. Zhang, Y.; Wang, W. Bidirectional regulation role of PARP-1 in high glucose-induced endothelial injury. *Exp. Cell Res.* **2022**, *421*, 113400. [CrossRef]
77. Wei, L.; Yang, C.; Li, K.Q.; Zhong, C.L.; Sun, Z.Y. 3-Aminobenzamide protects against cerebral artery injury and inflammation in rats with intracranial aneurysms. *Pharmazie* **2019**, *74*, 142–146. [CrossRef]
78. Daugherty, M.D.; Young, J.M.; Kerns, J.A.; Malik, H.S. Rapid evolution of PARP genes suggests a broad role for ADP-ribosylation in host-virus conflicts. *PLoS Genet.* **2014**, *10*, e1004403. [CrossRef]
79. Iwata, H.; Goettsch, C.; Sharma, A.; Ricchiuto, P.; Goh, W.W.; Halu, A.; Yamada, I.; Yoshida, H.; Hara, T.; Wei, M.; et al. PARP9 and PARP14 cross-regulate macrophage activation via STAT1 ADP-ribosylation. *Nat. Commun.* **2016**, *7*, 12849. [CrossRef]
80. Higashi, H.; Maejima, T.; Lee, L.H.; Yamazaki, Y.; Hottiger, M.O.; Singh, S.A.; Aikawa, M. A Study into the ADP-Ribosylome of IFN-gamma-Stimulated THP-1 Human Macrophage-like Cells Identifies ARTD8/PARP14 and ARTD9/PARP9 ADP-Ribosylation. *J. Proteome Res.* **2019**, *18*, 1607–1622. [CrossRef]
81. Jin, U.H.; Cheng, Y.; Park, H.; Davidson, L.A.; Callaway, E.S.; Chapkin, R.S.; Jayaraman, A.; Asante, A.; Allred, C.; Weaver, E.A.; et al. Short Chain Fatty Acids Enhance Aryl Hydrocarbon (Ah) Responsiveness in Mouse Colonocytes and Caco-2 Human Colon Cancer Cells. *Sci. Rep.* **2017**, *7*, 10163. [CrossRef]
82. Szanto, M.; Gupte, R.; Kraus, W.L.; Pacher, P.; Bai, P. PARPs in lipid metabolism and related diseases. *Prog. Lipid Res.* **2021**, *84*, 101117. [CrossRef] [PubMed]
83. Chen, Y.C.; Hsiao, C.C.; Chen, T.W.; Wu, C.C.; Chao, T.Y.; Leung, S.Y.; Eng, H.L.; Lee, C.P.; Wang, T.Y.; Lin, M.C. Whole Genome DNA Methylation Analysis of Active Pulmonary Tuberculosis Disease Identifies Novel Epigenotypes: PARP9/miR-505/RASGRP4/GNG12 Gene Methylation and Clinical Phenotypes. *Int. J. Mol. Sci.* **2020**, *21*, 3180. [CrossRef] [PubMed]
84. Thirunavukkarasu, S.; Ahmed, M.; Rosa, B.A.; Boothby, M.; Cho, S.H.; Rangel-Moreno, J.; Mbandi, S.K.; Schreiber, V.; Gupta, A.; Zuniga, J.; et al. Poly(ADP-ribose) polymerase 9 mediates early protection against Mycobacterium tuberculosis infection by regulating type I IFN production. *J. Clin. Investig.* **2023**, *133*, e158630. [CrossRef] [PubMed]
85. Xing, J.; Zhang, A.; Du, Y.; Fang, M.; Minze, L.J.; Liu, Y.J.; Li, X.C.; Zhang, Z. Identification of poly(ADP-ribose) polymerase 9 (PARP9) as a noncanonical sensor for RNA virus in dendritic cells. *Nat. Commun.* **2021**, *12*, 2681. [CrossRef]
86. Han, J.; Chen, C.; Wang, C.; Qin, N.; Huang, M.; Ma, Z.; Zhu, M.; Dai, J.; Jiang, Y.; Ma, H.; et al. Transcriptome-wide association study for persistent hepatitis B virus infection and related hepatocellular carcinoma. *Liver Int.* **2020**, *40*, 2117–2127. [CrossRef]
87. Liang, T.J. Hepatitis B: The virus and disease. *Hepatology* **2009**, *49*, S13–S21. [CrossRef]
88. Eckei, L.; Krieg, S.; Butepage, M.; Lehmann, A.; Gross, A.; Lippok, B.; Grimm, A.R.; Kummerer, B.M.; Rossetti, G.; Luscher, B.; et al. The conserved macrodomains of the non-structural proteins of Chikungunya virus and other pathogenic positive strand RNA viruses function as mono-ADP-ribosylhydrolases. *Sci. Rep.* **2017**, *7*, 41746. [CrossRef] [PubMed]
89. Fernandez, G.J.; Ramirez-Mejia, J.M.; Urcuqui-Inchima, S. Transcriptional and post-transcriptional mechanisms that regulate the genetic program in Zika virus-infected macrophages. *Int. J. Biochem. Cell Biol.* **2022**, *153*, 106312. [CrossRef] [PubMed]

90. Iqbal, M.B.; Johns, M.; Cao, J.; Liu, Y.; Yu, S.C.; Hyde, G.D.; Laffan, M.A.; Marchese, F.P.; Cho, S.H.; Clark, A.R.; et al. PARP-14 combines with tristetraprolin in the selective posttranscriptional control of macrophage tissue factor expression. *Blood* **2014**, *124*, 3646–3655. [CrossRef] [PubMed]
91. Mehrotra, P.; Riley, J.P.; Patel, R.; Li, F.; Voss, L.; Goenka, S. PARP-14 functions as a transcriptional switch for Stat6-dependent gene activation. *J. Biol. Chem.* **2011**, *286*, 1767–1776. [CrossRef] [PubMed]
92. Holbourn, K.P.; Shone, C.C.; Acharya, K.R. A family of killer toxins. Exploring the mechanism of ADP-ribosylating toxins. *FEBS J.* **2006**, *273*, 4579–4593. [CrossRef]
93. Atasheva, S.; Frolova, E.I.; Frolov, I. Interferon-stimulated poly(ADP-Ribose) polymerases are potent inhibitors of cellular translation and virus replication. *J. Virol.* **2014**, *88*, 2116–2130. [CrossRef] [PubMed]
94. Zhang, Y.; Mao, D.; Roswit, W.T.; Jin, X.; Patel, A.C.; Patel, D.A.; Agapov, E.; Wang, Z.; Tidwell, R.M.; Atkinson, J.J.; et al. PARP9-DTX3L ubiquitin ligase targets host histone H2BJ and viral 3C protease to enhance interferon signaling and control viral infection. *Nat. Immunol.* **2015**, *16*, 1215–1227. [CrossRef] [PubMed]
95. Gagne, J.P.; Isabelle, M.; Lo, K.S.; Bourassa, S.; Hendzel, M.J.; Dawson, V.L.; Dawson, T.M.; Poirier, G.G. Proteome-wide identification of poly(ADP-ribose) binding proteins and poly(ADP-ribose)-associated protein complexes. *Nucleic Acids Res.* **2008**, *36*, 6959–6976. [CrossRef]
96. Catara, G.; Grimaldi, G.; Schembri, L.; Spano, D.; Turacchio, G.; Lo Monte, M.; Beccari, A.R.; Valente, C.; Corda, D. PARP1-produced poly-ADP-ribose causes the PARP12 translocation to stress granules and impairment of Golgi complex functions. *Sci. Rep.* **2017**, *7*, 14035. [CrossRef] [PubMed]
97. Isabelle, M.; Gagne, J.P.; Gallouzi, I.E.; Poirier, G.G. Quantitative proteomics and dynamic imaging reveal that G3BP-mediated stress granule assembly is poly(ADP-ribose)-dependent following exposure to MNNG-induced DNA alkylation. *J. Cell Sci.* **2012**, *125*, 4555–4566. [CrossRef]
98. Leung, A.; Todorova, T.; Ando, Y.; Chang, P. Poly(ADP-ribose) regulates post-transcriptional gene regulation in the cytoplasm. *RNA Biol.* **2012**, *9*, 542–548. [CrossRef]
99. Karras, G.I.; Kustatscher, G.; Buhecha, H.R.; Allen, M.D.; Pugieux, C.; Sait, F.; Bycroft, M.; Ladurner, A.G. The macro domain is an ADP-ribose binding module. *EMBO J.* **2005**, *24*, 1911–1920. [CrossRef]
100. Egloff, M.P.; Malet, H.; Putics, A.; Heinonen, M.; Dutartre, H.; Frangeul, A.; Gruez, A.; Campanacci, V.; Cambillau, C.; Ziebuhr, J.; et al. Structural and functional basis for ADP-ribose and poly(ADP-ribose) binding by viral macro domains. *J. Virol.* **2006**, *80*, 8493–8502. [CrossRef]
101. Eriksson, K.K.; Cervantes-Barragan, L.; Ludewig, B.; Thiel, V. Mouse hepatitis virus liver pathology is dependent on ADP-ribose-1″-phosphatase, a viral function conserved in the alpha-like supergroup. *J. Virol.* **2008**, *82*, 12325–12334. [CrossRef] [PubMed]
102. Kuri, T.; Eriksson, K.K.; Putics, A.; Zust, R.; Snijder, E.J.; Davidson, A.D.; Siddell, S.G.; Thiel, V.; Ziebuhr, J.; Weber, F. The ADP-ribose-1″-monophosphatase domains of severe acute respiratory syndrome coronavirus and human coronavirus 229E mediate resistance to antiviral interferon responses. *J. Gen. Virol.* **2011**, *92*, 1899–1905. [CrossRef] [PubMed]
103. Fehr, A.R.; Athmer, J.; Channappanavar, R.; Phillips, J.M.; Meyerholz, D.K.; Perlman, S. The nsp3 macrodomain promotes virulence in mice with coronavirus-induced encephalitis. *J. Virol.* **2015**, *89*, 1523–1536. [CrossRef]
104. Voth, L.S.; O'Connor, J.J.; Kerr, C.M.; Doerger, E.; Schwarting, N.; Sperstad, P.; Johnson, D.K.; Fehr, A.R. Unique Mutations in the Murine Hepatitis Virus Macrodomain Differentially Attenuate Virus Replication, Indicating Multiple Roles for the Macrodomain in Coronavirus Replication. *J. Virol.* **2021**, *95*, e0076621. [CrossRef]
105. Grunewald, M.E.; Chen, Y.; Kuny, C.; Maejima, T.; Lease, R.; Ferraris, D.; Aikawa, M.; Sullivan, C.S.; Perlman, S.; Fehr, A.R. The coronavirus macrodomain is required to prevent PARP-mediated inhibition of virus replication and enhancement of IFN expression. *PLoS Pathog.* **2019**, *15*, e1007756. [CrossRef]
106. Fehr, A.R.; Channappanavar, R.; Jankevicius, G.; Fett, C.; Zhao, J.; Athmer, J.; Meyerholz, D.K.; Ahel, I.; Perlman, S. The Conserved Coronavirus Macrodomain Promotes Virulence and Suppresses the Innate Immune Response during Severe Acute Respiratory Syndrome Coronavirus Infection. *mBio* **2016**, *7*, e01721-16. [CrossRef]
107. Hoch, N.C. Host ADP-ribosylation and the SARS-CoV-2 macrodomain. *Biochem. Soc. Trans.* **2021**, *49*, 1711–1721. [CrossRef]
108. Cantini, F.; Banci, L.; Altincekic, N.; Bains, J.K.; Dhamotharan, K.; Fuks, C.; Furtig, B.; Gande, S.L.; Hargittay, B.; Hengesbach, M.; et al. (1)H, (13)C, and (15)N backbone chemical shift assignments of the apo and the ADP-ribose bound forms of the macrodomain of SARS-CoV-2 non-structural protein 3b. *Biomol. NMR Assign.* **2020**, *14*, 339–346. [CrossRef]
109. Michalska, K.; Kim, Y.; Jedrzejczak, R.; Maltseva, N.I.; Stols, L.; Endres, M.; Joachimiak, A. Crystal structures of SARS-CoV-2 ADP-ribose phosphatase: From the apo form to ligand complexes. *IUCrJ* **2020**, *7*, 814–824. [CrossRef] [PubMed]
110. Lin, M.H.; Chang, S.C.; Chiu, Y.C.; Jiang, B.C.; Wu, T.H.; Hsu, C.H. Structural, Biophysical, and Biochemical Elucidation of the SARS-CoV-2 Nonstructural Protein 3 Macro Domain. *ACS Infect. Dis.* **2020**, *6*, 2970–2978. [CrossRef]
111. Correy, G.J.; Kneller, D.W.; Phillips, G.; Pant, S.; Russi, S.; Cohen, A.E.; Meigs, G.; Holton, J.M.; Gahbauer, S.; Thompson, M.C.; et al. The mechanisms of catalysis and ligand binding for the SARS-CoV-2 NSP3 macrodomain from neutron and x-ray diffraction at room temperature. *Sci. Adv.* **2022**, *8*, eabo5083. [CrossRef]
112. Schuller, M.; Correy, G.J.; Gahbauer, S.; Fearon, D.; Wu, T.; Diaz, R.E.; Young, I.D.; Carvalho Martins, L.; Smith, D.H.; Schulze-Gahmen, U.; et al. Fragment binding to the Nsp3 macrodomain of SARS-CoV-2 identified through crystallographic screening and computational docking. *Sci. Adv.* **2021**, *7*, eabf8711. [CrossRef]

113. Claverie, J.M. A Putative Role of de-Mono-ADP-Ribosylation of STAT1 by the SARS-CoV-2 Nsp3 Protein in the Cytokine Storm Syndrome of COVID-19. *Viruses* **2020**, *12*, 646. [CrossRef] [PubMed]
114. Alhammad, Y.M.O.; Kashipathy, M.M.; Roy, A.; Gagne, J.P.; McDonald, P.; Gao, P.; Nonfoux, L.; Battaile, K.P.; Johnson, D.K.; Holmstrom, E.D.; et al. The SARS-CoV-2 Conserved Macrodomain Is a Mono-ADP-Ribosylhydrolase. *J. Virol.* **2021**, *95*, 10–1128. [CrossRef]
115. Brosey, C.A.; Houl, J.H.; Katsonis, P.; Balapiti-Modarage, L.P.F.; Bommagani, S.; Arvai, A.; Moiani, D.; Bacolla, A.; Link, T.; Warden, L.S.; et al. Targeting SARS-CoV-2 Nsp3 macrodomain structure with insights from human poly(ADP-ribose) glycohydrolase (PARG) structures with inhibitors. *Prog. Biophys. Mol. Biol.* **2021**, *163*, 171–186. [CrossRef] [PubMed]
116. Chea, C.; Lee, D.Y.; Kato, J.; Ishiwata-Endo, H.; Moss, J. Macrodomain Mac1 of SARS-CoV-2 Nonstructural Protein 3 Hydrolyzes Diverse ADP-ribosylated Substrates. *bioRxiv* **2023**. [CrossRef]
117. Russo, L.C.; Tomasin, R.; Matos, I.A.; Manucci, A.C.; Sowa, S.T.; Dale, K.; Caldecott, K.W.; Lehtio, L.; Schechtman, D.; Meotti, F.C.; et al. The SARS-CoV-2 Nsp3 macrodomain reverses PARP9/DTX3L-dependent ADP-ribosylation induced by interferon signaling. *J. Biol. Chem.* **2021**, *297*, 101041. [CrossRef]
118. Alhammad, Y.M.; Parthasarathy, S.; Ghimire, R.; O'Connor, J.J.; Kerr, C.M.; Pfannenstiel, J.J.; Chanda, D.; Miller, C.A.; Unckless, R.L.; Zuniga, S.; et al. SARS-CoV-2 Mac1 is required for IFN antagonism and efficient virus replication in mice. *bioRxiv* **2023**. [CrossRef]
119. Cohen, M.S. Interplay between compartmentalized NAD(+) synthesis and consumption: A focus on the PARP family. *Genes. Dev.* **2020**, *34*, 254–262. [CrossRef] [PubMed]
120. Gupte, R.; Liu, Z.; Kraus, W.L. PARPs and ADP-ribosylation: Recent advances linking molecular functions to biological outcomes. *Genes. Dev.* **2017**, *31*, 101–126. [CrossRef] [PubMed]
121. Dantoft, W.; Robertson, K.A.; Watkins, W.J.; Strobl, B.; Ghazal, P. Metabolic Regulators Nampt and Sirt6 Serially Participate in the Macrophage Interferon Antiviral Cascade. *Front. Microbiol.* **2019**, *10*, 355. [CrossRef] [PubMed]
122. Habeichi, N.J.; Tannous, C.; Yabluchanskiy, A.; Altara, R.; Mericskay, M.; Booz, G.W.; Zouein, F.A. Insights into the modulation of the interferon response and NAD(+) in the context of COVID-19. *Int. Rev. Immunol.* **2022**, *41*, 464–474. [CrossRef] [PubMed]
123. Block, T.; Kuo, J. Rationale for Nicotinamide Adenine Dinucleotide (NAD+) Metabolome Disruption as a Pathogenic Mechanism of Post-Acute COVID-19 Syndrome. *Clin. Pathol.* **2022**, *15*, 2632010X221106986. [CrossRef] [PubMed]
124. Jiang, Y.; Deng, Y.; Pang, H.; Ma, T.; Ye, Q.; Chen, Q.; Chen, H.; Hu, Z.; Qin, C.F.; Xu, Z. Treatment of SARS-CoV-2-induced pneumonia with NAD(+) and NMN in two mouse models. *Cell Discov.* **2022**, *8*, 38. [CrossRef]
125. Zhai, L.H.; Chen, K.F.; Hao, B.B.; Tan, M.J. Proteomic characterization of post-translational modifications in drug discovery. *Acta Pharmacol. Sin.* **2022**, *43*, 3112–3129. [CrossRef]
126. Gehrig, P.M.; Nowak, K.; Panse, C.; Leutert, M.; Grossmann, J.; Schlapbach, R.; Hottiger, M.O. Gas-Phase Fragmentation of ADP-Ribosylated Peptides: Arginine-Specific Side-Chain Losses and Their Implication in Database Searches. *J. Am. Soc. Mass. Spectrom.* **2021**, *32*, 157–168. [CrossRef]
127. Zee, B.M.; Garcia, B.A. Electron transfer dissociation facilitates sequencing of adenosine diphosphate-ribosylated peptides. *Anal. Chem.* **2010**, *82*, 28–31. [CrossRef]
128. Rosenthal, F.; Nanni, P.; Barkow-Oesterreicher, S.; Hottiger, M.O. Optimization of LTQ-Orbitrap Mass Spectrometer Parameters for the Identification of ADP-Ribosylation Sites. *J. Proteome Res.* **2015**, *14*, 4072–4079. [CrossRef]
129. Singh, S.A.; Kuraoka, S.; Pestana, D.V.S.; Nasir, W.; Delanghe, B.; Aikawa, M. The RiboMaP Spectral Annotation Method Applied to Various ADP-Ribosylome Studies Including INF-gamma-Stimulated Human Cells and Mouse Tissues. *Front. Cardiovasc. Med.* **2022**, *9*, 851351. [CrossRef]
130. Anagho, H.A.; Elsborg, J.D.; Hendriks, I.A.; Buch-Larsen, S.C.; Nielsen, M.L. Characterizing ADP-Ribosylation Sites Using Af1521 Enrichment Coupled to ETD-Based Mass Spectrometry. *Methods Mol. Biol.* **2023**, *2609*, 251–270. [CrossRef]
131. Buch-Larsen, S.C.; Hendriks, I.A.; Lodge, J.M.; Rykaer, M.; Furtwangler, B.; Shishkova, E.; Westphall, M.S.; Coon, J.J.; Nielsen, M.L. Mapping Physiological ADP-Ribosylation Using Activated Ion Electron Transfer Dissociation. *Cell Rep.* **2020**, *32*, 108176. [CrossRef] [PubMed]
132. Hendriks, I.A.; Larsen, S.C.; Nielsen, M.L. An Advanced Strategy for Comprehensive Profiling of ADP-ribosylation Sites Using Mass Spectrometry-based Proteomics. *Mol. Cell Proteom.* **2019**, *18*, 1010–1026. [CrossRef] [PubMed]
133. Daniels, C.M.; Ong, S.E.; Leung, A.K.L. ADP-Ribosylated Peptide Enrichment and Site Identification: The Phosphodiesterase-Based Method. *Methods Mol. Biol.* **2017**, *1608*, 79–93. [CrossRef]
134. Nowak, K.; Rosenthal, F.; Karlberg, T.; Butepage, M.; Thorsell, A.G.; Dreier, B.; Grossmann, J.; Sobek, J.; Imhof, R.; Luscher, B.; et al. Engineering Af1521 improves ADP-ribose binding and identification of ADP-ribosylated proteins. *Nat. Commun.* **2020**, *11*, 5199. [CrossRef] [PubMed]
135. Vivelo, C.A.; Wat, R.; Agrawal, C.; Tee, H.Y.; Leung, A.K. ADPriboDB: The database of ADP-ribosylated proteins. *Nucleic Acids Res.* **2017**, *45*, D204–D209. [CrossRef] [PubMed]
136. Ayyappan, V.; Wat, R.; Barber, C.; Vivelo, C.A.; Gauch, K.; Visanpattanasin, P.; Cook, G.; Sazeides, C.; Leung, A.K.L. ADPriboDB 2.0: An updated database of ADP-ribosylated proteins. *Nucleic Acids Res.* **2021**, *49*, D261–D265. [CrossRef] [PubMed]
137. Kuraoka, S.; Higashi, H.; Yanagihara, Y.; Sonawane, A.R.; Mukai, S.; Mlynarchik, A.K.; Whelan, M.C.; Hottiger, M.O.; Nasir, W.; Delanghe, B.; et al. A Novel Spectral Annotation Strategy Streamlines Reporting of Mono-ADP-ribosylated Peptides Derived from Mouse Liver and Spleen in Response to IFN-gamma. *Mol. Cell Proteomics* **2022**, *21*, 100153. [CrossRef]

138. Collier, R.J.; Cole, H.A. Diphtheria toxin subunit active in vitro. *Science* **1969**, *164*, 1179–1181. [CrossRef]
139. Hebert, A.S.; Prasad, S.; Belford, M.W.; Bailey, D.J.; McAlister, G.C.; Abbatiello, S.E.; Huguet, R.; Wouters, E.R.; Dunyach, J.J.; Brademan, D.R.; et al. Comprehensive Single-Shot Proteomics with FAIMS on a Hybrid Orbitrap Mass Spectrometer. *Anal. Chem.* **2018**, *90*, 9529–9537. [CrossRef]

pathogens

MDPI

Article

Characterization of an *Aedes* ADP-Ribosylation Protein Domain and Role of Post-Translational Modification during Chikungunya Virus Infection

Ramesh Kumar [1,2], Divya Mehta [1], Debasis Nayak [2,†] and Sujatha Sunil [1,*]

1 Vector Borne Diseases Group, International Centre for Genetic Engineering and Biotechnology, New Delhi 110067, India
2 Department of Biosciences and Biomedical Engineering, Indian Institute of Technology, Indore 453252, India
* Correspondence: sujatha@icgeb.res.in
† Current address: Department of Biological Sciences, Indian Institute of Science Education and Research Bhopal IISER, Bhopal 462066, India.

Abstract: Poly ADP-ribose polymerases (PARPs) catalyze ADP-ribosylation, a subclass of post-translational modification (PTM). Mono-ADP-ribose (MAR) moieties bind to target molecules such as proteins and nucleic acids, and are added as part of the process which also leads to formation of polymer chains of ADP-ribose. ADP-ribosylation is reversible; its removal is carried out by ribosyl hydrolases such as PARG (poly ADP-ribose glycohydrolase), TARG (terminal ADP-ribose protein glycohydrolase), macrodomain, etc. In this study, the catalytic domain of *Aedes aegypti* tankyrase was expressed in bacteria and purified. The tankyrase PARP catalytic domain was found to be enzymatically active, as demonstrated by an in vitro poly ADP-ribosylation (PARylation) experiment. Using in vitro ADP-ribosylation assay, we further demonstrate that the chikungunya virus (CHIKV) nsp3 (non-structural protein 3) macrodomain inhibits ADP-ribosylation in a time-dependent way. We have also demonstrated that transfection of the CHIKV nsP3 macrodomain increases the CHIKV viral titer in mosquito cells, suggesting that ADP-ribosylation may play a significant role in viral replication.

Keywords: poly ADP-ribosylation; *Aedes aegypti*; tankyrase; PARP; chikungunya virus (CHIKV)

Citation: Kumar, R.; Mehta, D.; Nayak, D.; Sunil, S. Characterization of an *Aedes* ADP-Ribosylation Protein Domain and Role of Post-Translational Modification during Chikungunya Virus Infection. *Pathogens* **2023**, *12*, 718. https://doi.org/10.3390/pathogens12050718

Academic Editors: Anthony K L Leung, Anthony Fehr and Rachy Abraham

Received: 21 February 2023
Revised: 27 April 2023
Accepted: 4 May 2023
Published: 16 May 2023

1. Introduction

ADP-ribosylation is a common modification that occurs in all life domains, including prokaryotic and eukaryotic organisms. It entails the attachment of mono or polymer units of ADP-ribose to target molecules, such as DNA, proteins, and RNA [1–3], and is known to play a role in a number of biological functions, including DNA damage repair, telomere maintenance, stress response, immunological response, cell signaling, and cell proliferation [4–6]. Three sets of proteins called writers, readers, and erasers control the process of ADP (Adenosine diphosphate)-ribosylation. The writers convert nicotinamide adenine dinucleotide (NAD^+) to nicotinamide (NAM) and ADP-ribose, then the latter attaches to target molecules. ADP-ribosyl transferases (ARTs) are commonly used for mono PARPs or Poly ADP-ribose polymerase (PARP), depending on whether they add only a single ADP-ribose (MARP) or multiple ADP-ribose units to target molecules [1,7]. Reader proteins, which contain one of the following domains such as macrodomain, WWE domain (named after three conserved single letter amino acid residues), PAR-binding motifs (PBMs), or PAR-binding zinc finger (PBZ) domain, are able to recognize ADP-ribosylation on target proteins [7]. These proteins have important roles in localization [8], DNA damage response [9], and ubiquitin-mediated proteasomal degradation [10,11] by interacting with ADP-ribosylated proteins. Eraser proteins bind to and remove ADP-ribosylation. They are divided into: terminal ADP-ribose protein glycohydrolase 1 (TARG1), poly ADP-ribose glycohydrolase (PARG), and ADP-ribosyl-acceptor hydrolases (ARH1 and ARH3). Macro

D1 and macro D2, Macro Ds, ARH1, ARH3, and TARG all eliminate mono-ADP-ribose residues, but PARG eliminates poly ADP-ribose chain [7,12].

In addition to glutamate and aspartate, additional acidic amino acid residues such as serine, arginine, and cysteine also act as acceptors for ADP-ribosylation by PARPs [2]. The next step is polymer extension, which entails repeatedly conjugating ADP-ribose from NAD+ to the previous ADP-ribose unit. This results in the construction of a linear ribose polymer chain (1″2′) made up of 2–200 ADP-ribose units, also known as a poly ADP-ribose (PAR) chain. Furthermore, branching is added to the PAR chains to boost both complexity and biological responsiveness [13–15]. By causing phase separation and encouraging protein–protein interactions, the PAR chains also aid the development of protein complexes [16].

The ADP-ribosylation process is also involved in the antiviral immune response [17–19], where it is known to activate different components in the immune pathway, such as ion channels [20], modulation of expression of genes involved in inflammation [21–23], and the RNAi (RNA interference) pathway [6], ultimately triggering the host defense mechanism against virus infection. It is well known that viruses of the *Coronaviridae*, *Togaviridae* and *Hepeviridae* families contain macrodomains, which hinder host-mediated immune response [24] by targeting stress granule formation [25], inhibiting the release of pro-inflammatory cytokines and interferons [26], enhancing pathogenesis [27], and promoting viral proliferation [28,29].

Aedes aegypti (*Ae. aegypti*) is an important vector for arboviruses, including dengue, Zika, chikungunya, yellow fever [30,31]. The effects of ADP-ribosylation during arboviral infections in the vector and function of the proteins implicated in the alteration in their survival are poorly understood. The non-structural protein 3 (nsP3) of the Chikungunya virus (CHIKV), an alphavirus, encodes a macrodomain that possesses ADP-ribosylhydrolase activity, which is important for virus replication and virulence [28,32]. In this study, we cloned and expressed the catalytic domain of *Ae. aegypti* tankyrase protein. The in vitro assay using the purified catalytic domain showed that the domain was able to auto-PARylate. When attached as monomer or polymer to target proteins, ADP-ribose moieties are known to alter their activity as well as have role in protein complex formation. The macrodomain is known to hydrolyze the ADP-ribose from the proteins and favor viral growth [25,33,34]. We found that PARylation was inhibited by the nsP3 protein, which is known to have a macrodomain, and its transfection into mosquito cells favored the CHIKV replication, indicating the crucial role of ADP-ribosylation and macrodomain play in deciding the outcome of mosquito-virus interaction.

2. Materials and Methods

2.1. Sequence Alignment and Phylogenetic Analysis

To identify PARP orthologs in *Ae. aegypti*, 17 human PARPs were blast aligned against known *Ae. aegypti* PARPs. Sequences were aligned and a phylogenetic tree was generated using MEGA 11 software [35]. The sequence alignment was done with MUSCLE algorithm. The sequences were then analyzed for phylogenetic analysis using the following method: statistical method: maximum likelihood, test of phylogeny: bootstrap method, no. of bootstrap replication: 1000, substitution model: Poisson model, ML heuristic method: nearest-neighbor-interchange (NNI), no. of threads: 5. Next, the *Aedes* PARP sequences were analyzed for domains present in them using ScanProsite (https://prosite.expasy.org/scanprosite/; accessed on 16 April 2022).

2.2. Cells, Virus, Infection and Transfection

Vero cells (ATCC® CCL-81™) were purchased from ATCC and *Ae. albopictus* (C6/36) cells were obtained from NCCS, Pune, India. *Ae. aegypti*-derived cells (Aag2) were a kind gift from Dr. Kevin Maringer, The Pirbright Institute, Surrey, UK. C6/36 and Vero cells were maintained in DMEM medium as mentioned previously [36]. Aag2 cells were grown in Leibovitz's L-15 Medium (Thermo Scientific Inc., Waltham, MA, USA, Cat. no. 11415064) and supplemented with 20% fetal bovine serum (Thermo Scientific Inc., Waltham, MA, USA, Cat. no. 10438018).

A lab-adapted CHIKV clinical strain, IND-2010#01 (Accession no. JF950631.1) was used to infect Aag2 cells [37], as mentioned in a previous study [36]. Vero cells were grown to full confluency and in Aag2 cells, 1×10^6 cells/well in a 12-well plate were seeded. The next day, the medium in the wells was changed with serum-supplemented medium absent of antibiotics. EGFP and macrodomain cloned pIB/V5-His plasmids (1.5 μg each) were suspended in 100 μL serum-free medium with 2 μL TransIT transfection reagent (Mirus Bio LLC, Madison, WI, USA) separately and kept at RT for 20 min (minutes). The mixture was added drop-wise to cells and after 24 h (hour) the cells were infected with CHIKV at MOI (multiplicity of infection) of 1. The cells/supernatant were collected at 24, 36, and 48 hpi (hours post infection).

2.3. Gene Cloning, and Expression

To clone the catalytic domain tankyrase of *Ae. aegypti*, the protocol uses RNA isolated from *Ae. aegypti* and PrimeScript One Step RT-PCR Kit (Takara Bio Inc., Japan, Cat RR055B) tankyrase-specific primers: forward primer 5′-ATAGGTACCAGCGGCACATCCATGGCCAACAG-3′ and reverse primer 5′-ATAGGATCCCTCGCTGGCTCCTGGGGCTAG-3′, with annealing at 65 °C and primer extension for 120 s (seconds). The PCR product was cloned into pET32a vector. The tankyrase-pET32a plasmid was transformed into *E. coli* CodonPlus cells, whereas CHIKV nsP3 cloned in pET29a was used from a previous study [38]. EGFP was also cloned using forward primer 5′-GGGGTACCATGCATCATCACCATCACCATCGGATGGTGAGCAAGGGCGAG-3′ and reverse primer 5′-ACCGGATCCCTTGTACAGCTCGTCCATGCCGAGAGTGATCCCG-3′, and CHIKV nsP3 macrodomain was cloned using forward primer 5′-ATAGGTACCATGGCACCGTCGTACCGGGTAAAACG-3′ and reverse primer 5′-ATAGGATCCGTCCGCATCTGTATGGCCTCAG-3′ into the pIB/V5-His vector.

2.4. Protein Purification

The tankyrase catalytic domain and CHIKV nsP3 protein were purified using a previously published protocol with slight modifications [38]. *E. coli* cultures having plasmid encoding *Ae. aegypti* tankyrase catalytic domain and CHIKV nsP3 were induced with 1mM IPTG for 16–20 h at 18 °C (degree Celsius). The cultures were pelleted and lysed in the lysis buffer (Tris-Cl (pH 8.0) 50 mM (milli Molar), NaCl 150 mM, EDTA 2 mM, glycerol 5%, β-mercaptoethanol 2 mM, and PMSF 1 mM) with lysozyme. This was followed by centrifugation. The clarified supernatant was mixed with freshly recharged Ni-NTA (Nickel-Nitrilotriacetic acid) agarose beads. The beads were eluted with lysis buffer containing 300 mM imidazole. The imidazole was removed by using dialysis membrane overnight with Tris-Cl (50 mM) pH 8.0, NaCl (150 mM), and DTT (2 mM) for further use of protein.

2.5. SDS-PAGE, Transfer, and Western Blotting

The SDS-PAGE and western blot were carried out according to prior methodology [36]. Briefly, Aag2 cells were lysed in RIPA buffer and 15–20 μg of each cell lysate or 4–6 μg of purified protein samples were resolved in SDS-PAGE gel and then transferred onto the nitrocellulose membrane (Bio-Rad Laboratories, Hercules, CA, USA). The membranes were probed with the following primary antibodies: anti-His HRP antibody (Santa Cruz, Dallas, TX, USA, Cat. no. sc-8036-HRP, 1:5000), anti-actin HRP (C4) (Santa Cruz, Dallas, USA, Cat. no. sc-47778 HRP, 1:6000), anti-pADPr antibody (10H) (Santa Cruz, Dallas, USA, Cat. no. sc-56198, 1:3000 dilution), and anti-V5 tag antibody (Thermo Fisher Scientific, Waltham, MA, USA, Cat. no. R960-25, 1:5000) and anti-CHIKV E1 (in house raised in mice, 1:3000). The blots probed with anti-pADPr antibody, anti-CHIKV E1 sera, and anti-V5 tag antibody were incubated with anti-mice IgG HRP antibody (Novus Biologicals, Colorado, USA, Cat. no. NB7539, 1:6000 dilution), washed with PBST, and then visualized using the Bio-Rad ChemiDoc MP System (Bio-Rad Laboratories, Hercules, CA, USA) after brief exposure to a chemiluminescent substrate. The uncropped images are included in supplementary information Figure S1.

2.6. Plaque Assay

The viral titration was done using plaque assay as per previously published protocol [39]. Briefly, the medium was replaced with serum-free medium 1 h prior to infection. The medium collected from CHIKV-infected Aag2 cells were initially diluted at 1:10 and added to the first well in triplicates, and then was diluted at 1:2 in the rest of the wells. The virus was allowed to bind to the cells for 90 min, and then the serum-supplemented medium was added. The wells were then added with 1% carboxymethyl cellulose (CMC) (Sigma–Aldrich, St. Louis, MO, USA, Cat. no. C4888) and plates were transferred back to the 5% CO_2 supplemented humidified incubator at 37 °C for 72 h. The cells were fixed with paraformaldehyde for 1 h. After fixing, crystal violet stain (0.25%) was added to the wells and incubated for 30 min. The stain solution was discarded and wells were rinsed with tap water. The plaques were calculated as plaque-forming units (pfu) = (number of plaques)/(dilution × volume of the virus).

2.7. In Vitro PARylation Assay and Co-Incubation Assay with CHIKV nsP3 Protein

The in vitro assay was performed following previous protocol [40]. The *E. coli* purified recombinant tankyrase protein (8 to 12 µg) was incubated for the specified time at 28 °C in PARP reaction buffer (50 mM Tris-Cl [pH 8.0], 4 mM $MgCl_2$, 0.2 mM DTT (dithiothreitol) containing 25 µM beta-Nicotinamide adenine dinucleotide sodium salt (NAD+) (Sigma–Aldrich, St. Louis, MO, USA, Cat. no. N0632-1G). The reactions were terminated by adding SDS loading buffer, and 4–6 µg of each protein sample was fractionated by 10% SDS-PAGE. The proteins were transferred onto a nitrocellulose membrane and probed with anti-pADPr antibody (10H) (Santa Cruz, USA, Cat. no. sc-56198 1:2000 dilution). The membrane was then probed with anti-mice IgG HRP antibody (Novus Biologicals, Centennial, CO, USA, Cat. no. NB7539) and visualized in a Bio-Rad ChemiDoc MP System after brief exposure to chemiluminescent substrate. Similar to the in vitro PARylation assay, the nsP3 co-incubation assay was carried out. The purified recombinant CHIKV nsP3 protein (8 to 12 µg) was added to the reaction mixture for the desired time and the reaction was stopped by adding SDS loading buffer. The 4–6 µg of each sample were then separated by SDS-PAGE, transferred onto the nitrocellulose membrane, and probed with anti-pADPr antibodies, followed by exposure to chemiluminescent substrate and visualization in ChemiDoc MP system.

2.8. In Vitro Transcription, RNA Isolation and Real-Time PCR

For double stranded RNA synthesis, T7 sequence (Tankyrase forward primer 5′-TAATACGACTCACTATAGGGTCACCGAACTGCTCATCAAG-3′, reverse primer 5′-TATCCGAAGCGAAAGCAAGTCCCTATAGTGAGTCGTATTA-3′, EGFP forward primer 5′-TAATACGACTCACTATAGGGATGGTGAGCAAGGGCGAGG-3′, and reverse primer 5′-TAATACGACTCACTATAGGGCTTGTACAGCTCGTCCATGCC-3′) was added to the primer and the desired product of around 400 bp was amplified using Dream Taq DNA polymerase (Thermo Fisher Scientific Inc., Waltham, MA, USA, Cat. no. EP0701) following the manufacturer's protocol. The PCR product was purified and an in vitro reaction was setup using MEGAscript T7 transcription kit (Thermo Fisher Scientific Inc., Waltham, MA, USA, Cat. no. AM1334) following the manufacturer's protocol. The dsRNA after in vitro transcription was treated with TURBO DNase. The RNA was purified using TRIzol reagent. Whole cell RNA isolation from Aag2 cells was done following previous protocol [41]. Total cellular RNA was isolated using TRIzol (Thermo Fisher Scientific Inc., Waltham, MA, USA). RNA was dissolved in DEPC treated water and quantified. One-step SYBR green real-time PCR was carried out on PIKOREAL 96 Well real-time PCR system (Thermo Fisher Scientific Inc., Waltham, MA, USA). A total of 300 ng total RNA per reaction was used with 0.3 µM of each primer with QuantiTect PCR kit (Qiagen, Hilden, Germany).

The RT-PCR conditions for the one-step RT-PCR consisted of a 30 min reverse transcription step at 50 °C and then 2 min of initial denaturation at 95 °C, followed by 40 cycles of PCR at 95 °C without holding time (denaturation), 60 °C for 30 s (annealing), and 72 °C for 30 s (extension). Small subunit ribosomal protein 7 (RPS7) was used as an internal control. Tankyrase real-time PCR sequence used forward primer 5′-GGTGAAGAACCTCGAGAAAGAA-3′ and reverse primer 5′-CAATAGCAGCAAAGCTGGAAC-3′ and RPS7 forward primer 5′-CCCGGTTGACGATGGATTT-3′ and reverse primer 5′-TCACGAAACCAGCGATCTTATT-3′.

2.9. Immunofluorescence Assay

Aag2 cells were cultured in 6 well plates containing sterile glass coverslips. Cells were infected with CHIKV at MOI 1 for 24 h and 48 h. The cells were then fixed with 4% paraformaldehyde for 30 min and then permeabilized using 0.1% Triton-X-100 for 30 min. Cells were then blocked with bovine serum and then incubated with anti-CHIKV nsP3 rabbit serum [38] at 1:200 dilution in PBS (phosphate buffer saline) + 2.5% BSA overnight. The following day, washing was done using PBST (PBS + 0.1% tween-20) 3 × 10 min. The cells were then added with a secondary antibody (anti-mice IgG Alexa 594) at 1:400 dilution. This was followed by washing with PBS + 0.1% tween-20 for 3 × 10 min. The cells were immersed in DAPI for a few minutes and then visualized in Nikon eclipse confocal microscope (Nikon Corp, Tokyo, Japan) with oil immersion for magnification.

2.10. Statistical Analysis and Software

Statistical analyses for plaque assay analysis and real-time PCR were performed using two-way ANOVA. The analyses were done using Graphpad prism software (version 9.1.1).

3. Results

3.1. Identification of ADP-Ribose Polymerases in Ae. aegypti

In order to compare the proteins in *Ae. aegypti* to the 17 human PARPs, sequence alignment was performed using blast tool (Blastp, NCBI). The results showed that *Ae. aegypti* encodes for three ADP-ribose polymerases, which are as follows: (1) tankyrase (NCBI accession: XP_021708496.1), (2) Poly ADP-ribose polymerase (PARP; NCBI accession: XP _001661932.1), and (3) Mono-ADP-ribose polymerase (MARP; NCBI accession: XP _001647568.1). The phylogenetic sequence analysis showed that the *Ae. aegypti* tankyrase protein was showing the highest sequence similarity to human PARP5a and PARP5b, and the *Ae. aegypti* PARP protein had the highest sequence similarity to human PARP1, while the *Ae. aegypti* MARP protein was showing the highest sequence similarity to human PARP16 (Figure 1A). These three ADP-ribose polymerases differed from one another in terms of the various sorts of domains. Here is the domain analysis for the three proteins:

1. **Tankyrase:** Tankyrase-1 (PARP5a) and tankyrase-2 (PARP5b) in human were found to be closest to *Ae.* tankyrase among the 17 human PARPs (Figure 1A). Three different types of domains were identified by the domain analysis: Ankyrin repeats, SAM domains, and PARP catalytic domains (Figure 1B). The 30–35 amino acid long motifs known as ankyrin repeats, which have a helix-turn-helix shape, are essential for protein–protein interactions [41,42]. Protein–protein interactions are mediated by another domain called the sterile alpha motif (SAM). These play a role in oligomerization as well as binding [43]. ADP-ribose is added by the third domain, called the PARP catalytic domain. Sequence alignment of tankyrase and PARP5b revealed an Ankyrin repeat region, which is crucial for protein–protein interaction and PARP catalytic domain, which is responsible for ADP-ribosylation activity, exhibited a higher region of similarity (Figure S2).
2. **PARP:** A poly ADP-ribose polymerase called PARP is the other protein found in *Ae. aegypti*. The most resemblance between *Ae. aegypti* PARP and human PARP1 was found during the phylogenetic analysis (Figure 1A). According to domain analysis, there are different types of domains: PARP Zn, BRCT, PARP alpha, and PARP catalytic domain (Figure 1B). A zinc finger domain, PARP Zn, included two copies. These

proteins, which typically reside in the nucleus, are implicated in DNA repair [44]. The BCRT (BRCA1 C-terminus) domain was the second domain from the protein's N-terminal. When the PARP alpha domain binds to the site of DNA damage, it transmits the activation signal [45]. The sequence alignment of PARP with PARP1 revealed several amino acid similarities between these two proteins, with the PARP catalytic domain showing the highest degree of similarity, indicating that this domain is mostly conserved in these animals (Figure S3). All of these facts suggest that the *Aedes* PARP protein is an enzyme that repairs DNA damage.

3. **MARP:** Mono-ADP-ribose polymerase (MARP) is responsible for adding mono-ADP-ribose units to proteins. These proteins cannot further connect ADP-ribose subunits to the terminals of those already attached [46]. According to the results of the phylogenetic research, the human MARP protein PARP16 and the *Ae. aegypti* MARP have the highest degree of similarity (Figure 1A). Proteins share comparable amino acids in the region responsible for catalytic activity of the protein, as seen by the sequence alignment of MARP and PARP16 (Figure S4). The MARP protein from *Ae. aegypti* is 362 amino acids long and only comprises a catalytic domain (Figure 1B), suggesting that it may be used for priming proteins or for MARylating proteins that are either activated or inactivated upon MARylation.

Based on analysis and evidence from the literature, we came to the conclusion that tankyrase is responsible for attaching the ADP-ribose chain to proteins, PARP is responsible for DNA repair, and MARP is responsible for intracellular signaling or priming. As the *Ae. aegypti* tankyrase protein contained a region implicated in protein–protein interaction, we moved forward with its cloning, production, purification, and characterization. These kinds of proteins are crucial for controlling cellular functions, and their discovery and characterization may shed light on the intricate mechanisms governing numerous biological processes. Full-length PARP proteins are required for the PARylation of target proteins in cells [47–49], but the catalytic domain alone is sufficient to create ADP-ribose chains on the proteins [49]. The catalytic domain of *Ae. aegypti* tankyrase protein (Figure 1B) was cloned into pET32a vector. The purified protein (of 60 kDa (kilo Dalton) size) was expressed in soluble form and was checked for purity using Coomassie stain and western blot (Figure 1C).

3.2. *In Vitro PARylation Assay of Catalytic Domain of Tankyrase Protein and Impact of nsP3 Macrodomain*

Each of the several domains that make up the PARP proteins is essential for their proper function in cells. Target proteins are added with long, variable-length ADP-ribose chains by these PARPs (Figure 2A) [15,16,50]. In this study, the capacity to add ADP-ribose subunits was initially assessed in the catalytic domain of tankyrase proteins. The tankyrase alone (Figure 3B, lane 1) and NAD^+ (Figure 2B, lane 2) as well as CHIKV capsid protein (Figure 2B, lane 3) were employed as a negative control (for a non-specific signal). The presence of tankyrase catalytic domain resulted in an intense band of higher molecular-weight proteins (Figure 2B, lane 4), indicating that tankyrase domain was using NAD^+ as a substrate to add ADP-ribose to itself via mechanism called auto-PARylation (Figure 3A). At both 30 min and 60 min, the PARylation assay revealed that the protein had undergone ADP-ribose modification with various lengths of the PAR chain (Figure 2C).

The in vitro PARylation assay showed that catalytic domain of tankyrase could add ADP-ribose units. Previous work from lab by Mathur et al. [51] has shown that CHIKV nsP3 macrodomain act as a viral suppressor of RNAi. CHIKV macrodomain is a mono-ADP-ribosylhydrolase and is crucial for the viral replication [28,52]. We were curious to find out if the CHIKV macrodomain, which is known to remove mono-ADP-ribose moieties [52], affected the poly ADP-ribosylation of proteins caused by *Ae. aegypti* PARPs. The bacterially purified CHIKV nsP3 protein (Figure S5) was added to the PARylation reaction mixture, and the reaction was run for 30 and 60 min in order to assess the function of the nsP3 macrodomain on ADP-ribosylation. Following the incubation of the nsP3 protein, it was found that the PARylation decreased as the incubation duration increased up to 60 min. In comparison to the 30 min and 60 min of the PARylated samples alone, the number or length of the PAR chains was lower at 30 min following the incubation of the nsP3 protein and dramatically decreased at 60 min (Figure 2D). The knockdown of tankyrase gene by dsRNA transfection resulted in increased titer of CHIKV (Figure 2E).

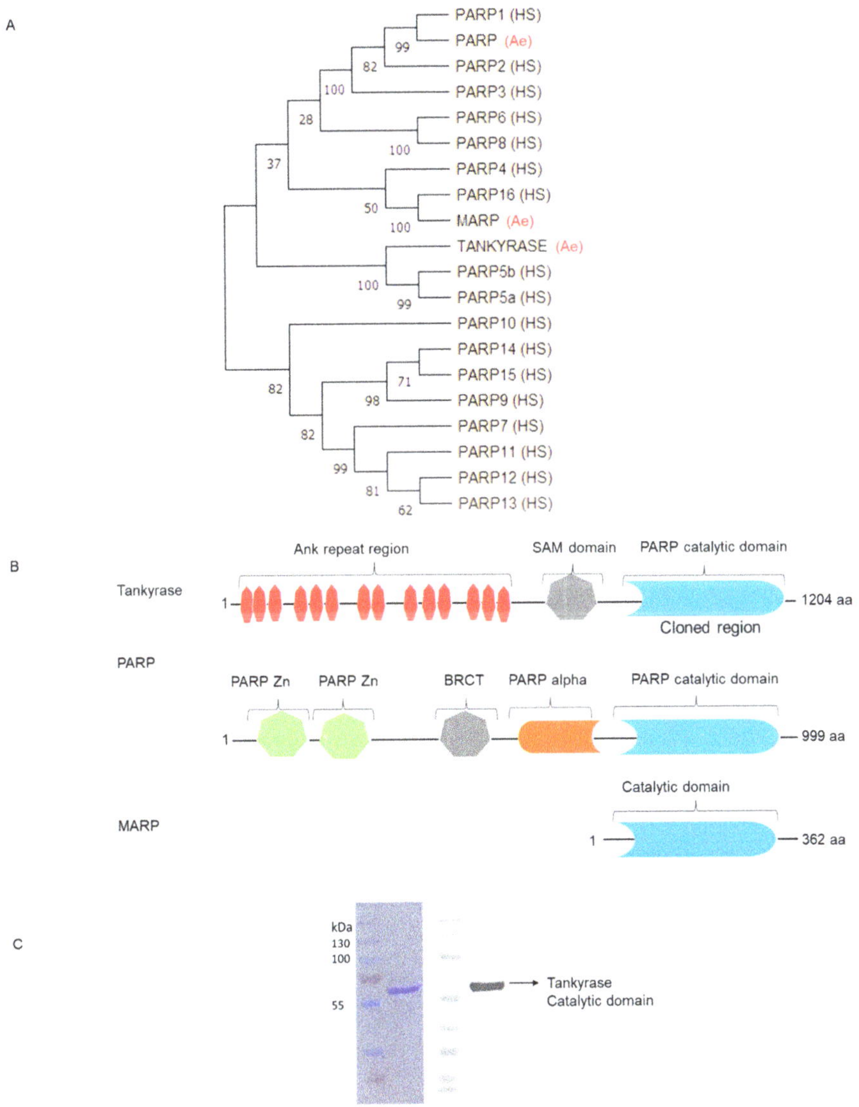

Figure 1. Analysis of ADP-ribose polymerase proteins in *Ae. aegypti*. (**A**) Phylogenetic comparison of 17 human PARP proteins with those from *Ae. aegypti*. *Aedes* PARPs are highlighted by red color. HS = *Homo sapiens*, Ae = *Ae. aegypti*; (**B**) Domain analysis of the ADP-ribose polymerase proteins from *Ae. aegypti*. Each color denotes a certain sort of domain, and the quantity of colored boxes indicates how many copies of that particular domain in the protein, and (**C**) Coomassie staining and western blot of purified recombinant tankyrase catalytic domain and western blot with anti-His HRP tagged antibody.

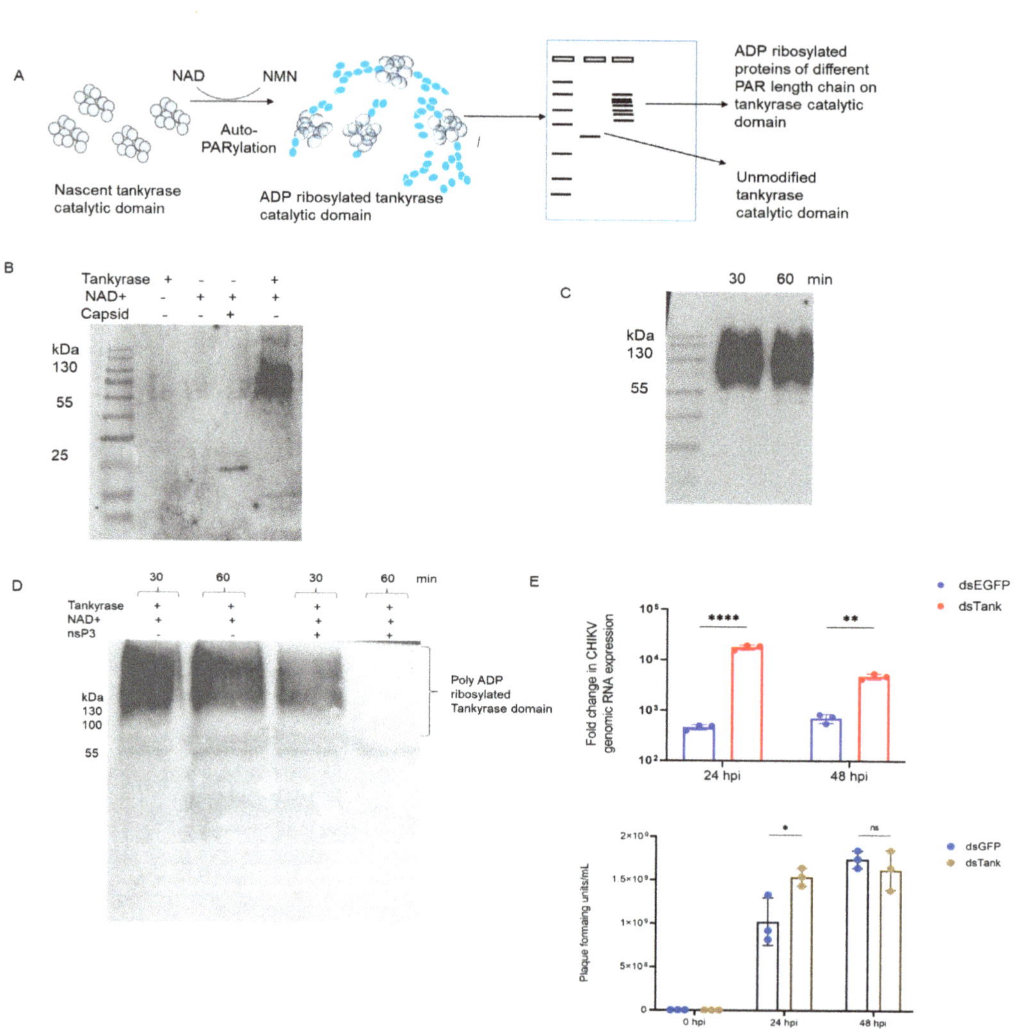

Figure 2. In vitro PARylation assay of tankyrase catalytic domain. (**A**) Schematic diagram of in vitro PARylation assay and sample (unmodified protein and ADP-ribose modified) protein separation on SDS-PAGE; (**B**) In vitro PARylation assay with tankyrase alone (negative control), NAD^+ alone (negative control), CHIKV capsid proteins with NAD^+ and tankyrase with NAD^+ for 30 min reaction (left image is the anti-PAR antibody blotted membrane and right image is the Ponceau-stained membrane before blocking and anti-PAR antibody exposure); (**C**) Western blot of time points (30 and 60 min) of in vitro PARylated tankyrase catalytic domain; (**D**) In vitro PARylation assay in the absence of CHIKV nsP3 protein for 30 and 60 min (lane 1 and lane 2 from the left side). The impact of nsP3 protein on PARylation was checked by co-incubation of nsP3 protein and PARylation buffer having tankyrase protein for 30 and 60 min (lane 3 and 4) and (**E**) dsRNA mediated knockdown of *Ae. aegypti* tankyrase transcript. Aag2 cells were transfected with dsRNA for tankyrase and EGFP (control) for 24 h and then infected with CHIKV at MOI of 1. Cells were collected at 24 hpi and 48 hpi and viral titer was quantified by CHIKV E1 specific primers using real-time PCR and plaque assay. ns- non-significant, * p-value < 0.05 ** p-value < 0.001 and **** p-value < 0.0001.

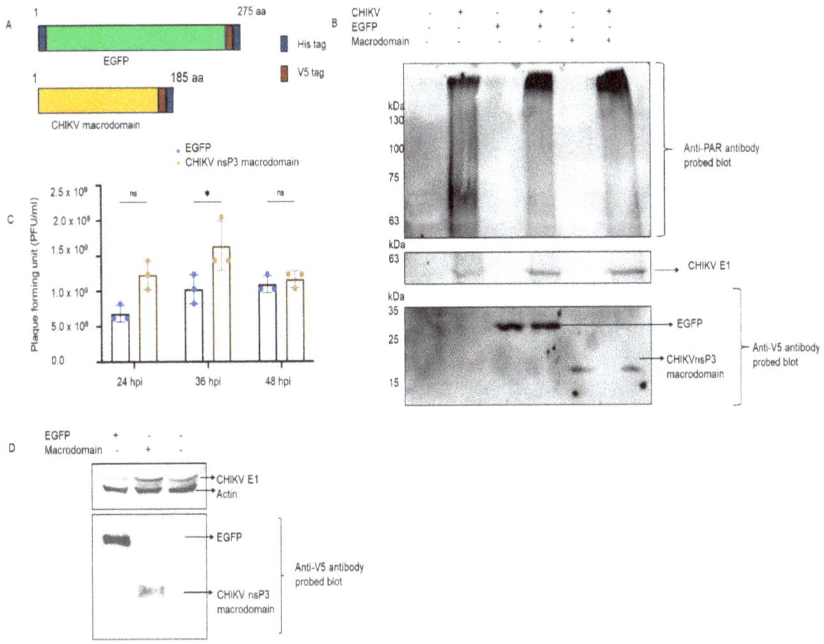

Figure 3. Effect of CHIKV nsP3 on PARylation. (**A**) Diagram of EGFP and CHIKV macrodomain cloned in insect vector (pIB/V5-His); (**B**) EGFP and CHIKV nsP3 macrodomain-pIB clones were transfected into Aag2 cells and then infected with CHIKV at MOI of 1 for 48 hpi. The lysates were blotted with mice anti-PAR antibody (for global PARylation level detection) and V5 (for detection of transfected EGFP and macrodomain in cells) and CHIKV E1 sera for different time points to observe the impact on viral growth (top), Ponceau-stained membrane sued to probing the antibodies (bottom); (**C**) Aag2 transfected with EGFP and macrodomain were infected with CHIKV at MOI of 1 for 24, 36, and 48 hpi. The medium was collected and used for viral titration via plaque assay, error bars represent standard deviation (sd). $n = 4$ (triplicates), and (**D**) Western blot of EGFP and macrodomain transfected cells infected with CHIKV at MOI of 1 after 48 hpi. The membrane was blotted with antibodies for CHIKV E1, actin, and V5 tag. ns- non-significant, * p-value < 0.05.

3.3. *Effect of CHIKV nsP3 Macrodomain on PARylation Activity of Tankyrase*

In CHIKV-infected Aag2 cells, the effect of the macrodomain alone on viral replication was also investigated. The nsP3 protein form discrete granules in cells, called replication complexes, and the number of cells increases with infection time (Figure S6), indicating that the nsP3 protein is not uniformly present in cells. To evaluate the impact of the macrodomain in viral kinetics, EGFP and the CHIKV macrodomain were cloned in the pIB/V5-His vector (Figure 3A) and transfected into Aag2 cells. The lysates were separated on SDS-PAGE gel and incubated with an anti-pADPr antibody. In all conditions of CHIKV infection (alone, with EGFP, or macrodomain transfected cells), PAR levels were increased compared to uninfected cells and transfected cells (Figure 3B). The global cellular PAR level difference between EGFP and macrodomain transfected cells was not significant. In plaque assay, we observed that the viral titer was high in macrodomain transfected cells compared to control (EGFP transfected) at 24 and 36 hpi, but at 48 hpi the difference between control and macrodomain was small (Figure 3C). At 24 hpi, western blot examination of CHIKV E1 protein revealed a similar pattern. CHIKV expression was lower in cells transfected with EGFP (lane 1) than it was in cells transfected with macrodomain (lane 2) (Figure 3D).

4. Discussion

ADP-ribosylation is an important PTM of proteins and nucleic acids that is mediated by PARPs. DNA repair, cell signaling, stress response, pathogen response, and gene control are just a few of the functions that PARPs are engaged in [49]. By targeting cellular transcripts, encouraging apoptosis [53], attenuating RISC (RNA induced silencing complex) mediated transcript silencing [54], inducing interferon-stimulated genes (ISGs), and degrading viral proteases, PARPs provide antiviral functions during viral infections [55]. In the current study, the sequence alignment with human PARPs led to the identification of three *Ae. aegypti* PARP proteins, including tankyrase (PARP5b), PARP (PARP1), and MARP (PARP16). Among these, tankyrase catalytic domain was cloned, expressed, and purified in a bacterial system. By using an in vitro PARylation assay, it was discovered that the tankyrase catalytic domain add PAR chains of variable length to its own molecules (auto-PARylation). The phosphate residues in the PAR chains imparts a negative charge, which interacts with the candidate proteins' PAR binding motif (PBM) [56]. The length and degree of branching impacts the propensity to create multimeric complexes [50,57] and also affects cellular systems [15]. Knockdown of tankyrase led to higher viral titer, indicating that there might be other PARP isoforms that are involved in the immunity against viruses, or tankyrase is involved in the inactivation of viral proteins, hence its knockdown increasing the viral titer. A recent study highlights that mono-ADP-ribosylation of viral protein (nsP2) by host PARP leads to inhibition of nsP2 enzymatic activity [34], raising the possibility that a similar protein might be playing a role in providing immunity to mosquito cells against viral infection.

The fact that the active macrodomain of an alphavirus is conserved in the active site region shows how crucial the active macrodomain is for viral life [32]. According to earlier research, mutation in the active areas influences viral replication [28,32]. The results of the current study demonstrated that CHIKV infection leads to increased PARylation of the cellular proteome, but nsP3 macrodomain transfection did not affect the global PARylation compared to the EGFP control. During infection, nsP3 protein is present as discrete granules in the cells, indicating that it may not be interacting with the whole host proteome but instead only a limited number of proteins. Transfection of nsP3 macrodomain significantly reduced viral titer, suggesting that the macrodomain prevents PAR-chain formation by hydrolyzing ADP-ribose. Based on our data, we proposed a model hypothesis that host PARP proteins are either ADP-ribosylate host or viral proteins which leads to the inhibition of viral replication (by activation of immune pathways or inhibition of crucial viral protein activity). To counter the host immune mechanism, the viral macrodomain removes ADP-ribose from the host or viral proteins, leading to an inactivation of host immune pathways or resumption of viral protein activity (Figure 4). Further in-depth studies are essential to identify and characterize other host targets and modes of action of macrodomain on modulating host/viral protein functions.

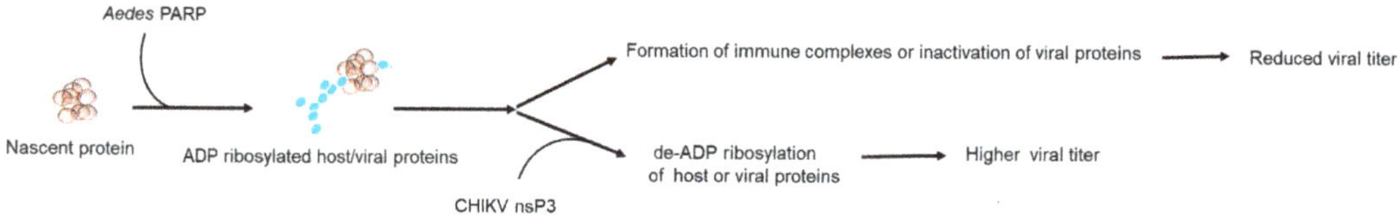

Figure 4. Proposed model of mechanism of tankyrase-mediated PARylation and CHIKV nsP3 macrodomain-mediated de-ADP-ribosylation of host/viral proteins. We hypothesize based on literature as well as current evidence that ADP-ribosylation inhibits viral growth by activating immune pathways or inactivating crucial viral proteins and thus reduces viral titer. CHIKV nsP3 has strong de-MARylation and weak de-PARylation activity. This indicates that nsp3 macrodomain either removes ADP-ribose from viral proteins, thereby preventing their inactivation or from host proteins thus inhibiting their role in immune pathways and crucial metabolic processes. This eventually leads to increased viral titer and compromised host immune system.

Supplementary Materials: The following supporting information can be downloaded at: https://www.mdpi.com/article/10.3390/pathogens12050718/s1, Figure S1: Uncropped images of SDS-PAGE gel, ponceau stained and chemiluminescent substrate exposed membranes used in the study.; Figure S2: Sequence alignment of *Ae. aegypti* Tankyrase proteins and human PARP5b protein.; Figure S3 Sequence alignment of *Ae. aegypti* PARP protein and human PARP1 protein.; Figure S4: Sequence alignment of *Ae. aegypti* MARP protein and human PARP16 protein.; Figure S5: Coomassie brilliant blue stained SDS-PAGE gel and western blot of bacterial purified recombinant nsP3 protein.; Figure S6: Immunofluorescence assay of CHIKV infected Aag2 cells.

Author Contributions: Conceptualization, S.S. and R.K.; methodology, R.K. and D.M.; validation, S.S., R.K. and D.M.; investigation, S.S.; resources, S.S.; writing—original draft preparation, R.K.; writing—review and editing, R.K., S.S. and D.N.; supervision, S.S. and D.N.; funding acquisition, S.S. All authors have read and agreed to the published version of the manuscript.

Funding: The work was supported by ICGEB core grant and Department of Biotechnology (BT/PR20554/MED/29/1107/2016). RK received Ph.D. funding from the Department of Biotechnology, India.

Institutional Review Board Statement: Not applicable.

Informed Consent Statement: Not applicable.

Data Availability Statement: Not applicable.

Conflicts of Interest: The authors declare no conflict of interest.

References

1. Cohen, M.S.; Chang, P. Insights into the biogenesis, function, and regulation of ADP-ribosylation. *Nat. Chem. Biol.* **2018**, *14*, 236–243. [CrossRef] [PubMed]
2. Hottiger, M.O.; Hassa, P.O.; Luscher, B.; Schuler, H.; Koch-Nolte, F. Toward a unified nomenclature for mammalian ADP-ribosyltransferases. *Trends Biochem. Sci.* **2010**, *35*, 208–219. [CrossRef] [PubMed]
3. Grunewald, M.E.; Fehr, A.R.; Athmer, J.; Perlman, S. The coronavirus nucleocapsid protein is ADP-ribosylated. *Virology* **2018**, *517*, 62–68. [CrossRef] [PubMed]
4. Liu, C.; Yu, X. ADP-ribosyltransferases and poly ADP-ribosylation. *Curr. Protein Pept. Sci.* **2015**, *16*, 491–501. [CrossRef]
5. Palazzo, L.; Mikolcevic, P.; Mikoc, A.; Ahel, I. ADP-ribosylation signalling and human disease. *Open Biol.* **2019**, *9*, 190041. [CrossRef]
6. Leung, A.K.; Vyas, S.; Rood, J.E.; Bhutkar, A.; Sharp, P.A.; Chang, P. Poly(ADP-ribose) regulates stress responses and microRNA activity in the cytoplasm. *Mol. Cell* **2011**, *42*, 489–499. [CrossRef]
7. Li, P.; Lei, Y.; Qi, J.; Liu, W.; Yao, K. Functional roles of ADP-ribosylation writers, readers and erasers. *Front Cell Dev. Biol.* **2022**, *10*, 941356. [CrossRef]
8. Osuagwu, N.; Dolle, C.; Tzoulis, C. Poly-ADP-ribose assisted protein localization resolves that DJ-1, but not LRRK2 or alpha-synuclein, is localized to the mitochondrial matrix. *PLoS ONE* **2019**, *14*, e0219909. [CrossRef]
9. Liu, C.; Vyas, A.; Kassab, M.A.; Singh, A.K.; Yu, X. The role of poly ADP-ribosylation in the first wave of DNA damage response. *Nucleic Acids Res.* **2017**, *45*, 8129–8141. [CrossRef]
10. Vivelo, C.A.; Ayyappan, V.; Leung, A.K.L. Poly(ADP-ribose)-dependent ubiquitination and its clinical implications. *Biochem. Pharmacol.* **2019**, *167*, 3–12. [CrossRef]
11. DaRosa, P.A.; Wang, Z.; Jiang, X.; Pruneda, J.N.; Cong, F.; Klevit, R.E.; Xu, W. Allosteric activation of the RNF146 ubiquitin ligase by a poly(ADP-ribosyl)ation signal. *Nature* **2015**, *517*, 223–226. [CrossRef] [PubMed]
12. O'Sullivan, J.; Tedim Ferreira, M.; Gagne, J.P.; Sharma, A.K.; Hendzel, M.J.; Masson, J.Y.; Poirier, G.G. Emerging roles of eraser enzymes in the dynamic control of protein ADP-ribosylation. *Nat. Commun.* **2019**, *10*, 1182. [CrossRef]
13. Rack, J.G.M.; Liu, Q.; Zorzini, V.; Voorneveld, J.; Ariza, A.; Honarmand Ebrahimi, K.; Reber, J.M.; Krassnig, S.C.; Ahel, D.; van der Marel, G.A.; et al. Mechanistic insights into the three steps of poly(ADP-ribosylation) reversal. *Nat. Commun.* **2021**, *12*, 4581. [CrossRef] [PubMed]
14. Chen, Q.; Kassab, M.A.; Dantzer, F.; Yu, X. PARP2 mediates branched poly ADP-ribosylation in response to DNA damage. *Nat. Commun.* **2018**, *9*, 3233. [CrossRef] [PubMed]
15. Aberle, L.; Kruger, A.; Reber, J.M.; Lippmann, M.; Hufnagel, M.; Schmalz, M.; Trussina, I.; Schlesiger, S.; Zubel, T.; Schutz, K.; et al. PARP1 catalytic variants reveal branching and chain length-specific functions of poly(ADP-ribose) in cellular physiology and stress response. *Nucleic Acids Res.* **2020**, *48*, 10015–10033. [CrossRef]
16. Jin, X.; Cao, X.; Liu, S.; Liu, B. Functional Roles of Poly(ADP-Ribose) in Stress Granule Formation and Dynamics. *Front Cell. Dev. Biol.* **2021**, *9*, 671780. [CrossRef]

17. Li, M.M.; Lau, Z.; Cheung, P.; Aguilar, E.G.; Schneider, W.M.; Bozzacco, L.; Molina, H.; Buehler, E.; Takaoka, A.; Rice, C.M.; et al. TRIM25 Enhances the Antiviral Action of Zinc-Finger Antiviral Protein (ZAP). *PLoS Pathog.* **2017**, *13*, e1006145. [CrossRef]
18. Fehr, A.R.; Singh, S.A.; Kerr, C.M.; Mukai, S.; Higashi, H.; Aikawa, M. The impact of PARPs and ADP-ribosylation on inflammation and host-pathogen interactions. *Genes Dev.* **2020**, *34*, 341–359. [CrossRef]
19. Chiu, H.P.; Chiu, H.; Yang, C.F.; Lee, Y.L.; Chiu, F.L.; Kuo, H.C.; Lin, R.J.; Lin, Y.L. Inhibition of Japanese encephalitis virus infection by the host zinc-finger antiviral protein. *PLoS Pathog.* **2018**, *14*, e1007166. [CrossRef]
20. Schwarz, N.; Drouot, L.; Nicke, A.; Fliegert, R.; Boyer, O.; Guse, A.H.; Haag, F.; Adriouch, S.; Koch-Nolte, F. Alternative splicing of the N-terminal cytosolic and transmembrane domains of P2X7 controls gating of the ion channel by ADP-ribosylation. *PLoS ONE* **2012**, *7*, e41269. [CrossRef]
21. Nie, Y.; Nirujogi, T.S.; Ranjan, R.; Reader, B.F.; Chung, S.; Ballinger, M.N.; Englert, J.A.; Christman, J.W.; Karpurapu, M. PolyADP-Ribosylation of NFATc3 and NF-kappaB Transcription Factors Modulate Macrophage Inflammatory Gene Expression in LPS-Induced Acute Lung Injury. *J. Innate Immun.* **2021**, *13*, 83–93. [CrossRef] [PubMed]
22. Nakajima, H.; Nagaso, H.; Kakui, N.; Ishikawa, M.; Hiranuma, T.; Hoshiko, S. Critical role of the automodification of poly(ADP-ribose) polymerase-1 in nuclear factor-kappaB-dependent gene expression in primary cultured mouse glial cells. *J. Biol. Chem.* **2004**, *279*, 42774–42786. [CrossRef] [PubMed]
23. Ke, Y.; Lv, X.; Fu, X.; Zhang, J.; Bohio, A.A.; Zeng, X.; Hao, W.; Wang, R.; Boldogh, I.; Ba, X. Poly(ADP-ribosyl)ation enhances HuR oligomerization and contributes to pro-inflammatory gene mRNA stabilization. *Cell Mol. Life Sci.* **2021**, *78*, 1817–1835. [CrossRef] [PubMed]
24. Alhammad, Y.M.O.; Fehr, A.R. The Viral Macrodomain Counters Host Antiviral ADP-Ribosylation. *Viruses* **2020**, *12*, 384. [CrossRef] [PubMed]
25. Jayabalan, A.K.; Adivarahan, S.; Koppula, A.; Abraham, R.; Batish, M.; Zenklusen, D.; Griffin, D.E.; Leung, A.K.L. Stress granule formation, disassembly, and composition are regulated by alphavirus ADP-ribosylhydrolase activity. *Proc. Natl. Acad. Sci. USA* **2021**, *118*, e2021719118. [CrossRef]
26. Fehr, A.R.; Channappanavar, R.; Jankevicius, G.; Fett, C.; Zhao, J.; Athmer, J.; Meyerholz, D.K.; Ahel, I.; Perlman, S. The Conserved Coronavirus Macrodomain Promotes Virulence and Suppresses the Innate Immune Response during Severe Acute Respiratory Syndrome Coronavirus Infection. *mBio* **2016**, *7*, e01721-16. [CrossRef]
27. Fehr, A.R.; Athmer, J.; Channappanavar, R.; Phillips, J.M.; Meyerholz, D.K.; Perlman, S. The nsp3 macrodomain promotes virulence in mice with coronavirus-induced encephalitis. *J. Virol.* **2015**, *89*, 1523–1536. [CrossRef]
28. McPherson, R.L.; Abraham, R.; Sreekumar, E.; Ong, S.E.; Cheng, S.J.; Baxter, V.K.; Kistemaker, H.A.; Filippov, D.V.; Griffin, D.E.; Leung, A.K. ADP-ribosylhydrolase activity of Chikungunya virus macrodomain is critical for virus replication and virulence. *Proc. Natl. Acad. Sci. USA* **2017**, *114*, 1666–1671. [CrossRef]
29. Parvez, M.K. The hepatitis E virus ORF1 'X-domain' residues form a putative macrodomain protein/Appr-1"-pase catalytic-site, critical for viral RNA replication. *Gene* **2015**, *566*, 47–53. [CrossRef]
30. Conway, M.J.; Colpitts, T.M.; Fikrig, E. Role of the Vector in Arbovirus Transmission. *Annu. Rev. Virol.* **2014**, *1*, 71–88. [CrossRef]
31. Souza-Neto, J.A.; Powell, J.R.; Bonizzoni, M. Aedes aegypti vector competence studies: A review. *Infect. Genet. Evol.* **2019**, *67*, 191–209. [CrossRef] [PubMed]
32. Abraham, R.; Hauer, D.; McPherson, R.L.; Utt, A.; Kirby, I.T.; Cohen, M.S.; Merits, A.; Leung, A.K.L.; Griffin, D.E. ADP-ribosyl-binding and hydrolase activities of the alphavirus nsP3 macrodomain are critical for initiation of virus replication. *Proc. Natl. Acad. Sci. USA* **2018**, *115*, E10457–E10466. [CrossRef] [PubMed]
33. Isabelle, M.; Gagne, J.P.; Gallouzi, I.E.; Poirier, G.G. Quantitative proteomics and dynamic imaging reveal that G3BP-mediated stress granule assembly is poly(ADP-ribose)-dependent following exposure to MNNG-induced DNA alkylation. *J. Cell. Sci.* **2012**, *125*, 4555–4566. [CrossRef] [PubMed]
34. Krieg, S.; Pott, F.; Potthoff, L.; Verheirstraeten, M.; Butepage, M.; Golzmann, A.; Lippok, B.; Goffinet, C.; Luscher, B.; Korn, P. Mono-ADP-ribosylation by PARP10 inhibits Chikungunya virus nsP2 proteolytic activity and viral replication. *Cell. Mol. Life Sci.* **2023**, *80*, 72. [CrossRef]
35. Tamura, K.; Stecher, G.; Kumar, S. MEGA11: Molecular Evolutionary Genetics Analysis Version 11. *Mol. Biol. Evol.* **2021**, *38*, 3022–3027. [CrossRef]
36. Kumar, R.; Mehta, D.; Chaudhary, S.; Nayak, D.; Sunil, S. Impact of CHIKV Replication on the Global Proteome of Aedes albopictus Cells. *Proteomes* **2022**, *10*, 38. [CrossRef] [PubMed]
37. Shrinet, J.; Jain, S.; Sharma, A.; Singh, S.S.; Mathur, K.; Rana, V.; Bhatnagar, R.K.; Gupta, B.; Gaind, R.; Deb, M.; et al. Genetic characterization of Chikungunya virus from New Delhi reveal emergence of a new molecular signature in Indian isolates. *Virol. J.* **2012**, *9*, 100. [CrossRef] [PubMed]
38. Kumar, R.; Srivastava, P.; Mathur, K.; Shrinet, J.; Dubey, S.K.; Chinnappan, M.; Kaur, I.; Nayak, D.; Chattopadhyay, S.; Bhatnagar, R.K.; et al. Chikungunya virus non-structural protein nsP3 interacts with Aedes aegypti DEAD-box helicase RM62F. *Virusdisease* **2021**, *32*, 657–665. [CrossRef]
39. Jain, J.; Kumar, A.; Narayanan, V.; Ramaswamy, R.S.; Sathiyarajeswaran, P.; Shree Devi, M.S.; Kannan, M.; Sunil, S. Antiviral activity of ethanolic extract of Nilavembu Kudineer against dengue and chikungunya virus through in vitro evaluation. *J. Ayurveda Integr. Med.* **2020**, *11*, 329–335. [CrossRef]

40. Bisht, K.K.; Dudognon, C.; Chang, W.G.; Sokol, E.S.; Ramirez, A.; Smith, S. GDP-mannose-4,6-dehydratase is a cytosolic partner of tankyrase 1 that inhibits its poly(ADP-ribose) polymerase activity. *Mol. Cell. Biol.* **2012**, *32*, 3044–3053. [CrossRef]
41. Li, J.; Mahajan, A.; Tsai, M.D. Ankyrin repeat: A unique motif mediating protein-protein interactions. *Biochemistry* **2006**, *45*, 15168–15178. [CrossRef] [PubMed]
42. Mosavi, L.K.; Cammett, T.J.; Desrosiers, D.C.; Peng, Z.Y. The ankyrin repeat as molecular architecture for protein recognition. *Protein Sci.* **2004**, *13*, 1435–1448. [CrossRef] [PubMed]
43. Slaughter, B.D.; Huff, J.M.; Wiegraebe, W.; Schwartz, J.W.; Li, R. SAM domain-based protein oligomerization observed by live-cell fluorescence fluctuation spectroscopy. *PLoS ONE* **2008**, *3*, e1931. [CrossRef] [PubMed]
44. Ali, A.A.E.; Timinszky, G.; Arribas-Bosacoma, R.; Kozlowski, M.; Hassa, P.O.; Hassler, M.; Ladurner, A.G.; Pearl, L.H.; Oliver, A.W. The zinc-finger domains of PARP1 cooperate to recognize DNA strand breaks. *Nat. Struct. Mol. Biol.* **2012**, *19*, 685–692. [CrossRef]
45. Ruf, A.; Mennissier de Murcia, J.; de Murcia, G.; Schulz, G.E. Structure of the catalytic fragment of poly(AD-ribose) polymerase from chicken. *Proc. Natl. Acad. Sci. USA* **1996**, *93*, 7481–7485. [CrossRef]
46. Corda, D.; Di Girolamo, M. Functional aspects of protein mono-ADP-ribosylation. *EMBO J.* **2003**, *22*, 1953–1958. [CrossRef]
47. Riffell, J.L.; Lord, C.J.; Ashworth, A. Tankyrase-targeted therapeutics: Expanding opportunities in the PARP family. *Nat. Rev. Drug Discov.* **2012**, *11*, 923–936. [CrossRef]
48. Bai, P. Biology of Poly(ADP-Ribose) Polymerases: The Factotums of Cell Maintenance. *Mol. Cell* **2015**, *58*, 947–958. [CrossRef]
49. Gupte, R.; Liu, Z.; Kraus, W.L. PARPs and ADP-ribosylation: Recent advances linking molecular functions to biological outcomes. *Genes Dev.* **2017**, *31*, 101–126. [CrossRef]
50. Fahrer, J.; Popp, O.; Malanga, M.; Beneke, S.; Markovitz, D.M.; Ferrando-May, E.; Burkle, A.; Kappes, F. High-affinity interaction of poly(ADP-ribose) and the human DEK oncoprotein depends upon chain length. *Biochemistry* **2010**, *49*, 7119–7130. [CrossRef]
51. Mathur, K.; Anand, A.; Dubey, S.K.; Sanan-Mishra, N.; Bhatnagar, R.K.; Sunil, S. Analysis of chikungunya virus proteins reveals that non-structural proteins nsP2 and nsP3 exhibit RNA interference (RNAi) suppressor activity. *Sci. Rep.* **2016**, *6*, 38065. [CrossRef] [PubMed]
52. Eckei, L.; Krieg, S.; Butepage, M.; Lehmann, A.; Gross, A.; Lippok, B.; Grimm, A.R.; Kummerer, B.M.; Rossetti, G.; Luscher, B.; et al. The conserved macrodomains of the non-structural proteins of Chikungunya virus and other pathogenic positive strand RNA viruses function as mono-ADP-ribosylhydrolases. *Sci. Rep.* **2017**, *7*, 41746. [CrossRef] [PubMed]
53. Todorova, T.; Bock, F.J.; Chang, P. PARP13 regulates cellular mRNA post-transcriptionally and functions as a pro-apoptotic factor by destabilizing TRAILR4 transcript. *Nat. Commun.* **2014**, *5*, 5362. [CrossRef]
54. Seo, G.J.; Kincaid, R.P.; Phanaksri, T.; Burke, J.M.; Pare, J.M.; Cox, J.E.; Hsiang, T.Y.; Krug, R.M.; Sullivan, C.S. Reciprocal inhibition between intracellular antiviral signaling and the RNAi machinery in mammalian cells. *Cell Host Microbe* **2013**, *14*, 435–445. [CrossRef]
55. Zhang, Y.; Mao, D.; Roswit, W.T.; Jin, X.; Patel, A.C.; Patel, D.A.; Agapov, E.; Wang, Z.; Tidwell, R.M.; Atkinson, J.J.; et al. PARP9-DTX3L ubiquitin ligase targets host histone H2BJ and viral 3C protease to enhance interferon signaling and control viral infection. *Nat. Immunol.* **2015**, *16*, 1215–1227. [CrossRef] [PubMed]
56. Gagne, J.P.; Isabelle, M.; Lo, K.S.; Bourassa, S.; Hendzel, M.J.; Dawson, V.L.; Dawson, T.M.; Poirier, G.G. Proteome-wide identification of poly(ADP-ribose) binding proteins and poly(ADP-ribose)-associated protein complexes. *Nucleic Acids Res.* **2008**, *36*, 6959–6976. [CrossRef]
57. Fahrer, J.; Kranaster, R.; Altmeyer, M.; Marx, A.; Burkle, A. Quantitative analysis of the binding affinity of poly(ADP-ribose) to specific binding proteins as a function of chain length. *Nucleic Acids Res.* **2007**, *35*, e143. [CrossRef]

pathogens

MDPI

Review

Roles of ADP-Ribosylation during Infection Establishment by *Trypanosomatidae* Parasites

Joshua Dowling and Craig L. Doig *

School of Science & Technology, Nottingham Trent University, Nottingham NG11 8NS, UK; joshua.dowling@ntu.ac.uk
* Correspondence: craig.doig@ntu.ac.uk

Abstract: ADP-ribosylation is a reversible post-translational protein modification, which is evolutionarily conserved in prokaryotic and eukaryotic organisms. It governs critical cellular functions, including, but not limited to cellular proliferation, differentiation, RNA translation, and genomic repair. The addition of one or multiple ADP-ribose moieties can be catalysed by poly(ADP-ribose) polymerase (PARP) enzymes, while in eukaryotic organisms, ADP-ribosylation can be reversed through the action of specific enzymes capable of ADP-ribose signalling regulation. In several lower eukaryotic organisms, including *Trypanosomatidae* parasites, ADP-ribosylation is thought to be important for infection establishment. *Trypanosomatidae* encompasses several human disease-causing pathogens, including *Trypanosoma cruzi*, *T. brucei*, and the *Leishmania* genus. These parasites are the etiological agents of Chagas disease, African trypanosomiasis (sleeping sickness), and leishmaniasis, respectively. Currently, licenced medications for these infections are outdated and often result in harmful side effects, and can be inaccessible to those carrying infections, due to them being classified as neglected tropical diseases (NTDs), meaning that many infected individuals will belong to already marginalised communities in countries already facing socioeconomic challenges. Consequently, funding to develop novel therapeutics for these infections is overlooked. As such, understanding the molecular mechanisms of infection, and how ADP-ribosylation facilitates infection establishment by these organisms may allow the identification of potential molecular interventions that would disrupt infection. In contrast to the complex ADP-ribosylation pathways in eukaryotes, the process of *Trypanosomatidae* is more linear, with the parasites only expressing one PARP enzyme, compared to the, at least, 17 genes that encode human PARP enzymes. If this simplified pathway can be understood and exploited, it may reveal new avenues for combatting *Trypanosomatidae* infection. This review will focus on the current state of knowledge on the importance of ADP-ribosylation in *Trypanosomatidae* during infection establishment in human hosts, and the potential therapeutic options that disrupting ADP-ribosylation may offer to combat *Trypanosomatidae*.

Keywords: ADP-ribosylation; PARP; PARG; *Trypanosoma*; *Leishmania*

Citation: Dowling, J.; Doig, C.L. Roles of ADP-Ribosylation during Infection Establishment by *Trypanosomatidae* Parasites. *Pathogens* **2023**, *12*, 708. https://doi.org/10.3390/pathogens12050708

Academic Editors: Anthony K L Leung, Anthony Fehr and Rachy Abraham

Received: 31 March 2023
Revised: 8 May 2023
Accepted: 9 May 2023
Published: 12 May 2023

1. ADP-Ribosylation in Infection

ADP-ribosylation is a fundamental post-translational protein modification in which single or several ADP-ribose units are covalently attached to proteins. The modification is commonly catalysed by members of the poly(ADP-ribose) polymerase (PARP) enzyme family. PARP enzymes attach ADP-ribose moieties to the aspartate, glutamate, lysine, arginine, cysteine, threonine, or serine residues, resulting in the creation of branched and linear polymers [1]. In addition to the actions of the PARP enzymes, ADP-ribosylation can also occur via the action of mono(ADP-ribosyl)transferases, which catalyse the attachment of ADP-ribose to arginine side chains via the activity of an essential and highly conserved R-S-EXE motif [2]. This motif is localised within a specialised loop used in target recognition for mono(ADP-ribosyl) transferases as well as for several other ADP-ribosyltransferases (ARTs), including human PARP-1.

1.1. ADP-Ribosylation in Viral Pathogens

PARP enzymes have long-documented actions in infection progression and protection against pathogens in humans [3]. The role of PARP-mediated protection, particularly against viral infection, has seen extensive study. Several human PARPs, including PARP 1, 5, 7, 9, 10, 12, and 13 are known antagonists of both DNA and RNA viruses [4–6]. Viral replication is disrupted by several PARP enzymes, predominantly targeting dysregulation in viral genomic translation and transcription, leading to inhibition or prevention of the completion of the viral life cycle. PARP13, in particular, demonstrates potent antiviral activity against many DNA and RNA viruses, including alphaviruses, influenza, filoviruses (including Ebola and Marburg viruses), herpes virus, HIV-1, coxsackie virus B3, hepatitis B, and Japanese encephalitis virus [7–10]. PARP13 utilizes multiple mechanisms in viral inhibition, binding viral RNA through its four zinc-finger motifs, to allow the inhibition of transcription and translation of the viral genome which disrupts the viral life cycle. PARP13 is, then, able to degrade the 5′ end of HIV-1 RNA via the recruitment of several degrading host factors, including poly(A)-specific ribonuclease (PARN) and the KHNYN endonuclease [11].

Viral replication is targeted by the host expressing PARP enzymes to defend against infection. For example, PARP1 and PARP5 serve as antagonists to critical binding sites used by viral proteins for genomic replication by Kaposi sarcoma-associated herpesvirus (KSHV) and Epstein–Barr virus (EBV), respectively [12]. PARP1 is also able to prevent transcriptional elongation of HIV-1 RNA through the binding of PARP1 to a TAR binding site, which is responsible for the binding of RNA elongation factor p-TEFb. PARP1 also binds with TAR more efficiently than p-TEFb, demonstrating its ability to inhibit viral replication through epigenetic modification [13]. Genome translation is essential for viral replication and a target for inhibition by PARPs. PARP13 has been shown to decrease the production of Nef, a protein that is present in HIV-1 and is critical for successful viral replication. PARP13 is also able to degrade viral RNA via exosome activation [14]. This is further evidence of the broad antiviral mechanisms exhibited by PARP13. Other PARPs demonstrate antiviral activity through the targeted degradation of essential viral proteins. PARP9 can form a protein-degrading complex with a ubiquitin ligase capable of *Picronoviridae* protein degradation [15]. PARP10 is capable of transfer to nuclei during avian influenza (AIV) viral infection to degrade the protein AIV NS1, which is important for AIV replication [16]. PARP12 can degrade Zika virus (ZIKV) proteins NS1 and NS3 via mon(ADP-ribosyl)ation, catalysed by PARP12 [17].

1.2. ADP-Ribosylation in Bacterial Pathogens

Studies examining the roles of PARPs during bacterial pathogen infection are not as extensive in comparison to those on PARPs in viral infections. Only a small number of bacterial species possess functional PARylation systems. Although the majority contain domains capable of PAR binding, in addition to PAR-degrading enzymes [18]. The majority of antibacterial studies on PARP enzymes have primarily focused on the use of PARP1. Studies have been performed on several notable human bacterial pathogens, including *Helicobacter*, *Salmonella*, *Escherichia coli*, *Pseudomonas aeruginosa*, *Streptococcus pneumonia*, *Streptococcus pyogenes*, and *Chlamydophila*. Experiments utilising these bacteria surmise a similar consensus, whereby shifts in PARP activity increase the difficulty of mounting an effective response in preventing damage from bacterial infection [19–23]. In particular, the depletion of PARP activity depresses the sufficient inflammatory response to bacterial infection. This is likely due to PARP1 regulation of NF-kB-mediated signalling and the activation of macrophage responses [24–26].

Several species of bacteria have also been found to possess PARG enzymes, including *Thermomonospora curvata* and *Herpetosiphon aurantiacus* [27]. In humans, PARG enzymes are a mechanism through which PAR can be removed from the cell via catabolism of poly(ADP-ribose), through hydrolysis of the ribose-ribose bonds. This prevents damage caused by excessive PAR accumulation in the cytoplasm [28]. PAR accumulation can lead to a PARP-mediated cell death pathway known as parthanatos, through which excessive PAR can lead to apoptosis via several mechanisms, including depletion of NAD and the PAR-mediated activation of an apoptosis-inducing factor (AIF) [29]. PAR can bind to AIF, which is followed by AIF translocation to the nucleus, resulting in extensive DNA fragmentation and chromatin damage [30]. PARG is the primary means through which excessive PAR is removed in human cells. Other human PAR hydrolases do exist, necessitated by PARG's inability to remove the most proximal ADP-ribose moieties [31], including ARH3, which is present during the removal of PAR from the mitochondria [32]. The network of PAR enzymes present in bacteria is much simpler in comparison. Human PARG enzymes act both as endo-glycohydrolases and exo-glcohydrolases, which leads to the production of free PAR moieties and mono-ADP ribose moieties, respectively, via the hydrolysis of ADP-ribose chains [28]. However, bacterial PARG enzymes are thought to be more limited in their roles, in that they are only capable of acting as exo-glcohydrolases due to the presence of a ribose cap located close to the C-terminus, which prevents PARG from binding efficiently to the internal binding sites necessary for endo-glycohydrolase activity [33]. Nevertheless, bacterial PARG enzymes are important to prevent PAR-mediated apoptosis, and as such, PARG inhibitors may present an interesting therapeutic option in the treatment of bacterial infections by limiting the pathogen's ability to remove harmful PAR accumulation within the bacterial cytoplasm.

There is much evidence to suggest the extensive roles of human PARP enzymes in immune protection during bacterial and viral infection. However, key questions remain for both. PARP activity is seemingly broader across enzymes in terms of antiviral activity (10 out of 17 human PARP enzymes have identified antiviral activity), potentially as a result of primarily cytoplasmic and nuclear localisation, which allows the PARP enzymes to interrupt viral replication cycles at several distinct stages [34]. Therefore, it seems that the expression of multiple cytoplasmic PARP enzymes, developed alongside the evolution of vertebrates, are seemingly as equally as important as the nuclear-localised PARPs in maintaining cellular health through the maintenance of essential processes and antimicrobial activity. Given the broad antiviral properties of the most studied PARPs, there is considerable potential to further explore the actions of the remaining PARPs to better understand their antiviral properties.

The use of broad-spectrum PARP agonists may present an attractive avenue to explore novel antiviral therapies. ADP-ribosyltransferase (ART) activity of PARP enzymes has been demonstrated in works examining bacterial infection [35,36]. Further investigation is required to fully understand the range of ART activity in inflammatory responses and infection. Interestingly, inhibition of PARP activity in animal models has displayed therapeutic benefits. Moreover, the acute septic shock has been resolved as a result of PARP modulation, likely due to a reduction in tissue damage, which usually results from enhanced PARP-mediated ART activity [35]. This poses an interesting approach to treating bacterial infection by finding the correct balance of PARP activity versus PARP inhibition. However, excessive inhibition of PARP-catalysed ART activity would lead to inefficient DNA repair during bacterial infection and restrict the positive impacts of the PARP activity. However, PARP inhibition does yield beneficial therapeutic effects in defined circumstances. Combinatorial therapy with appropriate antibiotics used in conjunction with a lowered dose of PARP inhibitors could theoretically allow the successful compromise of PARP activity. This would allow a limitation upon PARP-driven DNA damage, whilst retaining the benefits that PARP activity yields during the infection response. Nevertheless, this process would require significant clinical testing and optimisation for different bacterial species. Given the current dearth of effective treatment options for Trypanosomes, all options must be worthy of exploration.

2. Trypanosomatidae

Trypanosomatidae is an order of singular flagellate kinetoplastid parasites, the most relevant to human health being *Trypanosoma* and *Leishmania*. The parasites *T. cruzi*, *T. brucei*, and *Leishmania* are all responsible for infections categorised by the World Health Organisation (WHO) as neglected tropical diseases (NTDs) [37]. As such, these infections are responsible for profound impacts as they primarily arise in vulnerable people inhabiting developing countries. ADP-ribosylation and the associated enzymes, including PARP and PARG, play an essential role in the ability of *Trypanosomatidae* to establish a successful infection in the human host. Given the lack of safe and effective medications available to combat these parasites, and the parasites' reliance on the functioning ADP-ribosylation to maintain parasitic viability, the disruption of their ADP-ribosylation systems may present a novel and effective therapeutic option. However, a deep understanding of the molecular mechanisms underpinning the *Trypanosomatidae* ADP-ribosylation systems is required to accurately target it. As such, this review will offer insight into the current understanding of these systems and how they may be exploited to advance the options available for effective *Trypanosomatidae* treatment.

T. cruzi is the causative agent of Chagas disease, which is endemic to Central and South America. An initial infection during the acute stage results in flu-like systems. These symptoms mask the diagnosis of *T. cruzi* infection [38]. During chronic stages, an infection can lay dormant for decades, characterised by potentially minimal symptoms and low parasitemia, which causes issues in detecting the infection. The chronic infection eventually results in organ enlargement, with parasites primarily targeting the heart, leading to numerous cardiac complications and potential death. Chagas is a disease of poverty, meaning minimal financial gains in the development of tools can help to combat it, which is reflected in that only 0.67% of US funding for neglected diseases 2009–2019 was applied to Chagas [39]. These issues are compounded by the inefficacy of current treatments. Benznidazole and nifurtimox have limited cure rates and possess toxic side effects and the need for an effective alternative has been identified [40].

T. brucei is the etiological agent of African trypanosomiasis. Vectoral transmission is the most common route of infection, in which parasites enter the human host via bites inflicted by the primary tsetse fly host. Similar to Chagas disease, the initial stage of infection presents aspecific symptoms, followed by parasites eventually migrating to the brain, leading to neurological complications, and ultimately death, without proper treatment [41]. Available therapies for the neurological stage of infection are also limited, with Melarsoprol the only available drug, although this causes death in 5% of the patients who ingest it since Melarsoprol resistance is present in some strains [42]. The anti-parasitic Fexinidazole has shown activity against both the CNS and peripheral stages of African trypanosomiasis, although studies remain in clinical trials and effective drug alternatives are required to limit resistance [43].

Leishmaniasis is an umbrella term for three distinct diseases: visceral, cutaneous, and mucocutaneous leishmaniasis. Infections are predominantly found in Asia, the Middle East, and Northern Africa to differing degrees. The diseases are caused by several different *Leishmania* species and are transmitted commonly by bite wounds inflicted by an infected female phlebotomine sand fly. Visceral leishmaniasis is the most severe, as it causes a systemic infection that almost invariably is fatal without rapid treatment. The cutaneous infection leads to superficial skin lesions, whilst mucocutaneous infection leads to significant damage of the buccal and nasal cavities via the formation of damaging mucocutaneous lesions, which may disappear and reoccur repeatedly [44]. Given the severe nature of the infection, the majority of attention in the development of novel therapies for leishmaniasis has been focused on visceral leishmaniasis [45,46]. Currently, licenced therapies have several issues relating to their efficacy, dangerous side effects, and availability in endemic countries. Liposomal amphotericin B is often the first-choice drug, with miltefosine approved in 2014 by the FDA for oral treatment. Amphotericin B can cause significant side effects [47] and resistance against it has been documented, given that it is often a front-line drug for many fungal infections [48].

Consequently, given the scarcity of funding available to research the development of novel therapeutics to combat these NTDs and the lack of efficacy for existing treatments, the identification of novel ways to combat these infections is paramount. *T. cruzi*, *T. brucei*, and *Leishmania* all utilise ADP-ribosylation in several distinct ways to facilitate successful infection in the human host. Given the avenues with potential to be explored, whereby ADP-ribosylation manipulation can be harnessed to combat viral and bacterial infection, it is feasible that inhibiting or disrupting ADP-ribosylation in these parasites could lead to diminished parasite survival and proliferation.

In contrast to the extensive PARP network found in humans, the PARP system present in *Trypanosomatidae* is much simpler. *T. cruzi* and *T. brucei* both possess a singular PARP enzyme, designated TcPARP and TbPARP, respectively. Both parasites also utilise a sole poly(ADP-ribose) glycohydrolase (PARG) enzyme, which is used to reverse the action of PARP via the hydrolysis of ribose–ribose bonds present in PARP, which helps prevent extensive DNA damage by excessive PARP accumulation [49]. The primary ADP-ribosylation mechanisms within these parasites use polyADP-ribosylation, although there is also evidence of them using monoADP-ribosylation systems. Both *T. cruzi* and *T. brucei* possess proteins that are homologous to human MacroD1 and MacroD2, which are domains that hydrolyse and cleave ADP-ribose attachments to proteins in monoADP-ribosylation within human systems [50]. The homologous proteins present in *T. cruzi* and *T. brucei* (denoted TcMDO and TbMDO, respectively), both possess the core macrodomain fold present in human MacroD1/D2, although the N-terminus of the parasitic proteins differ heavily from the human MacroD1 and MacroD2, whereby they lack the adenine-binding region, indicating that TcMDO and TbMDO bind differently to the adenine moieties [51]. Given the nature of ADP-ribosylation in *Trypanosomes*, there is potential for the inhibition of TcMDO/TbMDO to have therapeutic penitential, yet there is a lack of inhibitors against macrodomain proteins [51].

3. ADP-Ribosylation in *Trypanosoma cruzi*

3.1. Use of ADP-Ribosylation and Associated PARP and PARG Enzymes in Trypanosoma cruzi

Trypanosoma cruzi expresses a sole PARP enzyme throughout its lifecycle, known as TcPARP (Figure 1). An initial study of TcPARP revealed several structural and molecular similarities to its human homologue, hPARP-1 [52]. Similar to hPARP-1, TcPARP is activated via DNA strand nicks, upon exposure of the parasite DNA to damaging agents such as H_2O_2. Initial studies on TcPARP revealed a highly evolutionarily conserved C-terminal catalytic domain that is homologous to *T. brucei* PARP, human PARP-1, and PARP-4. Furthermore, the catalytic triad of histidine, tyrosine, and glutamic acid is utilised for PAR elongation in hPARP-1, the closest human homologue to TcPARP is conserved within the parasite [53]. Though further structural similarities were identified between TcPARP and human PARPs (including the essential presence of glutamic acid to facilitate PARP activity) [54,55], TcPARP differentiates itself from human PARP through its method of DNA damage detection. Whilst DNA repair is essential for both humans and *Trypanosomatidae* to facilitate proper cellular function, human PARP enzymes utilise an N-terminal zinc-finger domain to bind DNA. This domain is seemingly absent in *T. cruzi*, which instead possesses an N-terminal region thought to be able to bind DNA through the large quantity of basic amino acid residues present in the domain (approximately 27% within the first 70 residues in TcPARP) [53], therefore, allowing *T. cruzi* to regulate DNA activity, which is corroborated by the ability of this region to bind hPARP-1 and hPARP-2 [56].

TcPARP activity is dependent upon the detection of DNA damage and is key in successful DNA repair and signalling. It serves roles in the ability of the parasite to differentiate into different life-cycle stages and establish infection in a human host. Consequently, the use of PARP inhibitors in the treatment of *T. cruzi* has been studied and has seen some success. Olaparib, a common PARP inhibitor, exerts strong inhibitory effects on hPARP-1 and TcPARP activities [52,57]. Olaparib has shown the ability to reduce amastigote (non-motile form lacking flagella) formation in a range of cell lines [52]. Another common hPARP inhibitor, EB-47 demonstrated an ability to prevent pADPr formation in human cell lines, although it failed to prevent pADPr formation in *T. cruzi* in vivo [58]. Villchez Larea and colleagues hypothesised that EB-47's inability to adequately prevent pADPr formation in vivo could be due to EB-47's comparatively large size and excessive polarity compared to other inhibitors. Significantly, Olaparib has been shown to reduce epimastigote growth in vitro by more than an 100× larger concentration of Nifurtimox (the established treatment modality for Chagas) [59]. Given the severity of the Nifurtimox side effects, the potential of PARP inhibitors as an alternative or combinatorial approach to Chagas therapy is plausible. However, specificity remains a challenge as TcPARP utilises the same nicotinamide binding site, to bind inhibitors, as the hPARP enzymes.

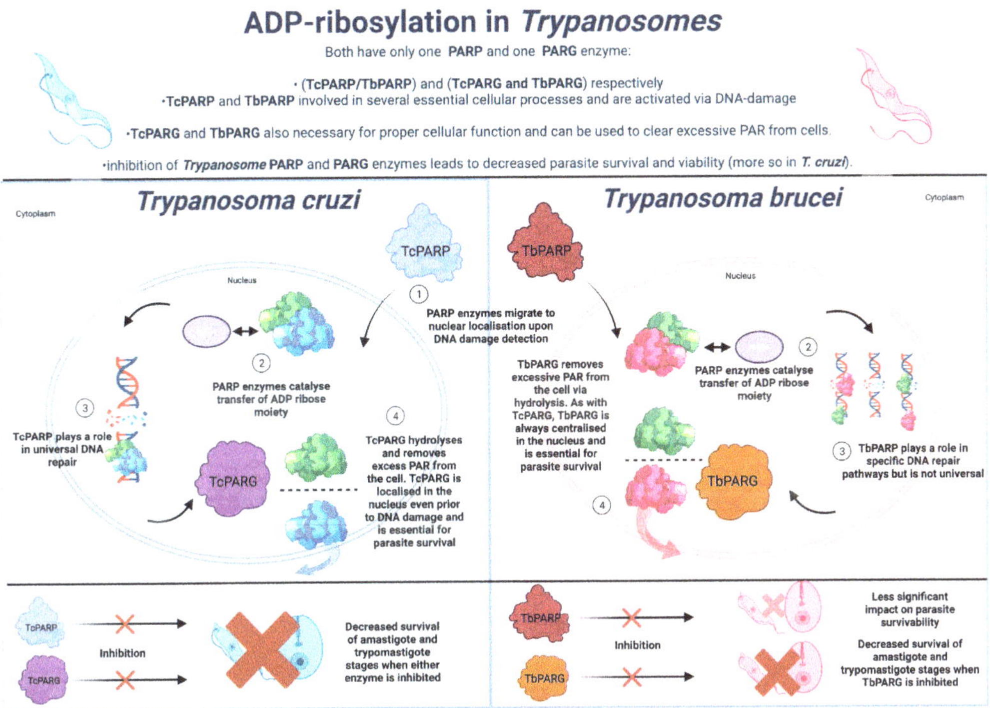

Figure 1. ADP-ribosylation in Trypanosomes. In contrast to pathways in eukaryotes, ADP-ribosylation in *Trypanosomatidae* is more linear, with the parasites only expressing one PARP enzyme. If this pathway can be understood and exploited, it may reveal new avenues for combatting *Trypanosomatidae* infection.

3.2. The Potential of ADP-Ribosylation Targeted Therapy for Trypanosoma cruzi

Intriguingly, there may be an interplay between hPARP-1 and TcPARP during infection establishment by *T. cruzi*. hPARP-1 silencing experiments in A549 cells demonstrated a significantly decreased amastigote number as well as a reduction in trypomastigotes in the culture media [52]. *T. cruzi* is dependent upon the ability of healthy trypomastigotes to establish a lasting infection and the subsequent differentiation and multiplication of amastigotes [60]. Therefore, successful inhibition and prevention of *T. cruzi* require the disruption of differentiation and growth. *T. cruzi* trypomastigote penetration of cardiac tissue in vitro leads to ROS generation, through activation of hPARP-1. ROS accumulation subsequently triggers the expression of NF-kB, which is a transcription factor that facilitates the transcription of multiple cytokines acting in concert to allow successful penetration of the parasite into cell lines in vitro [61,62]. The absence of cytokines TNF-α and IL-1β resulted in decreased levels of infection, demonstrating that successful infection by *T. cruzi* requires some activation of NF-kB [63]. Hence, this presents the interesting hypothesis that

T. cruzi relies not only on TcPARP but also the activation of cytokine factors, mediated by hPARP-1. Further studies on the relationship between hPARP-1 mediated persistence of *T. cruzi* are required. This offers evidence for the importance of both innate and host PARP enzymes in *T. cruzi*, furthering the case of PARP inhibitors as a treatment for Chagas. It is also clear that hPARP-1 is likely the critical enzyme to target, given that hPARP-1 silencing results in higher inhibitory effects on infection than Olaparib treatment. This suggests successful infection relies upon a mechanism involving hPARP-1 specifically, as opposed to other hPARP enzyme members.

However, chronic exposure to PARP inhibitors can be profoundly detrimental. Should PARP inhibitors be assessed regarding Chagas therapy, protection from DNA damage is a fundamental consideration. PARG enzymes can reverse the effects of PARP enzymes, via clearance of excessive PAR accumulation to circumvent any harm to cells [64]. The PARG enzyme present in *T. cruzi*, denoted TcPARG, has been demonstrated as essential for epimastigote growth and the infection cycle in vitro (Figure 1). TcPARG shares 46.5% sequence identity with human PARG and possesses a preserved domain, including the tyrosine residues utilised to bind PARG inhibitors [65]. Importantly, as with hPARP-1, *T. cruzi* seemingly uses host PARG during infection establishment. A significantly lower number of intracellular amastigotes and infected cells were observed in hPARG knockout experiments compared to PARG inhibitor experiments [66]. The generation of pARPr via the exo-glycosidase activity of human PARG is thought to regulate factors associated with protein binding and post-translational modification [66]. This work revealed that *T. cruzi* may rely on both innate and host PARG factors to allow successful infection in the human host. Given PARG's role in the removal of PARylation from cells, its inhibition would allow a longer and more impactful use of PARP inhibitors in Chagas treatment, but of course, the same considerations remain over limiting excess PARP-mediated DNA damage to the host [67]. As such, it seems apparent that there is a high level of interconnectivity between TcPARP, TcPARG, and hPARP/hPARG. Further investigation into this relationship may reveal the optimal avenues for PARP/PARG inhibition in treating Chagas infection, whilst preserving host cell homeostasis. However, *T. cruzi* relies on this relationship for infection establishment. The association between pathogen and human PARPs/PARG is open to intervention that may create new treatments to reduce infection and its associated clinical manifestations.

4. ADP-Ribosylation in *Trypanosoma brucei*

4.1. PARP Enzymes in Trypanosoma brucei and Potential Therapies

As in *T. cruzi*, *T. brucei* possesses one PARG and one PARP enzyme, denoted as TbPARG and TbPARP, respectively (Figure 1). TbPARP is highly conserved and similar to TcPARP and hPARP-1 in terms of its sequence, structure, and function [68,69]. A basic sequence of amino acids is likely responsible for DNA damage detection by TbPARP and the subsequent activity of the enzyme. Notably, TbPARP lacks a reliance on metal ions for TbPARP activity, both of which are characteristics that are shared with TcPARP. Metal ions (including Mn^{2+}, Ni^{2+}, and Zn^{2+}) can have an inhibitory effect on the activity of both TbPARP and TcPARP, which is likely due to the metal ions binding to sulfhydryl groups, thought to be necessary for disulphide bond formation and reduction reactions [68]. This is in contrast to the impacts that metal ions have on human PARP enzymes, with Ca^{2+} and Mn^{2+} able to increase the levels of PAR synthesis, with human PARP enzymes relying on an alternative mechanism for oxygen reduction and disulphide bond formation [70]. Studies on TbPARP structure have also elucidated N-terminus disorder, which is a characteristic shared by both hPARP2 and hPARP3 [24]. TbPARP, along with TcPARP, has also been shown to have a WGR binding domain composed of arginine, glycine, and tryptophan, which is present in several eukaryotic organisms, and has a demonstrated role in DNA-dependent PARP activity for PARP enzymes lacking the characteristic zinc-finger binding domain of hPARP-1 [71]. As such, this offers further insight into the mechanisms underpinning the activity of TbPARP, along with the potential role of the WGR domain in TbPARP activation.

The DNA-dependent activation of TbPARP specifically requires phosphorylated single-strand overhangs for the recognition of DNA nicks, with the WGR domain hypothesised to play a currently unknown role in this process [49].

Several studies have identified that the most efficacious inhibitors of TbPARP are the same as for TcPARP and hPARP-1 inhibition, most notably Olaparib, and EB-47 [71,72]. TbPARP shares the same nicotinamide-based NAD^+ binding site as TcPARP and hPARP-1, explaining the consistency in inhibitor potency. There are differences between some inhibitors, such as Rucaparib, which is unfavourable for TbPARP binding, thought to be due to the presence of a serine residue in place of the alanine present in the binding site of TcPARP and hPARP-1 [71]. Whilst this sequence difference means that some inhibitors may be less useful in *T. brucei* treatment than for *T. cruzi*, the use of the existing inhibitors is still a realistic treatment option for African trypanosomiasis, given the broad activity of PARP inhibitors on TbPARP and the role of TbPARP in parasite development and proliferation. Despite the promise shown in vitro, reducing PAR synthesis in vivo in *T. brucei* with conventional inhibitors remains a challenge. As in *T. cruzi*, for an inhibitor to exert an effect on PARP activity and PAR formation in the parasite, the inhibitor must be both small and polar enough to successfully cross parasite membranes.

TbPARP activation is dependent on the detection of DNA nicks and subsequent migration to the site of genomic damage in the nucleus, and TbPARP seemingly exerts similar protective effects to TcPARP in facilitating parasite growth and differentiation [73]. Moreover, excessive accumulation of PAR via TbPARP activity results in cellular damage and cell death [74].

4.2. PARG Enzymes in *Trypanosoma brucei* and Potential Therapies Targeting PARG Enzymes

The role of PARG-mediated suppression of PARP enzymes in *T. brucei* is unclear. However, a relationship between PARG and PARP enzymes exists within *T. brucei*. Therefore, although it is likely that a PARP/PARG-mediated treatment is closer for Chagas disease, a similar achievement could be attained for African trypanosomiasis. TbPARG shares 60% sequence similarity with human PARG, including the adenosine diphosphate hydroxymethyl pyrrolidinediol (ADP-HPD) binding site, with ADP-HPD commonly involved in PARG inhibition, meaning PARG inhibitors used in mammalian systems may offer relevance in *T. brucei* [65]. In *T. bruce*, the depletion of PARG has been shown to result in the increased nuclear accumulation of PAR in *Trypanosomes*, even in the absence of oxidative stress [75]. The duality of PARP and PARG enzymes likely exerts a similar effect in *T. brucei* as in *T. cruzi*, and their regulation strikes a fine balance between cellular protection and cell death. PARP enzymes exert protective effects against DNA damage in these parasites, but excessive PAR accumulation can lead to disrupted DNA repair, and NAD^+ and ATP depletion resulting in apoptosis. PARP and PARG enzymes remain credible targets for intervention in *T. brucei* as the parasite relies on both enzymes to establish infection. Nevertheless, the current state of understanding is poor in *T. brucei* in comparison to *T. cruzi*. Further work is required to elucidate the specific intricacies of the mechanisms involved in *T. brucei*, with therapies potentially requiring slight adjustments given the difference in the sequence of the nicotinamide binding site, of TbPARP specifically, compared to TcPARP and hPARP-1.

It has been hypothesised that TbPARP is less important to *T. brucei* parasite viability than TcPARP is to *T. cruzi* [68]. Nonhomologous end joining (NHEJ) repair of DNA breaks is not present in *Trypanosomatidae*, as it is in eukaryotes, and as such, double-strand repairs within these parasites rely on microhomology-mediated end joining (MMEJ) [76]. This DNA repair mechanism relies upon small homologous regions within the broken ends to align and repair the strands. This more commonly results in sequence deletions and other modifications than in NHEJ. MMEJ is often viewed as a less preferred alternative to NHEJ because of this, yet MMEJ is omnipresent in DNA repair for *T. brucei*, although is it not understood. Unlike the roles of TcPARP and hPARP-1, the role of TbPARP is likely more similar to hPARP2 and hPARP3, whereby it is used in specific DNA repair pathways and not as a universal DNA repair enzyme, evidenced by its specific role within MMEJ. Given the lack of necessity of TbPARP for *T. brucei* viability, further study is warranted to fully understand the role of TbPARP in DNA repair; however, currently, it would seem that attention may be better focused on exploiting PARG enzymes to treat African trypanosomiasis, given the strong effect PARG depletion has on decreasing parasite viability.

5. ADP-Ribosylation in *Leishmania*

5.1. LdARL-3A Ribosylation Factor as a Therapeutic Target in Leishmania

ADP-ribosylation in *Leishmania* is less studied and poorly understood compared to the *Trypanosomes*, though work has taken place to better understand the importance of ADP-ribosylation in relation to *Leishmania* viability and infection. LdARL-3A is an ADP-ribosylation factor identified in *Leishmania donovani*, which is expressed specifically during the promastigote stage, during the insect stage of the lifecycle. ARL (ADP ribosylation-like) enzymes are essential for numerous cellular processes, including trafficking, endocytosis, and other cell signalling pathways. LdARL-3A seemingly plays a role in flagellum formation and viability, as overexpression of a constitutively active LdARL-3A mutant led to a correlated decrease in flagellum length, with stronger levels of overexpression leading to larger decreases in flagellar length [77]. Therefore, if the function of LdARL-3A can be inhibited, the motility of the parasite in the insect host may be reduced before the infection of the human host. Further study of LdARL-3A revealed that LdARL-3A possesses two forms, a GDP-bound and a GTP-bound form [78]. The GDP-bound form is considered inactive and is the form in which overexpression led to parasite death in the stationary phase, whereas the GTP-bound form is active and overexpression led to diminished flagellar length. The disruption of LdARL-3A, with it switching between the inactive and active forms, is what leads to the reduction in flagellar length. Hence, this evidence suggests that LdARL-3A is a potential drug target, in the attempt to prevent transmission from the insect vector to the human host, by inhibiting the motility by decreasing the flagellar length.

5.2. LiSIR2RP1 as a Therapeutic Target in Leishmania

Leishmania infantum possesses a gene denoted LiSIR2RP1 that encodes the SIR2 protein, which is a deacetylase to several cellular substrates, including histone lysine residues [79]. This deacetylation activity is dependent upon the use of NAD^+. Disruption of LiSIR2RP1 has been shown to decrease the viability of *Leishmania infantum* amastigotes both in vivo and in vitro, with the close association of LiSIR2RP1 with the cytoskeletal structure of both *L. infantum* promastigote and amastigotes. LiSIR2RP1 can exert deacetylase activity upon tubulin, which is critical for parasite structural viability as well as the parasite's ability to interact with host cells [80]. *Leishmania* tubules and microtubules play an essential part in successful parasite division and structural integrity; thus, if LiSIR2RP1 can be further explored, it could yield a better understanding of the role LiSIR2RP1 plays in parasite integrity and remodelling and offer potential therapeutic alternatives via LiSIR2RP1 targeting.

5.3. Targeting NAD+ Salvage Pathways as a Therapeutic Option in Leishmania

Leishmania species are NAD^+ auxotrophs and require NAD^+ sequestered from host cells. *Leishmania* lacks both intrinsic de novo pathways that can be used for innate biosynthesis of NAD, which use either L-tryptophan or aspartic acid as a precursor (which are in eukaryotic and prokaryotic de novo pathways, respectively). *Leishmania* instead uses NR (nicotinamide riboside) as a precursor for NAD^+ production in salvage pathways, wherein NAD^+ is sequestered from the host [81]. The NAD nucelotidase (NadN) enzyme first described in *Haemophilus influenzae* was found to also be present in *Leishmania* [82]. NadN is a periplasmic enzyme thought to be involved in NAD^+ synthesis via NAD^+ hydrolysis into NR, adenosine, and phosphate, and finally, NR uptake across the inner membrane of the parasite and NadR catalysed resynthesis of NR into NAD^+ [83]. Given the high level of conformation change identified in the enzymatic pathway, it has been identified as a possible target for inhibition to prevent parasite replication, which is corroborated by

NadN knockout experiments in *Leishmania* resulting in a significant NAD^+ concentration decrease, which coincided with reduced parasite proliferation and virulence.

6. Summary

It is apparent that ADP-ribosylation plays a dynamic and important role in infection establishment for *Trypanosomatidae*, and there exists a considerable interplay between these parasites' innate ADP-ribosylation systems and the role of the host ADP-ribosylation systems, which exerts protective effects within the host to ward off infection. Given that *Trypanosomatidae* relies on ADP-ribosylation to maintain parasite viability, exploitation of these systems within the parasites offers an attractive avenue to explore novel therapies to prevent or treat the infection. This is especially important given the minimal current treatment options for all *Trypanosomatidae*.

T. cruzi, in particular, is reliant on the proper function of its innate PARP and PARG enzymes to be able to maintain an infection within a human host. Further evidence is required to be able to fully understand the interactions between humans and *T. cruzi* PARP and PARG enzymes. There exists the possibility of inhibiting both TcPARP and TcPARG to dysregulate *T. cruzi* function, as evidenced by the lack of parasite viability in TcPARP and TcPARG knockout experiments. Interestingly for *T. cruzi*, the knockout of hPARP-1 and hPARG also yielded lower parasite viability, in terms of lower numbers of amastigotes and trypomastigotes in culture, revealing that *T. cruzi* is also reliant on the action of hPARP-1 and hPARG, to some extent, highlighting the multifaceted way in which *T. cruzi* infection can potentially be disrupted. It is necessary to fully understand the mechanisms by which PARP and PARG may be used therapeutically, as the right balance needs to be struck for the host to benefit from this therapy without harm. Inhibiting PARP can disrupt infection, but PARP inhibition leads to detrimental impacts for the host; therefore, being able to clear the accumulation of PAR with PARG enzymes would be required. However, PARG enzymes could also be a target for inhibition themselves, given that TcPARG and hPARG are required for infection. TcPARP, TcPARG, hPARP, and hPARG are all credible targets for inhibition given their important roles in Chagas infection establishment, yet their close relationship necessitates fine-tuning of any potential PARP/PARG therapy against *T. cruzi*.

T. brucei similarly possesses innate PARP and PARG enzymes (TbPARP and TbPARG). However, unlike in *T. cruzi*, TbPARP is seemingly less essential for successful infection, though TbPARG maintains an essential role and is, therefore, likely the better candidate for targeted therapy to prevent successful *T. brucei* infection. Overall, the understanding of the interplay between the ADP-ribosylation factors of *T. brucei* is likely not as advanced as in *T. cruzi*. This highlights the need for subsequent research to better understand how ADP-ribosylation in *T. brucei* could be exploited therapeutically. TbPARG inhibition decreases *T. brucei* viability significantly through the reduced numbers of parasites, which makes TbPARG the leading candidate for therapy. TbPARP plays a less focal role in DNA repair than hPARP-1 and TcPARP and is, therefore, likely to not be the best candidate for therapy, although it may be used in a combinatorial approach when targeting TbPARG. Given the close genetics of *T. cruzi* and *T. brucei* and the similarities in their ADP-ribosylation systems, it may be worth exploring how or if *T. brucei* may compromise hPARP/hPARG enzymes to establish infection in a human host.

Given the status of *Leishmania* as an NAD^+ auxotroph, a therapeutic option for this genus is likely to be inhibiting their ability to harvest NAD^+ from the host via disruption of their NAD^+ salvage pathways. *Leishmania* uses NMN and NR as essential NAD^+ precursors, with the essential enzyme NadN being active in the pathway. Inhibition of NadN is another likely therapeutic target, given its specificity to *Leishmania* and other pathogens, and its critical role in allowing *Leishmania* to access NAD^+, necessary for several essential cellular processes that maintain parasite viability. Alternatively, LdARL3A and LiSIR2RP1 are essential genes for flagellar growth and parasite cytoskeletal maintenance, respectively. Knockout experiments of LdARL3A decreased flagellar growth to the extent that parasite movement could be prevented, which would aid in infection prevention as the parasite is unable to disseminate in the host. Knockout LiSIR2RP1 experiments decreased the number of viable parasites, likely due to the close relationship of LiSIR2RP1 to the cytoskeletal structure of *Leishmania infantum*. However, LdARL3A and LiSIR2RP1 are both specific to one species of *Leishmania* (*L. donovani* and *L. infantum*, respectively) and consequently, these approaches may not be universal for all *Leishmania* species, and so further research is needed to fully understand the conservancy of these genes across medically relevant *Leishmania* species, and if targeting these genes may require differing species-level approaches. Therefore, the universality of targeting NAD^+ salvaging for all *Leishmania* species likely presents the best current option for an ADP-ribosylation-related therapy in these parasites.

Author Contributions: Conceptualization, J.D. and C.L.D. writing—original draft preparation, J.D.; writing—review and editing, J.D. and C.L.D. All authors have read and agreed to the published version of the manuscript.

Funding: This research received no external funding.

Conflicts of Interest: The authors declare no conflict of interest.

References

1. Miwa, M.; Ishihara, M.; Takishima, S.; Takasuka, N.; Maeda, M.; Yamaizumi, Z.; Sugimura, T.; Yokoyama, S.; Miyazawa, T. The branching and linear portions of poly(adenosine diphosphate ribose) have the same alpha(1 leads to 2) ribose-ribose linkage. *J. Biol. Chem.* **1981**, *256*, 2916–2921. [CrossRef] [PubMed]
2. Laing, S.; Koch-Nolte, F.; Haag, F.; Buck, F. Strategies for the identification of arginine ADP-ribosylation sites. *J. Proteom.* **2011**, *75*, 169–176. [CrossRef] [PubMed]
3. Fehr, A.R.; Singh, S.A.; Kerr, C.M.; Mukai, S.; Higashi, H.; Aikawa, M. The impact of PARPs and ADP-ribosylation on inflammation and host–pathogen interactions. *Genes Dev.* **2020**, *34*, 341–359. [CrossRef]
4. Kuny, C.V.; Sullivan, C.S. Virus–Host Interactions and the ARTD/PARP Family of Enzymes. *PLoS Pathog.* **2016**, *12*, e1005453. [CrossRef] [PubMed]
5. Malgras, M.; Garcia, M.; Jousselin, C.; Bodet, C.; Lévêque, N. The Antiviral Activities of Poly-ADP-Ribose Polymerases. *Viruses* **2021**, *13*, 582. [CrossRef] [PubMed]
6. Abraham, R.; Hauer, D.; McPherson, R.L.; Utt, A.; Kirby, I.T.; Cohen, M.S.; Merits, A.; Leung, A.K.L.; Griffin, D.E. ADP-ribosyl–binding and hydrolase activities of the alphavirus nsP3 macrodomain are critical for initiation of virus replication. *Proc. Natl. Acad. Sci. USA* **2018**, *115*, E10457–E10466. [CrossRef]
7. Zhu, H.; Zheng, C. When PARPs Meet Antiviral Innate Immunity. *Trends Microbiol.* **2021**, *29*, 776–778. [CrossRef]
8. Chiu, H.-P.; Chiu, H.; Yang, C.-F.; Lee, Y.-L.; Chiu, F.-L.; Kuo, H.-C.; Lin, R.-J.; Lin, Y.-L. Inhibition of Japanese encephalitis virus infection by the host zinc-finger antiviral protein. *PLoS Pathog.* **2018**, *14*, e1007166. [CrossRef] [PubMed]
9. Liu, C.-H.; Zhou, L.; Chen, G.; Krug, R.M. Battle between influenza A virus and a newly identified antiviral activity of the PARP-containing ZAPL protein. *Proc. Natl. Acad. Sci. USA* **2015**, *112*, 14048–14053. [CrossRef] [PubMed]
10. Atasheva, S.; Akhrymuk, M.; Frolova, E.I.; Frolov, I. New PARP Gene with an Anti-Alphavirus Function. *J. Virol.* **2012**, *86*, 8147–8160. [CrossRef]
11. Todorova, T.; Bock, F.J.; Chang, P. PARP13 and RNA regulation in immunity and cancer. *Trends Mol. Med.* **2015**, *21*, 373–384. [CrossRef]
12. Deng, L.; Ammosova, T.; Pumfery, A.; Kashanchi, F.; Nekhai, S. HIV-1 Tat Interaction with RNA Polymerase II C-terminal Domain (CTD) and a Dynamic Association with CDK2 Induce CTD Phosphorylation and Transcription from HIV-1 Promoter. *J. Biol. Chem.* **2002**, *277*, 33922–33929. [CrossRef] [PubMed]
13. Parent, M.; Yung, T.M.C.; Rancourt, A.; Ho, E.L.Y.; Vispé, S.; Suzuki-Matsuda, F.; Uehara, A.; Wada, T.; Handa, H.; Satoh, M.S. Poly(ADP-ribose) Polymerase-1 Is a Negative Regulator of HIV-1 Transcription through Competitive Binding to TAR RNA with Tat·Positive Transcription Elongation Factor b (p-TEFb) Complex. *J. Biol. Chem.* **2005**, *280*, 448–457. [CrossRef] [PubMed]
14. Zhu, G.-D.; Gong, J.; Gandhi, V.B.; Liu, X.; Shi, Y.; Johnson, E.F.; Donawho, C.K.; Ellis, P.A.; Bouska, J.J.; Osterling, D.J.; et al. Discovery and SAR of orally efficacious tetrahydropyridopyridazinone PARP inhibitors for the treatment of cancer. *Bioorganic Med. Chem.* **2012**, *20*, 4635–4645. [CrossRef] [PubMed]
15. Xing, J.; Zhang, A.; Du, Y.; Fang, M.; Minze, L.J.; Liu, Y.-J.; Li, X.C.; Zhang, Z. Identification of poly(ADP-ribose) polymerase 9 (PARP9) as a noncanonical sensor for RNA virus in dendritic cells. *Nat. Commun.* **2021**, *12*, 2681. [CrossRef]
16. Yu, M.; Zhang, C.; Yang, Y.; Yang, Z.; Zhao, L.; Xu, L.; Wang, R.; Zhou, X.; Huang, P. The interaction between the PARP10 protein and the NS1 protein of H5N1 AIV and its effect on virus replication. *Virol. J.* **2011**, *8*, 546. [CrossRef]
17. Li, L.; Zhao, H.; Liu, P.; Li, C.; Quanquin, N.; Ji, X.; Sun, N.; Du, P.; Qin, C.-F.; Lu, N.; et al. *PARP12* suppresses Zika virus infection through PARP-dependent degradation of NS1 and NS3 viral proteins. *Sci. Signal.* **2018**, *11*, eaas9332. [CrossRef]
18. Miettinen, M.; Vedantham, M.; Pulliainen, A.T. Host poly(ADP-ribose) polymerases (PARPs) in acute and chronic bacterial infections. *Microbes Infect.* **2019**, *21*, 423–431. [CrossRef]

19. Soriano, F.G.; Liaudet, L.; Szabó, É.; Virág, L.; Mabley, J.G.; Pacher, P.; Szabó, C. Resistance to Acute Septic Peritonitis in Poly(ADP-ribose) Polymerase-1-Deficient Mice. *Shock* **2002**, *17*, 286–292. [CrossRef] [PubMed]
20. Murakami, K.; Enkhbaatar, P.; Shimoda, K.; Cox, R.A.; Burke, A.S.; Hawkins, H.K.; Traber, L.D.; Schmalstieg, F.C.; Salzman, A.L.; Mabley, J.G.; et al. Inhibition of Poly (ADP-ribose) Polymerase Attenuates Acute Lung Injury in an Ovine Model of Sepsis. *Shock* **2004**, *21*, 126–133. [CrossRef] [PubMed]
21. Yélamos, J.; Buendía, A.J.; Ortega, N.; Monreal, Y.; Gallego, M.C.; Sánchez, J.; Ramírez, P.; Parrilla, P.; Caro, M.R.; Aparicio, P.; et al. Genetic and pharmacological inhibition of poly(ADP-ribose) polymerase-1 interferes in the chlamydial life cycle. *Biochem. Biophys. Res. Commun.* **2004**, *324*, 840–848. [CrossRef] [PubMed]
22. Kunze, F.A.; Hottiger, M.O. Regulating Immunity via ADP-Ribosylation: Therapeutic Implications and Beyond. *Trends Immunol.* **2019**, *40*, 159–173. [CrossRef] [PubMed]
23. Chandrasekaran, S.; Caparon, M.G. The Streptococcus pyogenes NAD^+ glycohydrolase modulates epithelial cell PARylation and HMGB1 release. *Cell Microbiol.* **2015**, *17*, 1376–1390. [CrossRef] [PubMed]
24. Vyas, S.; Chesarone-Cataldo, M.; Todorova, T.; Huang, Y.-H.; Chang, P. A systematic analysis of the PARP protein family identifies new functions critical for cell physiology. *Nat. Commun.* **2013**, *4*, 2240. [CrossRef] [PubMed]
25. Oliver, F.J.; Ménissier-de Murcia, J.; Nacci, C.; Decker, P.; Andriantsitohaina, R.; Muller, S.; de la Rubia, G.; Stoclet, J.C.; de Murcia, G. Resistance to endotoxic shock as a consequence of defective NF-kappaB activation in poly (ADP-ribose) polymerase-1 deficient mice. *EMBO J.* **1999**, *18*, 4446–4454. [CrossRef]
26. Gupte, R.; Nandu, T.; Kraus, W.L. Nuclear ADP-ribosylation drives IFNγ-dependent STAT1α enhancer formation in macrophages. *Nat. Commun.* **2021**, *12*, 3931. [CrossRef] [PubMed]
27. Mikolčević, P.; Hloušek-Kasun, A.; Ahel, I.; Mikoč, A. ADP-ribosylation systems in bacteria and viruses. *Comput. Struct. Biotechnol. J.* **2021**, *19*, 2366–2383. [CrossRef] [PubMed]
28. Harrision, D.; Gravells, P.; Thompson, R.; Bryant, H.E. Poly(ADP-Ribose) Glycohydrolase (PARG) vs. Poly(ADP-Ribose) Polymerase (PARP)—Function in Genome Maintenance and Relevance of Inhibitors for Anti-cancer Therapy. *Front. Mol. Biosci.* **2020**, *7*, 191. [CrossRef]
29. Andrabi, S.A.; Kim, N.S.; Yu, S.-W.; Wang, H.; Koh, D.W.; Sasaki, M.; Klaus, J.A.; Otsuka, T.; Zhang, Z.; Koehler, R.C.; et al. Poly(ADP-ribose) (PAR) polymer is a death signal. *Proc. Natl. Acad. Sci. USA* **2006**, *103*, 18308–18313. [CrossRef] [PubMed]
30. Wang, Y.; Kim, N.S.; Haince, J.-F.; Kang, H.C.; David, K.K.; Andrabi, S.A.; Poirier, G.G.; Dawson, V.L.; Dawson, T.M. Poly(ADP-Ribose) (PAR) Binding to Apoptosis-Inducing Factor Is Critical for PAR Polymerase-1–Dependent Cell Death (Parthanatos). *Sci. Signal.* **2011**, *4*, ra20. [CrossRef]
31. Barkauskaite, E.; Brassington, A.; Tan, E.S.; Warwicker, J.; Dunstan, M.S.; Banos, B.; Lafite, P.; Ahel, M.; Mitchison, T.J.; Ahel, I.; et al. Visualization of poly(ADP-ribose) bound to PARG reveals inherent balance between exo- and endo-glycohydrolase activities. *Nat. Commun.* **2013**, *4*, 2164. [CrossRef]
32. O'Sullivan, J.; Ferreira, M.T.; Gagné, J.-P.; Sharma, A.K.; Hendzel, M.J.; Masson, J.-Y.; Poirier, G.G. Emerging roles of eraser enzymes in the dynamic control of protein ADP-ribosylation. *Nat. Commun.* **2019**, *10*, 1182. [CrossRef] [PubMed]
33. Pourfarjam, Y.; Kasson, S.; Tran, L.; Ho, C.; Lim, S.; Kim, I.-K. PARG has a robust endo-glycohydrolase activity that releases protein-free poly(ADP-ribose) chains. *Biochem. Biophys. Res. Commun.* **2020**, *527*, 818–823. [CrossRef] [PubMed]
34. Sethi, G.S.; Dharwal, V.; Naura, A.S. Poly(ADP-Ribose)Polymerase-1 in Lung Inflammatory Disorders: A Review. *Front. Immunol.* **2017**, *8*, 1172. [CrossRef] [PubMed]
35. Wasyluk, W.; Zwolak, A. PARP Inhibitors: An Innovative Approach to the Treatment of Inflammation and Metabolic Disorders in Sepsis. *J. Inflamm. Res.* **2021**, *14*, 1827–1844. [CrossRef] [PubMed]
36. Caprara, G.; Prosperini, E.; Piccolo, V.; Sigismondo, G.; Melacarne, A.; Cuomo, A.; Boothby, M.; Rescigno, M.; Bonaldi, T.; Natoli, G. PARP14 Controls the Nuclear Accumulation of a Subset of Type I IFN–Inducible Proteins. *J. Immunol.* **2018**, *200*, 2439–2454. [CrossRef]
37. Ackley, C.; Elsheikh, M.; Zaman, S. Scoping review of Neglected Tropical Disease Interventions and Health Promotion: A framework for successful NTD interventions as evidenced by the literature. *PLoS Negl. Trop. Dis.* **2021**, *15*, e0009278. [CrossRef] [PubMed]
38. Echeverria, L.E.; Morillo, C.A. American Trypanosomiasis (Chagas Disease). *Infect. Dis. Clin. N. Am.* **2019**, *33*, 119–134. [CrossRef]
39. Sangenito, L.S.; Branquinha, M.H.; Santos, A.L.S. Funding for Chagas Disease: A 10-Year (2009–2018) Survey. *Trop. Med. Infect. Dis.* **2020**, *5*, 88. [CrossRef] [PubMed]
40. Bustamante, J.M.; Tarleton, R.L. Potential new clinical therapies for Chagas disease. *Expert Rev. Clin. Pharmacol.* **2014**, *7*, 317–325. [CrossRef] [PubMed]
41. Büscher, P.; Cecchi, G.; Jamonneau, V.; Priotto, G. Human African trypanosomiasis. *Lancet* **2017**, *390*, 2397–2409. [CrossRef]
42. Kennedy, P.G. Clinical features, diagnosis, and treatment of human African trypanosomiasis (sleeping sickness). *Lancet Neurol.* **2013**, *12*, 186–194. [CrossRef]
43. Baker, N.; de Koning, H.P.; Mäser, P.; Horn, D. Drug resistance in African trypanosomiasis: The melarsoprol and pentamidine story. *Trends Parasitol.* **2013**, *29*, 110–118. [CrossRef] [PubMed]
44. Desjeux, P. The increase in risk factors for leishmaniasis worldwide. *Trans. R. Soc. Trop. Med. Hyg.* **2001**, *95*, 239–243. [CrossRef]
45. Zulfiqar, B.; Shelper, T.B.; Avery, V.M. Leishmaniasis drug discovery: Recent progress and challenges in assay development. *Drug Discov. Today* **2017**, *22*, 1516–1531. [CrossRef] [PubMed]

46. Jain, V.; Jain, K. Molecular targets and pathways for the treatment of visceral leishmaniasis. *Drug Discov. Today* **2018**, *23*, 161–170. [CrossRef] [PubMed]
47. Oliveira, S.S.; Marques, C.S.; de Sousa, D.P.; Andrade, L.N.; Fricks, A.T.; Jain, S.; Branquinha, M.H.; Souto, E.B.; Santos, A.L.; Severino, P. Analysis of the mechanisms of action of isopentenyl caffeate against Leishmania. *Biochimie* **2021**, *189*, 158–167. [CrossRef]
48. Laniado-Laborín, R.; Cabrales-Vargas, M.N. Amphotericin B: Side effects and toxicity. *Rev. Iberoam. Micol.* **2009**, *26*, 223–227. [CrossRef]
49. Schlesinger, M.; Larrea, S.C.V.; Haikarainen, T.; Narwal, M.; Venkannagari, H.; Flawiá, M.M.; Lehtiö, L.; Villamil, S.H.F. Disrupted ADP-ribose metabolism with nuclear Poly (ADP-ribose) accumulation leads to different cell death pathways in presence of hydrogen peroxide in procyclic Trypanosoma brucei. *Parasites Vectors* **2016**, *9*, 173. [CrossRef]
50. Agnew, T.; Munnur, D.; Crawford, K.; Palazzo, L.; Mikoč, A.; Ahel, I. MacroD1 Is a Promiscuous ADP-Ribosyl Hydrolase Localized to Mitochondria. *Front. Microbiol.* **2018**, *9*, 20. [CrossRef]
51. Haikarainen, T.; Lehtiö, L. Proximal ADP-ribose Hydrolysis in Trypanosomatids is Catalyzed by a Macrodomain. *Sci. Rep.* **2016**, *6*, 24213. [CrossRef] [PubMed]
52. Larrea, S.C.V.; Alonso, G.D.; Schlesinger, M.; Torres, H.N.; Flawiá, M.M.; Villamil, S.H.F. Poly(ADP-ribose) polymerase plays a differential role in DNA damage-response and cell death pathways in Trypanosoma cruzi. *Int. J. Parasitol.* **2011**, *41*, 405–416. [CrossRef]
53. Fernández Villamil, S.H.; Vilchez Larrea, S.C. Poly(ADP-ribose) metabolism in human parasitic protozoa. *Acta Trop.* **2020**, *208*, 105499. [CrossRef]
54. Otto, H.; Reche, P.A.; Bazan, F.; Dittmar, K.; Haag, F.; Koch-Nolte, F. In silico characterization of the family of PARP-like poly(ADP-ribosyl)transferases (pARTs). *BMC Genom.* **2005**, *6*, 139. [CrossRef]
55. Villamil, S.H.F.; Baltanás, R.; Alonso, G.D.; Larrea, S.C.V.; Torres, H.N.; Flawiá, M.M. TcPARP: A DNA damage-dependent poly(ADP-ribose) polymerase from Trypanosoma cruzi. *Int. J. Parasitol.* **2008**, *38*, 277–287. [CrossRef] [PubMed]
56. Oliver, A.W.; Amé, J.C.; Roe, S.M.; Good, V.; de Murcia, G.; Pearl, L.H. Crystal structure of the catalytic fragment of murine poly(ADP-ribose) polymerase-2. *Nucleic Acids Res.* **2004**, *32*, 456–464. [CrossRef]
57. Gonçalves, V.M.; Matteucci, K.C.; Buzzo, C.L.; Miollo, B.H.; Ferrante, D.; Torrecilhas, A.C.; Rodrigues, M.M.; Alvarez, J.M.; Bortoluci, K.R. NLRP3 Controls Trypanosoma cruzi Infection through a Caspase-1-Dependent IL-1R-Independent NO Production. *PLoS Negl. Trop. Dis.* **2013**, *7*, e2469. [CrossRef]
58. Stahl, P.; Ruppert, V.; Meyer, T.; Schmidt, J.; Campos, M.A.; Gazzinelli, R.T.; Maisch, B.; Schwarz, R.T.; Debierre-Grockiego, F. Trypomastigotes and amastigotes of Trypanosoma cruzi induce apoptosis and STAT3 activation in cardiomyocytes in vitro. *Apoptosis* **2013**, *18*, 653–663. [CrossRef]
59. Priotto, S.; Sartori, M.; Repossi, G.; Valentich, M. Trypanosoma cruzi: Participation of cholesterol and placental alkaline phosphatase in the host cell invasion. *Exp. Parasitol.* **2009**, *122*, 70–73. [CrossRef]
60. Nogueira, N.P.; Saraiva, F.M.S.; Sultano, P.E.; Cunha, P.R.B.B.; Laranja, G.A.T.; Justo, G.A.; Sabino, K.C.C.; Coelho, M.G.P.; Rossini, A.; Atella, G.C.; et al. Proliferation and Differentiation of Trypanosoma cruzi inside Its Vector Have a New Trigger: Redox Status. *PLoS ONE* **2015**, *10*, e0116712. [CrossRef] [PubMed]
61. Poveda, C.; Fresno, M.; Gironès, N.; Martins-Filho, O.A.; Ramírez, J.D.; Santi-Rocca, J.; Marin-Neto, J.A.; Morillo, C.; Rosas, F.; Guhl, F. Cytokine Profiling in Chagas Disease: Towards Understanding the Association with Infecting Trypanosoma cruzi Discrete Typing Units (A BENEFIT TRIAL Sub-Study). *PLoS ONE* **2014**, *9*, e91154. [CrossRef]
62. Ba, X.; Gupta, S.; Davidson, M.; Garg, N.J. Trypanosoma cruzi Induces the Reactive Oxygen Species-PARP-1-RelA Pathway for Up-regulation of Cytokine Expression in Cardiomyocytes. *J. Biol. Chem.* **2010**, *285*, 11596–11606. [CrossRef]
63. Pinto, A.M.T.; Sales, P.C.M.; Camargos, E.R.S.; Silva, A.M. Tumour necrosis factor (TNF)-mediated NF-κB activation facilitates cellular invasion of non-professional phagocytic epithelial cell lines by Trypanosoma cruzi. *Cell. Microbiol.* **2011**, *13*, 1518–1529. [CrossRef] [PubMed]
64. Luo, X.; Kraus, W.L. On PAR with PARP: Cellular stress signaling through poly(ADP-ribose) and PARP-1. *Genes Dev.* **2012**, *26*, 417–432. [CrossRef]
65. Alonso, G.D.; Vilchez Larrea, S.C.; Fernández Villamil, S.H. Metabolism of poly-ADP-ribose in trypanosomatids. In *Parasitology Research Trends*; Le Novo Science Publishers Inc.: Buenos Aires, Argentina, 2010.
66. Larrea, S.C.V.; Schlesinger, M.; Kevorkian, M.L.; Flawiá, M.M.; Alonso, G.D.; Villamil, S.H.F. Host Cell Poly(ADP-Ribose) Glycohydrolase Is Crucial for Trypanosoma cruzi Infection Cycle. *PLoS ONE* **2013**, *8*, e67356. [CrossRef]
67. Wen, J.J.; Yin, Y.W.; Garg, N.J. PARP1 depletion improves mitochondrial and heart function in Chagas disease: Effects on POLG dependent mtDNA maintenance. *PLoS Pathog.* **2018**, *14*, e1007065. [CrossRef]
68. Haikarainen, T.; Schlesinger, M.; Obaji, E.; Villamil, S.H.F.; Lehtiö, L. Structural and Biochemical Characterization of Poly-ADP-ribose Polymerase from Trypanosoma brucei. *Sci. Rep.* **2017**, *7*, 3642. [CrossRef]
69. Niemirowicz, G.T.; Cazzulo, J.J.; Álvarez, V.E.; Bouvier, L.A. Simplified inducible system for Trypanosoma brucei. *PLoS ONE* **2018**, *13*, e0205527. [CrossRef]
70. Kun, E.; Kirsten, E.; Mendeleyev, J.; Ordahl, C.P. Regulation of the Enzymatic Catalysis of Poly(ADP-ribose) Polymerase by dsDNA, Polyamines, Mg^{2+}, Ca^{2+}, Histones H_1 and H_3, and ATP. *Biochemistry* **2003**, *43*, 210–216. [CrossRef] [PubMed]

71. van Beek, L.; McClay, É.; Patel, S.; Schimpl, M.; Spagnolo, L.; Maia de Oliveira, T. Parp power: A structural perspective on parp1, parp2, and parp3 in dna damage repair and nucleosome remodelling. *Int. J. Mol. Sci.* **2021**, *2*, 5112. [CrossRef]
72. Klebanov-Akopyan, O.; Mishra, A.; Glousker, G.; Tzfati, Y.; Shlomai, J. Trypanosoma brucei UMSBP2 is a single-stranded telomeric DNA binding protein essential for chromosome end protection. *Nucleic Acids Res.* **2018**, *46*, 7757–7771. [CrossRef]
73. Glover, L.; McCulloch, R.; Horn, D. Sequence homology and microhomology dominate chromosomal double-strand break repair in African trypanosomes. *Nucleic Acids Res.* **2008**, *36*, 2608–2618. [CrossRef]
74. Welburn, S.C.; Maudlin, I. Control of Trypanosoma brucei brucei infections in tsetse, Glossina morsitans. *Med. Veter. Entomol.* **1997**, *11*, 286–289. [CrossRef] [PubMed]
75. Nepomuceno-Mejía, T.; Lara-Martínez, R.; Hernández, R.; De, M.; Segura-Valdez, L.; Jiménez-García, L.F. Nucleologenesis in *Trypanosoma cruzi*. *Microsc. Microanal.* **2016**, *22*, 621–629. [CrossRef]
76. Nenarokova, A.; Záhonová, K.; Krasilnikova, M.; Gahura, O.; McCulloch, R.; Zíková, A.; Yurchenko, V.; Lukeš, J. Causes and Effects of Loss of Classical Nonhomologous End Joining Pathway in Parasitic Eukaryotes. *mBio* **2019**, *10*, e01541-19. [CrossRef]
77. Cuvillier, A.; Miranda, J.C.; Ambit, A.; Barral, A.; Merlin, G. Abortive infection of Lutzomyia longipalpis insect vectors by aflagellated LdARL-3A-Q70L overexpressing Leishmania amazonensis parasites. *Cell. Microbiol.* **2003**, *5*, 717–728. [CrossRef] [PubMed]
78. Sahin, A.; Espiau, B.; Tetaud, E.; Cuvillier, A.; Lartigue, L.; Ambit, A.; Robinson, D.R.; Merlin, G. The *Leishmania* ARL-1 and Golgi Traffic. *PLoS ONE* **2008**, *3*, e1620. [CrossRef]
79. Tavares, J.; Ouaissi, A.; Kongâthooâlin, P.; Loureiro, I.; Kaur, S.; Roy, N.; Cordeiro-Da-Silva, A. Bisnaphthalimidopropyl Derivatives as Inhibitors of *Leishmania* SIR_2 Related Protein 1. *ChemBioChem* **2009**, *5*, 140–147. [CrossRef]
80. Ronin, C.; Costa, D.M.; Tavares, J.; Faria, J.; Ciesielski, F.; Ciapetti, P.; Smith, T.K.; MacDougall, J.; Cordeiro-Da-Silva, A.; Pemberton, I.K. The crystal structure of the Leishmania infantum Silent Information Regulator 2 related protein 1: Implications to protein function and drug design. *PLoS ONE* **2018**, *13*, e0193602. [CrossRef] [PubMed]
81. Gazanion, E.; Garcia, D.; Silvestre, R.; Gérard, C.; Guichou, J.F.; Labesse, G.; Seveno, M.; Cordeiro-Da-Silva, A.; Ouaissi, A.; Sereno, D.; et al. The Leishmania nicotinamidase is essential for NAD+ production and parasite proliferation. *Mol. Microbiol.* **2011**, *82*, 21–38. [CrossRef]
82. Reidl, J.; Schlör, S.; Kraiß, A.; Schmidt-Brauns, J.; Kemmer, G.; Soleva, E. NADP and NAD utilization in Haemophilus influenzae. *Mol. Microbiol.* **2000**, *35*, 1573–1581. [CrossRef] [PubMed]
83. Moreira, D.; Rodrigues, V.; Abengozar, M.; Rivas, L.; Rial, E.; Laforge, M.; Li, X.; Foretz, M.; Viollet, B.; Estaquier, J.; et al. Leishmania infantum Modulates Host Macrophage Mitochondrial Metabolism by Hijacking the SIRT1-AMPK Axis. *PLoS Pathog.* **2015**, *11*, e1004684. [CrossRef] [PubMed]

pathogens

MDPI

Article

Recurrent Loss of Macrodomain Activity in Host Immunity and Viral Proteins

Sofia E. Delgado-Rodriguez, Andrew P. Ryan and Matthew D. Daugherty *

Department of Molecular Biology, School of Biological Sciences, University of California—San Diego, La Jolla, CA 92093, USA
* Correspondence: mddaugherty@ucsd.edu

Abstract: Protein post-translational modifications (PTMs) are an important battleground in the evolutionary arms races that are waged between the host innate immune system and viruses. One such PTM, ADP-ribosylation, has recently emerged as an important mediator of host antiviral immunity. Important for the host–virus conflict over this PTM is the addition of ADP-ribose by PARP proteins and removal of ADP-ribose by macrodomain-containing proteins. Interestingly, several host proteins, known as macroPARPs, contain macrodomains as well as a PARP domain, and these proteins are both important for the host antiviral immune response and evolving under very strong positive (diversifying) evolutionary selection. In addition, several viruses, including alphaviruses and coronaviruses, encode one or more macrodomains. Despite the presence of the conserved macrodomain fold, the enzymatic activity of many of these proteins has not been characterized. Here, we perform evolutionary and functional analyses to characterize the activity of macroPARP and viral macrodomains. We trace the evolutionary history of macroPARPs in metazoans and show that PARP9 and PARP14 contain a single active macrodomain, whereas PARP15 contains none. Interestingly, we also reveal several independent losses of macrodomain enzymatic activity within mammalian PARP14, including in the bat, ungulate, and carnivore lineages. Similar to macroPARPs, coronaviruses contain up to three macrodomains, with only the first displaying catalytic activity. Intriguingly, we also reveal the recurrent loss of macrodomain activity within the alphavirus group of viruses, including enzymatic loss in insect-specific alphaviruses as well as independent enzymatic losses in two human-infecting viruses. Together, our evolutionary and functional data reveal an unexpected turnover in macrodomain activity in both host antiviral proteins and viral proteins.

Keywords: ADP-ribosylation; macrodomain; PARP; host–virus evolution; phylogenetics; alphaviruses; coronaviruses

Citation: Delgado-Rodriguez, S.E.; Ryan, A.P.; Daugherty, M.D. Recurrent Loss of Macrodomain Activity in Host Immunity and Viral Proteins. *Pathogens* **2023**, *12*, 674. https://doi.org/10.3390/pathogens12050674

Academic Editors: Anthony K L Leung, Anthony Fehr and Rachy Abraham

Received: 25 March 2023
Revised: 29 April 2023
Accepted: 30 April 2023
Published: 3 May 2023

1. Introduction

ADP-ribosylation is a reversible post-translational modification (PTM) of proteins that is widely found in bacteria, eukaryotes, and viruses [1–4]. The PTM is catalyzed by diverse ADP-ribosyltransferases, including the family of PARP enzymes in eukaryotes [3]. Completing the cycle of PTM addition and removal, a variety of enzymatic domains can catalyze the removal of ADP-ribose from proteins [5]. Primary among these ADP-ribosylhydrolases is the macrodomain, which is a structurally conserved 120-200 amino acid module that can both recognize ('read") and reverse ('erase") ADP-ribosylation of proteins [5–7].

Macrodomains are found in a wide variety of eukaryotic, bacterial, and viral proteins. Of particular note are several metazoan proteins known as macroPARPs, which contain both a PARP domain (a "writer") and two or more macrodomains ("readers" and "erasers"). Interestingly, mammalian macroPARPs, which include human PARP9, PARP14, and PARP15, are highly upregulated in response to the immune signaling molecule interferon (IFN), and have evolved under very strong positive (diversifying) selection [8], both of which are characteristic of genes that are engaged in host–pathogen evolutionary "arms races" [9–11]. Such data prompted us to propose that macroPARPs may be involved in a molecular and genetic conflict with viruses over ADP-ribosylation addition and removal [8]. Indeed, several papers have now revealed important roles for macroPARPs in directly or indirectly potentiating the host antiviral immune response, including evidence that PARP9 and PARP14 regulate the antiviral IFN response and other immune signaling pathways, and that PARP14 inhibits coronavirus replication [12–16]. However, the importance of macrodomains in these innate immune functions of macroPARPs has not been determined.

On the other side of the host–virus conflict surrounding ADP-ribosylation are viral proteins that contain macrodomains. Several groups of positive-sense single-stranded RNA (+ssRNA) viruses contain macrodomains embedded within non-structural proteins, including alphaviruses (e.g., chikungunya and equine encephalitis viruses), hepeviruses (e.g., hepatitis E virus), and coronaviruses (e.g., SARS-CoV-2) [17,18]. Notably, viral macrodomains have been shown to be critical for not only viral replication, but also for virulence and evasion of the IFN-mediated antiviral immune response [12,17,19–26]. In many cases, mutation of key catalytic residues in the viral macrodomain results in viral attenuation or increased sensitivity to antiviral immunity, suggesting that macrodomain ADP-ribosylhydrolase activity is a critical viral function.

These results position macrodomains and ADP-ribosylhydrolase activity at the center of a conflict between host antiviral immunity and viruses. As such, one expectation might be that macrodomain enzymatic activity would be well conserved throughout host and viral evolution. However, the degree to which macrodomains and ADP-ribosylhydrolase activity is conserved or divergent has not been analyzed in many cases. Here we analyze both host macroPARPs and viral macrodomain-containing proteins for conservation of key catalytic residues required for ADP-ribosylhydrolase activity. Strikingly, we find that key residues have been mutated in several independent lineages of host macroPARPs, as well as independent lineages of alphaviruses. Using an enzymatic assay for ADP-ribosylhydrolase activity in human cells, we confirm the loss of macrodomain activity consistent with the observed sequence changes. These results reveal at least three mammalian lineages in bats, ungulates, and carnivores that lack PARP14 macrodomain activity. Moreover, we find that macrodomains from several alphaviruses, including a human alphavirus and insect-specific alphaviruses, lack enzymatic activity. Together, our evolutionary and functional data reveal an unexpected turnover in macrodomain activity in both host antiviral proteins and viral proteins, shedding further light on the dynamic evolution of this critical PTM.

2. Materials and Methods

2.1. MacroPARP Homology Searches

A portion of human PARP14 (accession NP_060024.2) spanning the three macrodomains (residues 791-1387) was used to query the NCBI RefSeq protein database (including "metazoans (taxid:33208)") using BLASTP [27] with an e-value cutoff of 1×10^{-20} and a query coverage cutoff of 40%. Using only the tandem macrodomain region as a search eliminated results from PARP proteins that lack macrodomains. The resulting 1846 sequences were downloaded as complete protein sequences and aligned using Clustal Omega [28]. Sequences that lacked a complete PARP domain were eliminated from further analyses, as were other incomplete sequences and poorly aligning proteins, resulting in 1091 "full length" macroPARP sequences. To eliminate closely related sequences and reduce total sequence number, sequences with >95% identity were reduced to a single unique sequence using CD-HIT with a 0.95 sequence identity cutoff [29]. The resulting 741 sequences are listed in Supplementary Material File S1. For genomes shown in Figure S1, the absence of PARP9 or PARP14 proteins was confirmed by performing a BLASTP search of the indicated genome with an e-value cutoff 0.05 and using the HMMER webserver [30] to search the indicated genomes with an e-value cutoff of 0.05. In all cases, and as expected, macrodomain-containing proteins were identified with these searches. However, all proteins that had both a macrodomain and a PARP domain that were identified in the *Petromyzon marinus*, *Asterias rubens*, *Crassostrea gigas*, and *Stylophora pistillata* genomes had a domain structure that resembled PARP14 rather than PARP9 or PARP15. Moreover, these searches yielded no protein in the *Drosophila melanogaster* and *Caenorhabditis elegans* genomes that contained both a macrodomain and a PARP domain, consistent with the conclusion that these genomes lack macroPARPs, based on previous iterative PSI-BLAST searches [31].

2.2. MacroPARP Phylogenetic Analyses

All homologs shown in Supplemental File S1 were aligned using Clustal Omega using two iterations of refinement. For full-length macroPARP analyses, such as the one shown in Figure 1B, the alignment was trimmed to only the region spanning from the macrodomains to the PARP domain (corresponding to residues 791-1801 of human PARP14). Maximum likelihood phylogenetic trees were generated using IQ-TREE [32]. IQ-TREE phylogenies were generated using the "-bb 1000-alrt 1000" commands for generation of 1000 ultrafast bootstrap [33] and SH-aLRT support values. The best-fitting substitution model was determined by ModelFinder [34] using the "-m AUTO" command. For macrodomain analyses, such as the one shown in Figure 1C, the individual macrodomains were extracted from the full-length macroPARP alignment described above. Human PARP14 macrodomain boundaries were used: Macrodomain1–residues 791-978, Macrodomain1–residues 1003-1190, and Macrodomain3–residues 1216-1387. Individual extracted macrodomain alignments, along with macrodomains from human MACROD1 (accession NP_054786.2, residues 140-324), MACROD2 (accession NP_542407.2, residues 59-243), and GDAP2 (accession NP_060156.1, residues 43-226), were realigned using Clustal Omega with two rounds of refinement, and maximum likelihood phylogenetic trees were generated with IQ-TREE as described above. All phylogenetic trees were visualized using FigTree (http://tree.bio.ed.ac.uk/software/figtree/, accessed on 1 July 2022). All consensus logos were visualized using Geneious Prime 2022.1.1 (https://www.geneious.com/, accessed on 1 July 2022).

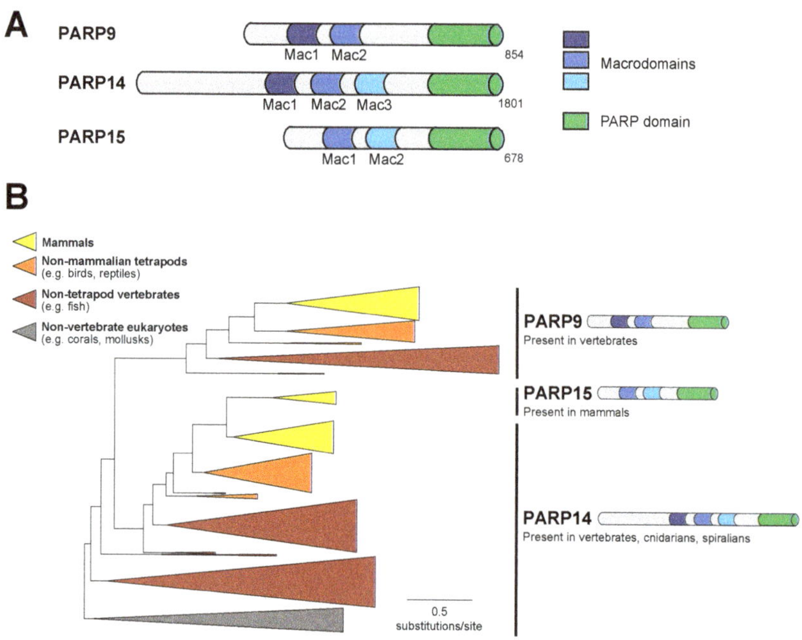

Figure 1. *Cont.*

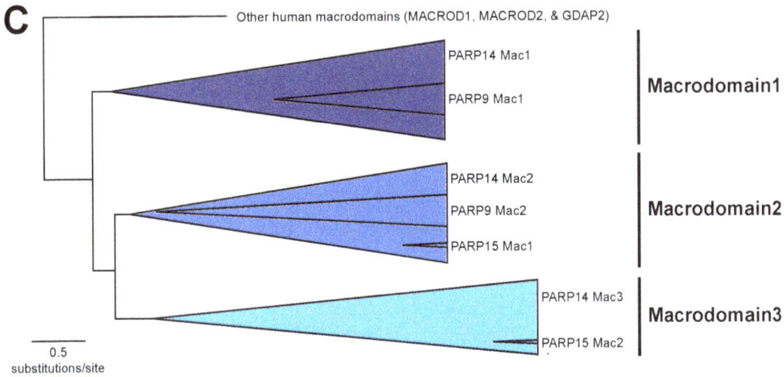

Figure 1. Evolution of macroPARP macrodomains within metazoans. (**A**) Domain structure of the three human macroPARP proteins, PARP9, PARP14, and PARP15. Macrodomains and PARP domains are shown, as is the total amino acid length of each protein. For simplicity, other domains within macroPARPs are not displayed. (**B**) Phylogenetic tree of metazoan macroPARP proteins. Clades of proteins with PARP9-like, PARP14-like, and PARP15-like domain architectures are indicated on the right. Colors represent groups of species as indicated in the key. (**C**) Phylogenetic tree of individual macroPARP macrodomains along with other indicated human macrodomains. There are three clear macroPARP macrodomain clades, corresponding to Macrodomains 1–3. As indicated, each large clade comprises two or three individual macrodomains from the metazoan macroPARPs.

2.3. Coronavirus Macrodomain Homology Searches and Phylogenetic Analyses

The nsP3 protein from SARS-CoV-2 (accession YP_009724389.1) was used to query the NCBI RefSeq protein database (including "viruses (taxid:10239)") using BLASTP with a query coverage cutoff of 25%. Resulting sequences were aligned and curated as for macroPARPs. Identical sequences were removed, but no CD-HIT removal of near-identical sequences was performed. Resulting sequences are listed in Supplemental File S2. Sequences were aligned using Clustal Omega with two rounds of refinement and maximum likelihood phylogenetic trees were generated using IQ-TREE.

2.4. Alphavirus Macrodomain Homology Searches and Phylogenetic Analyses

The non-structural polyprotein from Sindbis virus (accession NP_062888.1) was used to query the NCBI RefSeq protein database (including "viruses (taxid:10239)") using BLASTP with a query coverage cutoff of 25%. Resulting sequences were aligned and curated as for macroPARPs. Identical sequences were removed, but no CD-HIT removal of near-identical sequences was performed. Resulting sequences are listed in Supplemental File S3. Sequences were aligned using Clustal Omega with two rounds of refinement and maximum likelihood phylogenetic trees were generated using IQ-TREE.

2.5. SARS-CoV-2 Macrodomain2 and Macrodomain3 Structure Prediction

Sequences for Macrodomain2 (residues 415–541) and Macrodomain3 (residues 549–676) were extracted from the SARS-CoV-2 nsp3 region of the ORF1ab polyprotein (accession YP_009724389.1). Structural models for these domains were predicted using AlphaFold2 via the ColabFold package [35] with default parameters. Although the sequence similarity is low, there was a clear overall fold similarity of Macrodomain2 and Macrodomain3 to the experimentally determined SARS-CoV-2 Macrodomain1 (PDB code: 6WEY [36]) structure in terms of a core of beta strands (β1 through β5) with stereotypical interruption by α-helices. Using this, it was possible to determine the bounds of loop 1 (between β3 and the proximal downstream α-helix) and loop 2 (between β4 and the proximal downstream α-helix). Whereas the exact sequence alignment between these loop residues may not be precise, based on the fact that the sequences are so divergent, we are able to infer from those loop sequences that Macrodomain2 and Macrodomain3 lack the full repertoire of catalytic residues that would be expected to confer ADP-ribosylhydrolase enzymatic activity. Predicted structures, as well as experimentally determined structures for PARP14 Macrodomain1 bound to ADP-ribose (PDB code: 3Q6Z [37]) and SARS-CoV-2 Macrodomain1 (PDB code: 6WEY [36]) were displayed using PyMol (The PyMOL Molecular Graphics System, Version 2.1 Schrödinger, LLC. New York, NY, USA).

2.6. Plasmids and Constructs

For PARP10 overexpression, the coding sequence of human PARP10 (accession NP_116178.2) was cloned into the pcDNA5/FRT/TO backbone with an N-terminal mCherry, P2A linker, and 3×FLAG epitope tag. For macrodomain overexpression, codon-optimized sequences (see Supplemental File S4) were synthesized by Twist Biosciences (San Francisco, CA, USA) and cloned into pCMV-Twist with an N-terminal HA tag.

2.7. Cell Culture and Transient Transfection

Cell lines (HEK293T, obtained from ATCC (Manassas, VA, USA)), were routinely tested for mycoplasma infection using a PCR kit and kept at a low passage number. Cells were grown in complete media using DMEM (Gibco, Billings, MT, USA) with 10% FBS (Peak Serum, Wellington, CO, USA) and 1% Pen/Strep (Invitrogen, Carlsbad, CA, USA). Cells were seeded a day before transfection in a 24-well plate with 500 uL of media per well such that they would be at 60% confluent the following day for transfection. Cells were transfected using 500 ng of total plasmid DNA with 1.5 uL Transit-X2 transfection reagent (Mirus Bio, Madison, WI, USA) in 100 uL of OptiMEM (Invitrogen, Carlsbad, CA, USA) per well. In all assays, 100 ng of the plasmid expressed mCherry-P2A-3xFlag-PARP10. Except for the case shown in Figure S2, 250 ng of the HA-tagged macrodomain plasmid was used. In the case of Figure S2, either 25 ng, 100 ng, or 400 ng of HA-tagged macrodomain was transfected. In all transfections, the total amount of DNA added was supplemented to 500 ng with an empty cloning vector, pQCXIP (Clontech, Mountain View, CA, USA). Detection of ADP-ribosylation has been shown to be highly dependent on sample conditions [38], and we have observed that the edges of multiwell plates give less consistent signal than the middle of plates. As a result, only the central eight wells of any given plate were used for transfection.

2.8. Sample Preparation, Immunoblotting, and Antibodies

Cells that had been transfected with plasmids as described above were harvested 20 h post transfection. One hour prior to harvest, veliparib (VWR, Radnor, PA, USA), a selective PARP1/PARP2 inhibitor [39,40], was added to culture media to a final concentration of 1 μM to inhibit PARP1 activity as has been previously used [41]. At the time of harvest, media was aspirated, PBS was added to cells and aspirated, and then plates were frozen at −80 °C. After at least 1 h at −80 °C, plates were thawed on ice for 10 min and 75 uL of ADPr lysis buffer (50 mM Tris (pH 7.4), 150 mM NaCl, 1 mM $MgCl_2$, 1% triton-X-100, 1X protease inhibitor, 1 μM PDD00017273 (PARG inhibitor, Sigma-Aldrich, St. Louis, MO, USA)), 1 μM veliparib, 1 mM DTT) was added to each well. After a 10 min incubation on ice, lysates were transferred and centrifuged at 10,000× g at 4 °C for 5 min. The resulting supernatant was transferred to a new tube and 4× NuPAGE LDS sample buffer (Invitrogen) containing 5% β-mercaptoethanol (VWR) was added. Samples were boiled at 95 °C for 10 min and briefly centrifuged before being loaded onto a 4–12% Bis-Tris SDS-PAGE gel (Invitrogen) and run in 1X MOPS (Invitrogen) running buffer. Samples were then wet transferred onto nitrocellulose membrane and blocked with PBS-T containing 5% bovine serum albumin for 1 h. This was followed by incubation overnight at 4 °C with primary antibodies for mono/poly ADPr (anti-poly/mono-ADP-ribose antibody, E6F6A [42], Cell Signaling Technology, Danvers, MA, USA), anti-FLAG M2 (Sigma-Aldrich), anti-HA (Sigma-Aldrich), or anti-GAPDH (Cell Signaling Technology, Danvers, MA, USA). Membranes were then rinsed in PBS-T three times then incubated with the appropriate HRP-conjugated secondary antibodies (Fisher Scientific, Pittsburg, PA, USA). Membranes were then rinsed in PBS-T three times, and developed with SuperSignal West Pico PLUS Chemiluminescent Substrate (Fisher Scientific, Pittsburg, PA, USA), and imaged on a BioRad GelDoc (BioRad, Hercules, CA, USA).

3. Results

3.1. A Single Macrodomain in Human PARP9 and PARP14 Contains ADP-Ribosylhydrolase Activity

Among human PARP proteins, only PARP9, PARP14, and PARP15 contain a combination of macrodomains and a PARP domain (Figure 1A). To understand the distribution of macroPARPs within metazoans, we performed a phylogenetic analysis of homologs of PARP9, PARP14, and PARP15, characterizing them as either PARP9-like, PARP14-like, or PARP15-like based on their domain architecture and position within the protein phylogeny (Figure 1B). As previously observed, PARP15-like proteins only exist in mammalian species [8]. In contrast, we observed PARP9 homologs in jawed vertebrate species, including fish, reptiles, birds, and mammals, but lacking in the jawless sea lamprey, *Petromyzon marinus*, and non-vertebrate metazoans (Figure S1). PARP14 is the most broadly distributed in metazoans, with homologs in cnidarians (corals), spiralians (mollusks), and vertebrates, but noticeably absent in arthropods and nematodes (Figures 1B and S1). From this, we conclude that the PARP14 domain structure of three tandem macrodomains and a PARP domain is the most ancestral form of macroPARP, with PARP9 and PARP15 arising in the vertebrate and mammalian lineages, respectively, as the result of partial duplication of PARP14. These date the existence of different macroPARPs in metazoans to >700 million years old for PARP14, ~500 million years old for PARP9, and ~100 million years old for PARP15 [43,44].

To further characterize the macrodomains present within metazoan macroPARPs, we extracted individual macrodomain sequences from each macroPARP and performed additional phylogenetic analyses. As shown in Figure 1C, the two macrodomains of PARP9 group phylogenetically with Macrodomain1 and Macrodomain2 of PARP14, respectively, whereas the two macrodomains of PARP15 correspond to Macrodomain2 and Macrodomain3 of PARP14, respectively. These data further support the model that PARP9 and PARP15 were partial duplications of the ancestral three macrodomain PARP14 architectures.

We next wished to ask which of the human macroPARP macrodomains display ADP-ribosylhydrolase activity. Several papers have described sequence characteristics that are important for host and viral macrodomain catalytic activity [18,45–48]. In particular, these analyses have focused on the importance of Asn/Ser and Gly residues flanking "loop 1" in the N-terminal region of the macrodomain and a hydrophobic (e.g., Ala, Thr, Ile, Val, or Leu) residue followed by an aromatic (e.g., Tyr or Phe) residue within "loop 2" in the C-terminal end of the protein (Figure 2A). Based on these features, we observed the presence of all of these key catalytic residues only in macrodomain1 of PARP9 and PARP14. As such, only Macrodomain1 would be expected to have catalytic activity, whereas the other macrodomains have sequence characteristics that would be predicted to inactivate ADP-ribosylhydrolase activity (Figure 2A).

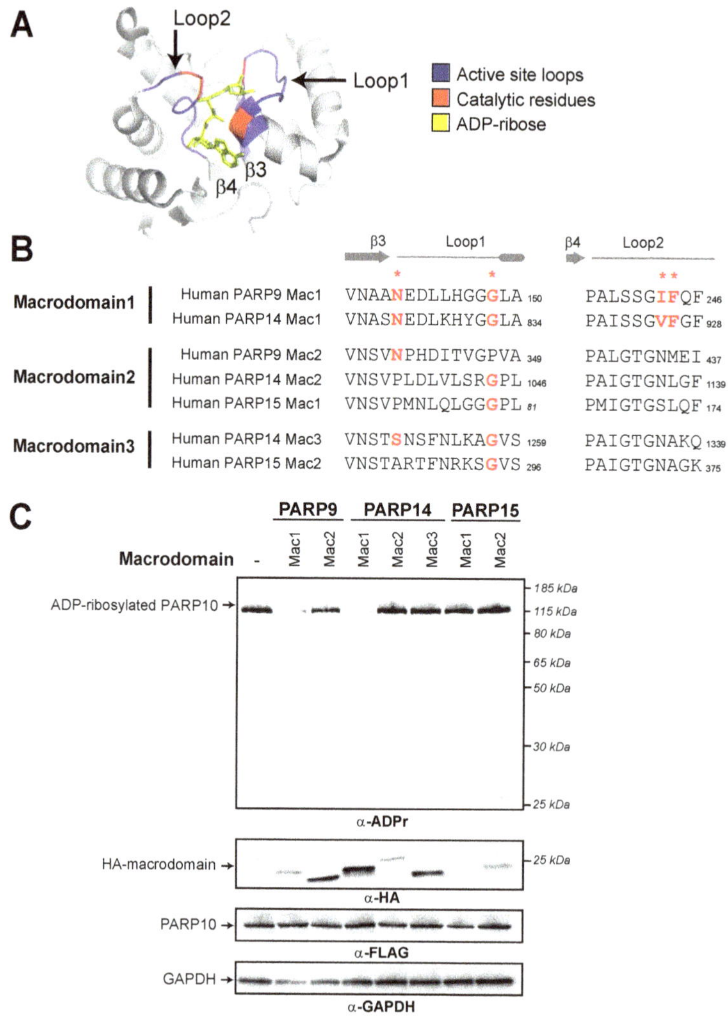

Figure 2. Presence of catalytic residues and enzymatic activity within individual macroPARP macrodomains. (**A**) Critical structural features and catalytic residues are mapped on the structure of PARP14 macrodomain1 solved in complex with ADP-ribose (PDB code: 3Q6Z [37]). Loop1 and Loop2, colored blue, are named in accordance with [46]. Important residues for ADP-ribosylhydrolase activity have been identified in several publications (see for example [18,45–48]) and are colored red. (**B**) Positions of catalytic residues (red asterisks) in Loop1 and Loop2 in human macroPARP macrodomains. Amino acids that are predicted to be compatible with enzymatic activity are shown in bold red. Residue number of the C-terminal residue in each motif is shown. (**C**) Enzymatic assay for ADP-ribosylhydrolase activity by transient overexpression of the indicated human macroPARP macrodomain with human PARP10 in human (HEK293T) cells. In the absence of any macrodomain, PARP10 (100 ng plasmid) is auto-ADP-ribosylated, resulting in a single band as detected by an anti-ADP-ribose antibody. A decrease in band intensity indicates that the indicated macrodomain (250 ng plasmid) is enzymatically active as an ADP-ribosylhydrolase. Anti-FLAG (PARP10) and anti-HA (macrodomain) blots are shown, as is an anti-GAPDH loading control. Expected positions of indicated proteins are shown, as are positions of molecular weight markers. Detailed information about the experimental protocol is found in the Section 2.

To test these functional hypotheses, we expressed individual human macroPARP macrodomains with PARP10 and monitored auto-ADP-ribosylation of PARP10. We used ADP-riboslyation levels of PARP10 as a readout for ADP-ribosylhydrolase activity, since this is a commonly used substrate in the field [20,23,24,45–47]. In the absence of any macrodomain, PARP10 is robustly ADP-ribosylated as measured using an antibody that detects ADP-ribosylated proteins (Cell Signaling Technology anti-poly/mono-ADP-ribose antibody, E6F6A) [42]. As shown in Figure 2B, and confirming our bioinformatic predictions here and elsewhere [8], we only observed macrodomain ADP-ribosylhydrolase activity with Macrodomain1 of human PARP9 and human PARP14. Our results showing that Macrodomain1 of human PARP14 is an active ADP-ribosylhydrolase contrasts with a previous report that mouse PARP14 Macrodomain1 is enzymatically inactive [47]. The source of this discrepancy is unclear, but it should be noted that there are substantial differences in the methods used; whereas we assayed for activity from human cells in which macrodomains and PARP10 were overexpressed, the previous study used purified recombinant macrodomains and tested them against purified ADP-ribosylated PARP10 [47]. Beyond PARP9 and PARP14 Macrodomain1s, and again consistent with our bioinformatic predictions, other human macrodomains showed no obvious ADP-ribosylhydrolase activity, although the first macrodomain of PARP15 expresses poorly, so it is difficult to confirm a lack of enzymatic activity. Together, our bioinformatic and functional results indicate that two human macroPARP macrodomains are catalytically active, whereas the other five macrodomains found in human macroPARPs lack ADP-ribosylhydrolase activity.

3.2. Recurrent Loss of Macrodomain Enzymatic Activity in Mammalian PARP14s

We next sought to determine whether the existence of enzymatic activity within a given macroPARP macrodomain is conserved across species. We were particularly interested in this question as we had previously observed that all three macroPARPs are evolving under very strong positive selection in primates, with a large number of amino acid changes occurring in the macrodomains of each macroPARP [8]. We therefore considered the possibility that whereas human macroPARPs have catalytic activity in the Macrodomain1 of PARP14 and PARP9, other species may have a different constellation of macrodomains with enzymatic activity.

To first ask this question, we returned to our macrodomain alignments shown in Figure 1C and looked for conserved sequence features that might suggest gain or loss of enzymatic activity. Based on the sequence logos shown in Figure 3A–C, we predicted that only Macrodomain1, which is present in PARP9 and PARP14 but not PARP15, would be an enzymatically active ADP-ribosylhydrolase. Specifically, we found that sequence features that are required for catalytic activity are broadly conserved in Macrodomain1 from diverse species including cnidarians, spiralians, and vertebrates (Figure 3A). This includes our observation that all key catalytic residues are present in PARP14 Macrodomain1 from the hood coral, *Stylophora pistillata*, which is a cnidarian species and is therefore representative of a PARP14 macrodomain that diverged from mammalian PARP14 > 700 million years ago [43]. In contrast, using the same groups of species, we observed poor conservation of many of the key catalytic residues in Macrodomain2 (Figure 3B) and Macrodomain3 (Figure 3C). These results suggest that across metazoan macroPARPs, the ancestral state of PARP14 contained a catalytically active Macrodomain1, whereas Macrodomain2 and Macrodomain3 lacked catalytic activity.

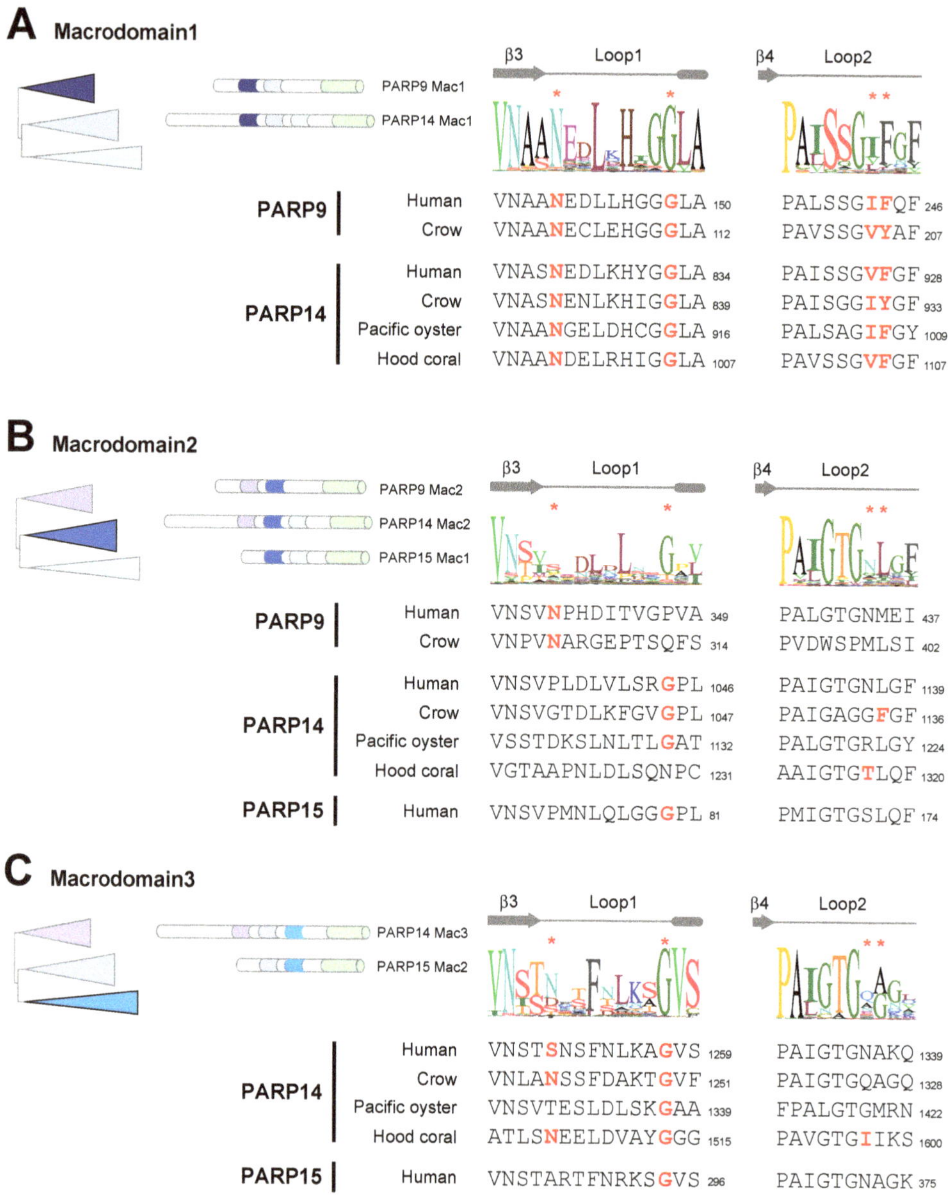

Figure 3. Catalytic residues are well conserved in macroPARP Macrodomain1 but not Macrodomain2 or Macrodomain3. (**A**) Cartoon of the phylogenetic position and protein position of Macrodomain1 as in Figure 1. A consensus logo of Loop1 and Loop2 across all analyzed Macrodomain1 sequences is shown, with critical residue positions indicated by red asterisks. Below are individual sequences from Macrodomain1s from vertebrates (human (*Homo sapiens*) and crow (*Corvus hawaiiensis*)), a spiralian (Pacific oyster, *Crassostrea gigas*) and a cnidarian (hood coral, *Stylophora pistillata*). Amino acids that are predicted to be compatible with enzymatic activity are shown in bold red. Residue number of the C-terminal residue in each motif is shown. (**B**) Same as A, except for Macrodomain2. (**C**) Same as A, except for Macrodomain3.

Interestingly, we did note that several species of mammals had mutations in the Macrodomain1 of PARP14 that disrupt critical residues for ADP-ribosylhydrolase activity. For instance, key residues have been mutated independently in P14 Macrodomain1 from little brown bat (*Myotis lucifugus*), cow (*Bos taurus*), and polar bear (*Ursus maritimus*) (Figure 4A). These data suggest that while Macrodomain1 has broadly retained enzymatic activity, several individual mammalian lineages have independently lost catalytic activity.

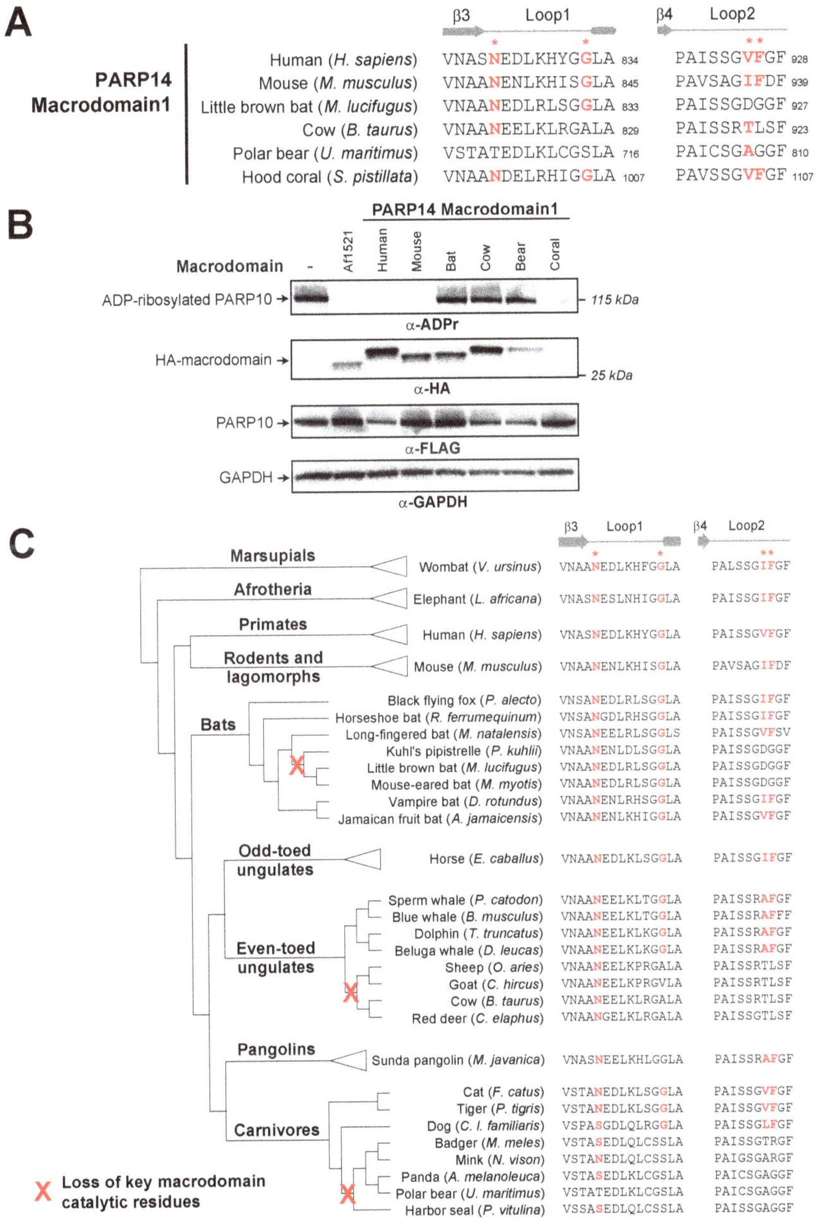

Figure 4. Recurrent loss of macrodomain enzymatic activity in mammalian PARP14. (**A**) Sequences of PARP14 Macrodomain1 from several metazoan species for Loop1 and Loop2. Critical residue

positions indicated by red asterisks. Amino acids that are predicted to be compatible with enzymatic activity are shown in bold red. Residue number of the C-terminal residue in each motif is shown. (**B**) Enzymatic assay for ADP-ribosylhydrolase activity by transient overexpression of the indicated PARP14 Macrodomain1 with human PARP10 in human (HEK293T) cells as in Figure 2C. Expected positions of indicated proteins are shown, as are positions of molecular weight markers. As a positive control for ADP-ribosylhydrolase activity, the well-characterized macrodomain from *Archaeoglobus fulgidus* (Af1521) was included. Detailed information about the experimental protocol is found in the Section 2. (**C**) An expanded view of the phylogenetic tree for mammalian PARP14 from Figure 1B, with major mammalian clades and example species shown. To the right are sequences for each indicated species in Loop1 and Loop2, with red bold letters indicating presence of residues that are predicted to confer catalytic activity. Red "X's" on the transformed phylogenetic tree indicate the inferred branch along which ADP-ribosylhydrolase activity was lost.

To test the hypothesis that P14 Macrodomain1 from individual mammalian species has lost catalytic activity, we compared the activity of PARP14 Macrodomain1 from species we predicted would have catalytic activity to those that we predicted had lost catalytic activity (Figure 4B). As a positive control for ADP-ribosylhydrolase activity, we used the well-established macrodomain from the archael species *Archaeoglobus fulgidus* (Af1521) [49]. Consistent with our evolutionary model, we observed no catalytic activity for the PARP14 Macrodomain1 from cow, polar bear, and little brown bat. In contrast, PARP14 Macrodomain1 from humans and mice are enzymatically active, with robust ADP-ribosylhydrolase activity against PARP10 (Figure 4B,C). In addition, we observed that the cnidarian *S. pistillata* PARP14 Macrodomain1 also has robust catalytic activity, indicative of the ancient presence of ADP-ribosylhydrolase activity in macroPARP proteins (Figure 4B).

To further characterize the evolutionary origins of the mutations to the key catalytic residues, we analyzed the PARP14 Macrodomain1 sequences from species closely related to those that had lost catalytic activity (Figure 4D). For instance, based on available bat PARP14 sequences, we infer that inactivating mutations that disrupt catalytic activity only arose in the vespertine microbats, including species in the *Myotis* and *Pipistrellus* genera. In contrast, other microbat species, including horseshoe bat (*Rhinolophus ferrumequinum*) and vampire bat (*Desmodus* rotundus), as well as megabats such as the black flying fox (*Pteroptus alecto*), retain all residues of the ancestral enzymatically active macrodomain. Likewise, within the even-toed ungulates (*Artiodactyla*), our data suggest that Macrodomain1 catalytic residues were disrupted in the ruminant lineage, including cow, sheep, goat, and deer, but are retained in cetaceans such as dolphins and whales. Finally, within carnivores, feline and canine species retain the indicated residues required for catalytic activity, whereas most other carnivores lack these critical residues. Together, our analyses shown in Figure 4B indicate at least three independent instances of loss of critical catalytic residues across the mammalian phylogeny. Based on estimates of divergence times of internal nodes in the mammalian phylogeny [50,51], all three of these independent losses of catalytic residues occurred between 25 and 65 million years ago during the mammalian diversification that followed the Cretaceous–Paleogene (KPg) mass extinction.

3.3. *Tandem Macrodomain Orientation Is Shared between MacroPARPs and Coronaviruses*

Having analyzed macrodomain activity in host macroPARPs, we next turned our attention to viral macrodomains. Coronaviruses have a conserved macrodomain that has been the target of substantial interest, as it is required for antagonizing the host immune response [12,17,20–22]. Interestingly, several coronaviruses, including SARS-CoV-2, encode tandem macrodomains, as is seen in host macroPARPs (Figure 5A,B). We therefore wished to determine whether the catalytic residues that are required for ADP-ribosylhydrolase activity are conserved in one or all coronavirus macrodomains. As with macroPARPs, we found that Macrodomain1 contains conserved residues that are predicted to be consistent with catalytic activity (Figure 5C,D), although in this case, we observed no cases in which the catalytic residues were mutated in any coronavirus. However, it was difficult to reliably align Macrodomain2 and Macrodomain3 to Macrodomain1 based on primary sequence alone. As a result, we performed structural predictions using AlphaFold [35] to identify residues in positions that are analogous to Loop1 and Loop2 in Macrodomain1 (Figure 5C,E). These structure-based homology models allowed us to predict the absence of catalytic residues in the SARS-CoV-2 Macrodomain2 and Macrodomain3 (Figure 5F).

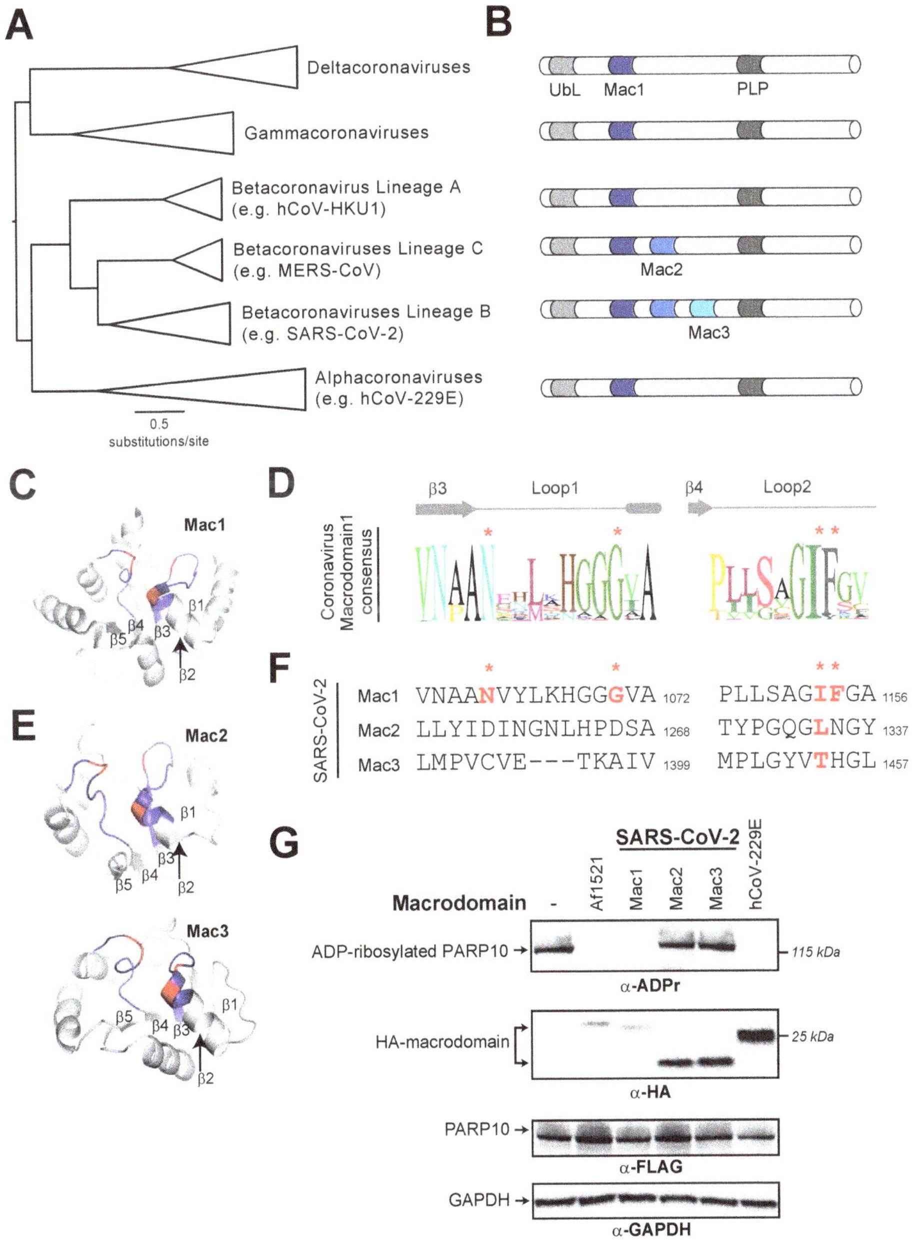

Figure 5. Distribution of active and inactive macrodomains in coronaviruses. (**A**) Phylogenetic tree of nsP3 proteins from diverse coronaviruses. Major coronavirus clades are shown. (**B**) Domain

cartoon of nsP3 proteins from the indicated coronavirus clades. Flanking nsP3 macrodomains are a ubiquitin-like (UbL) domain and a papain-like protease (PLP). The number of macrodomains found in each protein is shown. For clarity, other domains are not shown. (**C**) Critical structural features and catalytic residues are mapped on the structure of the SARS-CoV-2 nsP3 macrodomain1 (PDB code: 6WEY [36]). Loop1 and Loop2, colored blue, and important residues for ADP-ribosylhydrolase activity, colored red, as indicated as in Figure 2A. The first five beta-strands in the structure are also labeled. (**D**) Consensus logo of Loop1 and Loop2 across all analyzed coronavirus Macrodomain1 sequences. Key residue positions are marked by red asterisks. (**E**) Structural models for SARS-CoV-2 nsP3 Macrodomain2 and Macrodomain3 were predicted using AlphaFold2 via the ColabFold package [35]. Loops, positions of important residues, and beta-strands are marked as in part C. (**F**) Sequences of Loop1 and Loop2 from the indicated SARS-CoV-2 macrodomains. Although there is little sequence similarity to other viral or host macrodomains, the sequences of Loop1 and Loop2 in Macrodomain2 and Macrodomain3 were identified using the structural models shown in panel (**E**) (see Materials and Methods for additional explanation). Red bold letters indicate presence of residues that are predicted to confer catalytic activity. Residue number of the C-terminal residue in each motif relative to the start of the viral ORF1ab polyprotein is shown. (**G**) Enzymatic assay for ADP-ribosylhydrolase activity by transient overexpression of the indicated coronavirus macrodomain with human PARP10 in human (HEK293T) cells as in Figure 2C. Expected positions of indicated proteins are shown, as are positions of molecular weight markers. Detailed information about the experimental protocol is found in the Section 2.

We next wished to test the hypothesis that, like macroPARPs, only the first macrodomain of the SARS-CoV-2 tandem macrodomains is enzymatically active. We therefore cloned and expressed individual macrodomains from SARS-CoV-2 as an example of a three-macrodomain viral protein and from hCoV-229E as an example of a one-macrodomain viral protein. As with PARP14 and PARP9, we observed that Macrodomain1 of each virus had measurable ADP-ribosylhydrolase activity, which is consistent with several previous studies [12,18,20,21] (Figure 5G). In contrast, we observed no activity from Macrodomain2 or Macrodomain3 from SARS-CoV-2, consistent with the absence of residues required for catalytic activity and with prior observations that Macrodomain2 and Macrodomain3 of SARS-CoV specifically bind nucleic acids [52].

3.4. Recurrent Loss of Macrodomain Activity in Alphaviruses

We finally wished to analyze the macrodomain activity within the alphavirus genus of *Togaviridae*. Alphaviruses encode a single macrodomain within the nsP3 protein that has important roles in tissue tropism, viral replication and virulence [23–26]. Indeed, previous macrodomain mutations have been shown to prevent replication in both mammalian and mosquito cells [24]. Based on this functional importance, as well as the observed conservation of macrodomains across diverse coronaviruses (Figure 5D and [18]), we therefore expected strong conservation of macrodomain sequences and catalytic activity within the alphaviruses.

We first generated a phylogenetic tree of nonstructural polyproteins from diverse alphaviruses (Figure 6A). We then extracted macrodomain sequences from these viruses. Similar to the macroPARP Macrodomain1 alignment (Figure 3A), we observed a consensus sequence that contained residues shown to be important for catalysis, but also observed that these residues were not perfectly conserved (Figure 6B). We therefore investigated whether any alphaviruses might lack residues important for ADP-ribosylhydrolase activity. Consistent with previous observations [18], we noted that the insect-specific alphaviruses, including Eilat virus (EILV) [53] and Tai Forest virus [54], lack residues expected to confer enzymatic activity (Figure 6C). More surprisingly, we also found two separate additional cases of alphavirus macrodomains that lacked one or more of the catalytic residues. The first occurs in Middelburg virus (MIDV), a virus which has been isolated from cerebrospinal fluid and blood from humans [55] and is associated with severe disease in horses [56]. Importantly, phylogenetic analyses indicate that MIDV is more closely related to several human alphaviruses than it is to the insect-specific alphaviruses (Figure 6A), suggesting this was an independent loss of macrodomain catalytic activity. Finally, we observed another loss of catalytic residues in a recently discovered virus known as Caaingua virus (CAAV), which was isolated from mosquitoes but could be cultured in human mononuclear cells [57]. Consistent with prior phylogenetic analyses [57], we found Caaingua groups with other vertebrate-infecting viruses, rather than with the insect-specific viruses (Figure 6A).

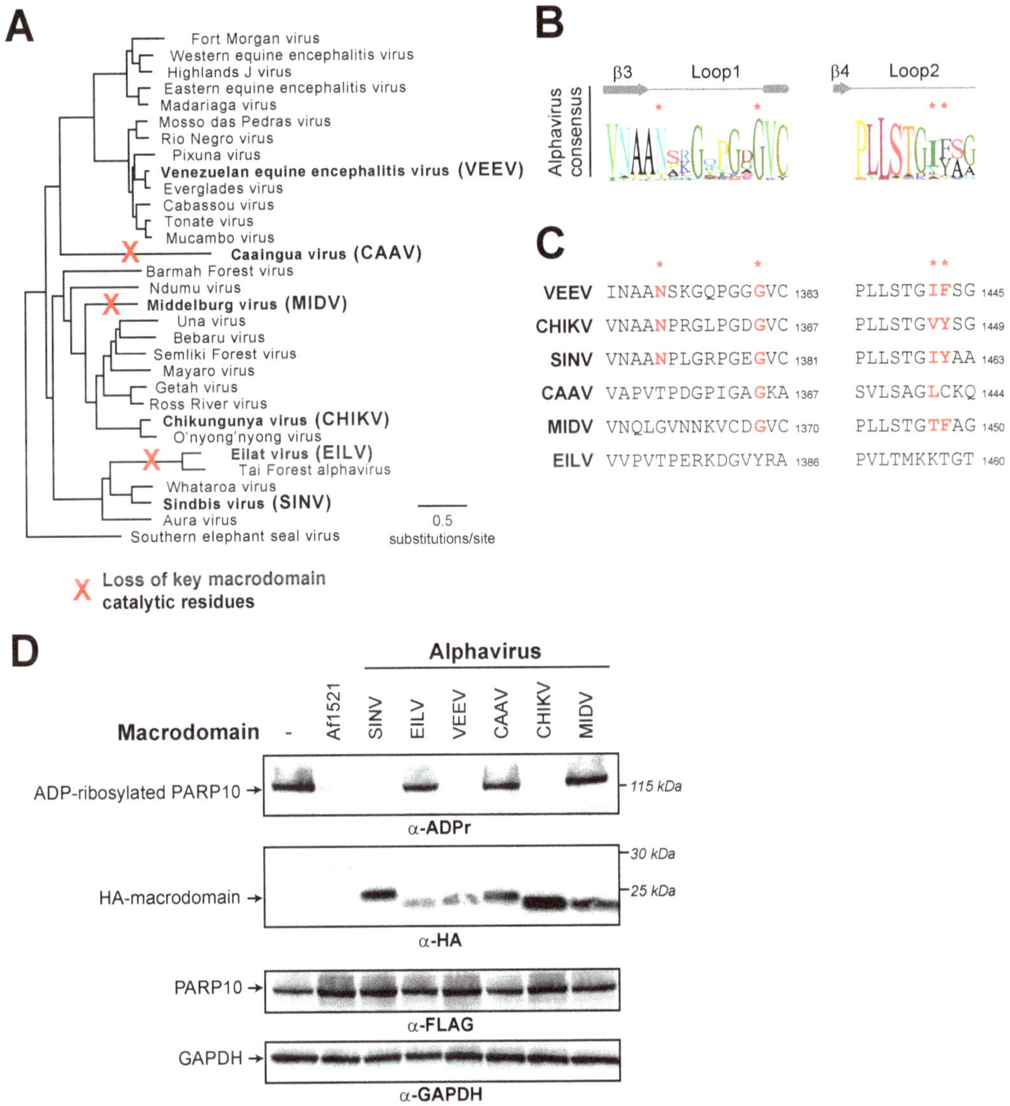

Figure 6. Recurrent loss of macrodomain enzymatic activity in alphaviruses. (**A**) Phylogenetic tree of nonstructural polyproteins from diverse alphaviruses. Species in bold are those that are shown in panels (**C**,**D**). Red "X's" indicate the inferred branch along which ADP-ribosylhydrolase activity was lost based on data shown in panel (**C**,**D**). (**B**) Consensus logo of Loop1 and Loop2 across all analyzed alphavirus macrodomain sequences. Key residue positions are marked by red asterisks.

(**C**) Sequences of Loop1 and Loop2 from the indicated alphavirus macrodomains. Red bold letters indicate presence of residues that are predicted to confer catalytic activity. Residue number of the C-terminal residue in each motif relative to the start of the viral nsP1-4 polyprotein is shown. (**D**) Enzymatic assay for ADP-ribosylhydrolase activity by transient overexpression of the indicated alphavirus macrodomain with human PARP10 in human (HEK293T) cells as in Figure 2C. Expected positions of indicated proteins are shown, as are positions of molecular weight markers. Detailed information about the experimental protocol is found in the Section 2.

To test the hypothesis that macrodomains from EILV, MIDV, and CAAV are enzymatically inactive, we cloned and expressed each macrodomain in the presence of human PARP10. In addition to the panel of alphavirus macrodomains we predicted would be inactive, we also tested several diverse alphavirus macrodomains that we predicted would be active, including those from Sindbis virus (SINV), chikungunya virus (CHIKV), and Venezuelan equine encephalitis virus (VEEV). Consistent with our predictions from sequence data, we observed no ADP-ribosylhydrolase activity against PARP10 with EILV, MIDV, and CAAV, whereas we observed robust activity with the other alphavirus macrodomains (Figure 6D). These data indicate that ADP-ribosylhydrolase activity is not absolutely essential for a macrodomain in a vertebrate-infecting alphavirus. In addition, when placed in the context of our phylogenetic analyses (Figure 6C), these data suggest that, like PARP14, macrodomain activity has been independently lost several times in the alphaviruses.

4. Discussion

Addition, recognition, and removal of ADP-ribosylation has emerged as an important battleground between the host antiviral immune response and viruses. Although the molecular targets of ADP-ribosylation, and the mechanistic consequences of that ADP-ribosylation, are mostly uncharacterized, there is clear function of host PARPs in the antiviral immune response and a clear role of viral macrodomains in antagonizing the host immune response [17,58–61]. Our evolutionary and functional data suggest that an important consideration is the degree to which the ADP-ribosylhydrolase activity encoded by some macrodomains is conserved amongst host antiviral macroPARPs and is conserved amongst viral macrodomains.

In addition to ADP-ribosylhydrolase activity, macrodomains have been observed to have several other functions. In particular, some macrodomains that lack the ability to remove ('erase") ADP-ribosylation still retain the ability to recognize ('read") ADP-ribosylated proteins [5–7]. In most of the cases we describe, individual substitutions in catalytic residues would likely still retain this "reader" function, potentially allowing these proteins to still function in some aspects of ADP-ribose biology. In addition, macrodomains have been shown to function in processes not directly related to protein ADP-ribosylation, including binding nucleic acids and catalyzing degradation of a product of tRNA splicing [5–7], and it is unknown whether the enzymatically inactive macrodomains may participate in these functions. By sampling macrodomain diversity found throughout metazoan macroPARPs and viruses, additional insights into the many functions of macrodomains may emerge.

Although the recurrent loss of enzymatic activity in an important host–virus battleground is seemingly paradoxical, it is reminiscent of the observation that two interferon-stimulated antiviral PARPs lack the ADP-ribosyltransferase activity that all of the other mammalian PARPs display. PARP13, also known as zinc-finger antiviral protein (ZAP), lacks ADP-ribosyltransferase activity but is a potent antiviral factor against a wide range of viruses, including retroviruses and alphaviruses [62]. In addition, PARP9 is a macroPARP that can potentiate the IFN response [16], but lacks the canonical activity of other PARPs. Previously characterized as a catalytically inactive ADP-ribosyltransferase like PARP13 due to a lack of conserved catalytic residues [3], PARP9 has been implicated in specific ADP-ribosylation of ubiquitin when in complex with its E3-ubiquitin ligase binding partner, DTX3L [63]. Regardless of potential residual enzymatic activity in PARP9, it is striking that the only two PARPs that lack canonical ADP-ribosyltransferase activity are also upregulated by the antiviral cytokine IFN, have antiviral function, and are evolving under strong positive selection indicative of a host–virus genetic conflict [8]. Whether the loss of enzymatic activity in those antiviral PARPs, or whether the loss of macrodomain enzymatic activity in the antiviral host protein PARP14, is adaptive or confers some additional function to these proteins remains to be determined.

In addition, the loss of macrodomain activity in several alphavirus lineages is surprising. Whereas the loss of macrodomain activity in insect-specific alphaviruses may be rationalized by the observation that insects lack most PARP proteins found in vertebrates, the loss of catalytic activity from Caaingua and especially Middelburg viruses is more difficult to reconcile. It will be interesting to determine how these viruses can infect human cells while lacking ADP-ribosylhydrolase activity that has been shown to be critical for other alphaviruses.

In sum, our results highlight an unexpected but recurrent loss of enzymatic activity in host and viral macrodomains in a way that will fundamentally affect their interactions with protein ADP-ribosylation. Such observations go against the assumption that macrodomain activity will be broadly conserved, especially in viruses. These results indicate that there remain many aspects of ADP-ribosylation, especially at the interface between the host antiviral immune response and viruses, that remain to be fully understood.

Supplementary Materials: The following supporting information can be downloaded at: https://www.mdpi.com/article/10.3390/pathogens12050674/s1, File S1: Accession numbers and species names of metazoan macroPARPs used for phylogenetic analyses; File S2: Accession numbers and species names of coronavirus macrodomain-containing proteins used for phylogenetic analyses; File S3: Accession numbers and species names of alphavirus nsP3 macrodomain-containing proteins used for phylogenetic analyses; File S4: Macrodomain sequences used for functional analyses; Figure S1: Phylogenomic distribution of macroPARPs in metazoans; Figure S2: Dose response of human and bat (M. lucifugus) PARP14.

Author Contributions: Conceptualization, S.E.D.-R., A.P.R. and M.D.D.; Data curation, M.D.D.; Formal analysis, M.D.D.; Funding acquisition, M.D.D.; Investigation, S.E.D.-R., A.P.R. and M.D.D.; Methodology, A.P.R. and M.D.D.; Project administration, M.D.D.; Resources, M.D.D.; Supervision, M.D.D.; Validation, S.E.D.-R., A.P.R. and M.D.D.; Visualization, S.E.D.-R. and M.D.D.; Writing–original draft, M.D.D.; Writing–review and editing, S.E.D.-R., A.P.R. and M.D.D. All authors have read and agreed to the published version of the manuscript.

Funding: This work was supported by the National Institutes of Health (R35 GM133633), the Pew Biomedical Scholars program and the Burroughs Wellcome Fund Investigators in the Pathogenesis of Infectious Disease program to M.D.D., and NIH T32 GM007240 to A.P.R.

Institutional Review Board Statement: Not applicable.

Informed Consent Statement: Not applicable.

Data Availability Statement: All data are contained within the article or supplementary material. All sequences used for analysis are available in the NCBI protein database (https://www.ncbi.nlm.nih.gov/protein/, accessed on 1 January 2023) using the accession numbers shown in the supplementary material.

Acknowledgments: We thank members of the Daugherty laboratory for their suggestions and manuscript comments.

Conflicts of Interest: The authors declare no conflict of interest.

References

1. Hoch, N.C.; Polo, L.M. ADP-ribosylation: From molecular mechanisms to human disease. *Genet. Mol. Biol.* **2019**, *43*, e20190075. [CrossRef] [PubMed]
2. Cohen, M.S.; Chang, P. Insights into the biogenesis, function, and regulation of ADP-ribosylation. *Nat. Chem. Biol.* **2018**, *14*, 236–243. [CrossRef] [PubMed]
3. Luscher, B.; Ahel, I.; Altmeyer, M.; Ashworth, A.; Bai, P.; Chang, P.; Cohen, M.; Corda, D.; Dantzer, F.; Daugherty, M.D.; et al. ADP-ribosyltransferases, an update on function and nomenclature. *FEBS J.* **2022**, *289*, 7399–7410. [CrossRef] [PubMed]
4. Luscher, B.; Butepage, M.; Eckei, L.; Krieg, S.; Verheugd, P.; Shilton, B.H. ADP-Ribosylation, a Multifaceted Posttranslational Modification Involved in the Control of Cell Physiology in Health and Disease. *Chem. Rev.* **2018**, *118*, 1092–1136. [CrossRef]
5. Rack, J.G.M.; Palazzo, L.; Ahel, I. (ADP-ribosyl)hydrolases: Structure, function, and biology. *Genes Dev.* **2020**, *34*, 263–284. [CrossRef]
6. Rack, J.G.; Perina, D.; Ahel, I. Macrodomains: Structure, Function, Evolution, and Catalytic Activities. *Annu. Rev. Biochem.* **2016**, *85*, 431–454. [CrossRef]
7. Feijs, K.L.; Forst, A.H.; Verheugd, P.; Luscher, B. Macrodomain-containing proteins: Regulating new intracellular functions of mono(ADP-ribosyl)ation. *Nat. Rev. Mol. Cell Biol.* **2013**, *14*, 443–451. [CrossRef]
8. Daugherty, M.D.; Young, J.M.; Kerns, J.A.; Malik, H.S. Rapid evolution of PARP genes suggests a broad role for ADP-ribosylation in host-virus conflicts. *PLoS Genet.* **2014**, *10*, e1004403. [CrossRef]
9. Daugherty, M.D.; Malik, H.S. Rules of engagement: Molecular insights from host-virus arms races. *Annu. Rev. Genet.* **2012**, *46*, 677–700. [CrossRef]
10. Duggal, N.K.; Emerman, M. Evolutionary conflicts between viruses and restriction factors shape immunity. *Nat. Rev. Immunol.* **2012**, *12*, 687–695. [CrossRef]
11. Sironi, M.; Cagliani, R.; Forni, D.; Clerici, M. Evolutionary insights into host-pathogen interactions from mammalian sequence data. *Nat. Rev. Genet.* **2015**, *16*, 224–236. [CrossRef] [PubMed]
12. Grunewald, M.E.; Chen, Y.; Kuny, C.; Maejima, T.; Lease, R.; Ferraris, D.; Aikawa, M.; Sullivan, C.S.; Perlman, S.; Fehr, A.R. The coronavirus macrodomain is required to prevent PARP-mediated inhibition of virus replication and enhancement of IFN expression. *PLoS Pathog.* **2019**, *15*, e1007756. [CrossRef] [PubMed]
13. Caprara, G.; Prosperini, E.; Piccolo, V.; Sigismondo, G.; Melacarne, A.; Cuomo, A.; Boothby, M.; Rescigno, M.; Bonaldi, T.; Natoli, G. PARP14 Controls the Nuclear Accumulation of a Subset of Type I IFN-Inducible Proteins. *J. Immunol.* **2018**, *200*, 2439–2454. [CrossRef] [PubMed]
14. Iwata, H.; Goettsch, C.; Sharma, A.; Ricchiuto, P.; Goh, W.W.; Halu, A.; Yamada, I.; Yoshida, H.; Hara, T.; Wei, M.; et al. PARP9 and PARP14 cross-regulate macrophage activation via STAT1 ADP-ribosylation. *Nat. Commun.* **2016**, *7*, 12849. [CrossRef] [PubMed]
15. Xing, J.; Zhang, A.; Du, Y.; Fang, M.; Minze, L.J.; Liu, Y.J.; Li, X.C.; Zhang, Z. Identification of poly(ADP-ribose) polymerase 9 (PARP9) as a noncanonical sensor for RNA virus in dendritic cells. *Nat. Commun.* **2021**, *12*, 2681. [CrossRef] [PubMed]
16. Zhang, Y.; Mao, D.; Roswit, W.T.; Jin, X.; Patel, A.C.; Patel, D.A.; Agapov, E.; Wang, Z.; Tidwell, R.M.; Atkinson, J.J.; et al. PARP9-DTX3L ubiquitin ligase targets host histone H2BJ and viral 3C protease to enhance interferon signaling and control viral infection. *Nat. Immunol.* **2015**, *16*, 1215–1227. [CrossRef] [PubMed]
17. Leung, A.K.L.; Griffin, D.E.; Bosch, J.; Fehr, A.R. The Conserved Macrodomain Is a Potential Therapeutic Target for Coronaviruses and Alphaviruses. *Pathogens* **2022**, *11*, 94. [CrossRef] [PubMed]
18. Rack, J.G.M.; Zorzini, V.; Zhu, Z.; Schuller, M.; Ahel, D.; Ahel, I. Viral macrodomains: A structural and evolutionary assessment of the pharmacological potential. *Open Biol.* **2020**, *10*, 200237. [CrossRef]
19. Alhammad, Y.M.O.; Fehr, A.R. The Viral Macrodomain Counters Host Antiviral ADP-Ribosylation. *Viruses* **2020**, *12*, 384. [CrossRef]
20. Alhammad, Y.M.O.; Kashipathy, M.M.; Roy, A.; Gagne, J.P.; McDonald, P.; Gao, P.; Nonfoux, L.; Battaile, K.P.; Johnson, D.K.; Holmstrom, E.D.; et al. The SARS-CoV-2 Conserved Macrodomain Is a Mono-ADP-Ribosylhydrolase. *J. Virol.* **2021**, *95*, e01969-20. [CrossRef]
21. Fehr, A.R.; Channappanavar, R.; Jankevicius, G.; Fett, C.; Zhao, J.; Athmer, J.; Meyerholz, D.K.; Ahel, I.; Perlman, S. The Conserved Coronavirus Macrodomain Promotes Virulence and Suppresses the Innate Immune Response during Severe Acute Respiratory Syndrome Coronavirus Infection. *mBio* **2016**, *7*, e01721-16. [CrossRef]
22. Kuri, T.; Eriksson, K.K.; Putics, A.; Zust, R.; Snijder, E.J.; Davidson, A.D.; Siddell, S.G.; Thiel, V.; Ziebuhr, J.; Weber, F. The ADP-ribose-1″-monophosphatase domains of severe acute respiratory syndrome coronavirus and human coronavirus 229E mediate resistance to antiviral interferon responses. *J. Gen. Virol.* **2011**, *92*, 1899–1905. [CrossRef]
23. Abraham, R.; Hauer, D.; McPherson, R.L.; Utt, A.; Kirby, I.T.; Cohen, M.S.; Merits, A.; Leung, A.K.L.; Griffin, D.E. ADP-ribosyl-binding and hydrolase activities of the alphavirus nsP3 macrodomain are critical for initiation of virus replication. *Proc. Natl. Acad. Sci. USA* **2018**, *115*, E10457–E10466. [CrossRef] [PubMed]
24. McPherson, R.L.; Abraham, R.; Sreekumar, E.; Ong, S.E.; Cheng, S.J.; Baxter, V.K.; Kistemaker, H.A.; Filippov, D.V.; Griffin, D.E.; Leung, A.K. ADP-ribosylhydrolase activity of Chikungunya virus macrodomain is critical for virus replication and virulence. *Proc. Natl. Acad. Sci. USA* **2017**, *114*, 1666–1671. [CrossRef]

25. Aguilar, E.G.; Paniccia, G.; Adura, C.; Singer, Z.S.; Ashbrook, A.W.; Razooky, B.S.; Rice, C.M.; MacDonald, M.R. Sindbis Macrodomain Poly-ADP-Ribose Hydrolase Activity Is Important for Viral RNA Synthesis. *J. Virol.* **2022**, *96*, e0151621. [CrossRef] [PubMed]
26. Park, E.; Griffin, D.E. The nsP3 macro domain is important for Sindbis virus replication in neurons and neurovirulence in mice. *Virology* **2009**, *388*, 305–314. [CrossRef]
27. Altschul, S.F.; Madden, T.L.; Schäffer, A.A.; Zhang, J.; Zhang, Z.; Miller, W.; Lipman, D.J. Gapped BLAST and PSI-BLAST: A new generation of protein database search programs. *Nucleic Acids Res.* **1997**, *25*, 3389–3402. [CrossRef]
28. Sievers, F.; Wilm, A.; Dineen, D.; Gibson, T.J.; Karplus, K.; Li, W.; Lopez, R.; McWilliam, H.; Remmert, M.; Söding, J.; et al. Fast, scalable generation of high-quality protein multiple sequence alignments using Clustal Omega. *Mol. Syst. Biol.* **2011**, *7*, 539. [CrossRef]
29. Li, W.; Godzik, A. Cd-hit: A fast program for clustering and comparing large sets of protein or nucleotide sequences. *Bioinformatics* **2006**, *22*, 1658–1659. [CrossRef] [PubMed]
30. Potter, S.C.; Luciani, A.; Eddy, S.R.; Park, Y.; Lopez, R.; Finn, R.D. HMMER web server: 2018 update. *Nucleic Acids Res.* **2018**, *46*, W200–W204. [CrossRef] [PubMed]
31. Otto, H.; Reche, P.A.; Bazan, F.; Dittmar, K.; Haag, F.; Koch-Nolte, F. In silico characterization of the family of PARP-like poly(ADP-ribosyl)transferases (pARTs). *BMC Genom.* **2005**, *6*, 139. [CrossRef]
32. Nguyen, L.-T.; Schmidt, H.A.; von Haeseler, A.; Minh, B.Q. IQ-TREE: A fast and effective stochastic algorithm for estimating maximum-likelihood phylogenies. *Mol. Biol. Evol.* **2015**, *32*, 268–274. [CrossRef]
33. Hoang, D.T.; Chernomor, O.; von Haeseler, A.; Minh, B.Q.; Vinh, L.S. UFBoot2: Improving the Ultrafast Bootstrap Approximation. *Mol. Biol. Evol.* **2018**, *35*, 518–522. [CrossRef] [PubMed]
34. Kalyaanamoorthy, S.; Minh, B.Q.; Wong, T.K.F.; von Haeseler, A.; Jermiin, L.S. ModelFinder: Fast model selection for accurate phylogenetic estimates. *Nat. Methods* **2017**, *14*, 587–589. [CrossRef] [PubMed]
35. Mirdita, M.; Schutze, K.; Moriwaki, Y.; Heo, L.; Ovchinnikov, S.; Steinegger, M. ColabFold: Making protein folding accessible to all. *Nat. Methods* **2022**, *19*, 679–682. [CrossRef] [PubMed]
36. Frick, D.N.; Virdi, R.S.; Vuksanovic, N.; Dahal, N.; Silvaggi, N.R. Molecular Basis for ADP-Ribose Binding to the Mac1 Domain of SARS-CoV-2 nsp3. *Biochemistry* **2020**, *59*, 2608–2615. [CrossRef]
37. Forst, A.H.; Karlberg, T.; Herzog, N.; Thorsell, A.G.; Gross, A.; Feijs, K.L.; Verheugd, P.; Kursula, P.; Nijmeijer, B.; Kremmer, E.; et al. Recognition of mono-ADP-ribosylated ARTD10 substrates by ARTD8 macrodomains. *Structure* **2013**, *21*, 462–475. [CrossRef]
38. Weixler, L.; Ikenga, N.J.; Voorneveld, J.; Aydin, G.; Bolte, T.M.; Momoh, J.; Butepage, M.; Golzmann, A.; Luscher, B.; Filippov, D.V.; et al. Protein and RNA ADP-ribosylation detection is influenced by sample preparation and reagents used. *Life Sci. Alliance* **2023**, *6*, e202201455. [CrossRef]
39. Donawho, C.K.; Luo, Y.; Luo, Y.; Penning, T.D.; Bauch, J.L.; Bouska, J.J.; Bontcheva-Diaz, V.D.; Cox, B.F.; DeWeese, T.L.; Dillehay, L.E.; et al. ABT-888, an orally active poly(ADP-ribose) polymerase inhibitor that potentiates DNA-damaging agents in preclinical tumor models. *Clin. Cancer Res.* **2007**, *13*, 2728–2737. [CrossRef]
40. Thorsell, A.G.; Ekblad, T.; Karlberg, T.; Low, M.; Pinto, A.F.; Tresaugues, L.; Moche, M.; Cohen, M.S.; Schuler, H. Structural Basis for Potency and Promiscuity in Poly(ADP-ribose) Polymerase (PARP) and Tankyrase Inhibitors. *J. Med. Chem.* **2017**, *60*, 1262–1271. [CrossRef]
41. Rodriguez, K.M.; Buch-Larsen, S.C.; Kirby, I.T.; Siordia, I.R.; Hutin, D.; Rasmussen, M.; Grant, D.M.; David, L.L.; Matthews, J.; Nielsen, M.L.; et al. Chemical genetics and proteome-wide site mapping reveal cysteine MARylation by PARP-7 on immune-relevant protein targets. *Elife* **2021**, *10*, e60480. [CrossRef] [PubMed]
42. Lu, A.Z.; Abo, R.; Ren, Y.; Gui, B.; Mo, J.R.; Blackwell, D.; Wigle, T.; Keilhack, H.; Niepel, M. Enabling drug discovery for the PARP protein family through the detection of mono-ADP-ribosylation. *Biochem. Pharm.* **2019**, *167*, 97–106. [CrossRef]
43. Dohrmann, M.; Worheide, G. Dating early animal evolution using phylogenomic data. *Sci. Rep.* **2017**, *7*, 3599. [CrossRef]
44. Kumar, S.; Hedges, S.B. A molecular timescale for vertebrate evolution. *Nature* **1998**, *392*, 917–920. [CrossRef] [PubMed]
45. Li, C.; Debing, Y.; Jankevicius, G.; Neyts, J.; Ahel, I.; Coutard, B.; Canard, B. Viral Macro Domains Reverse Protein ADP-Ribosylation. *J. Virol.* **2016**, *90*, 8478–8486. [CrossRef]
46. Jankevicius, G.; Hassler, M.; Golia, B.; Rybin, V.; Zacharias, M.; Timinszky, G.; Ladurner, A.G. A family of macrodomain proteins reverses cellular mono-ADP-ribosylation. *Nat. Struct. Mol. Biol.* **2013**, *20*, 508–514. [CrossRef]
47. Rosenthal, F.; Feijs, K.L.; Frugier, E.; Bonalli, M.; Forst, A.H.; Imhof, R.; Winkler, H.C.; Fischer, D.; Caflisch, A.; Hassa, P.O.; et al. Macrodomain-containing proteins are new mono-ADP-ribosylhydrolases. *Nat. Struct. Mol. Biol.* **2013**, *20*, 502–507. [CrossRef]
48. Eckei, L.; Krieg, S.; Butepage, M.; Lehmann, A.; Gross, A.; Lippok, B.; Grimm, A.R.; Kummerer, B.M.; Rossetti, G.; Luscher, B.; et al. The conserved macrodomains of the non-structural proteins of Chikungunya virus and other pathogenic positive strand RNA viruses function as mono-ADP-ribosylhydrolases. *Sci. Rep.* **2017**, *7*, 41746. [CrossRef]
49. Karras, G.I.; Kustatscher, G.; Buhecha, H.R.; Allen, M.D.; Pugieux, C.; Sait, F.; Bycroft, M.; Ladurner, A.G. The macro domain is an ADP-ribose binding module. *EMBO J.* **2005**, *24*, 1911–1920. [CrossRef]
50. Meredith, R.W.; Janecka, J.E.; Gatesy, J.; Ryder, O.A.; Fisher, C.A.; Teeling, E.C.; Goodbla, A.; Eizirik, E.; Simao, T.L.; Stadler, T.; et al. Impacts of the Cretaceous Terrestrial Revolution and KPg extinction on mammal diversification. *Science* **2011**, *334*, 521–524. [CrossRef]

51. Upham, N.S.; Esselstyn, J.A.; Jetz, W. Inferring the mammal tree: Species-level sets of phylogenies for questions in ecology, evolution, and conservation. *PLoS Biol.* **2019**, *17*, e3000494. [CrossRef] [PubMed]
52. Tan, J.; Vonrhein, C.; Smart, O.S.; Bricogne, G.; Bollati, M.; Kusov, Y.; Hansen, G.; Mesters, J.R.; Schmidt, C.L.; Hilgenfeld, R. The SARS-unique domain (SUD) of SARS coronavirus contains two macrodomains that bind G-quadruplexes. *PLoS Pathog.* **2009**, *5*, e1000428. [CrossRef] [PubMed]
53. Nasar, F.; Palacios, G.; Gorchakov, R.V.; Guzman, H.; Da Rosa, A.P.; Savji, N.; Popov, V.L.; Sherman, M.B.; Lipkin, W.I.; Tesh, R.B.; et al. Eilat virus, a unique alphavirus with host range restricted to insects by RNA replication. *Proc. Natl. Acad. Sci. USA* **2012**, *109*, 14622–14627. [CrossRef] [PubMed]
54. Hermanns, K.; Zirkel, F.; Kopp, A.; Marklewitz, M.; Rwego, I.B.; Estrada, A.; Gillespie, T.R.; Drosten, C.; Junglen, S. Discovery of a novel alphavirus related to Eilat virus. *J. Gen. Virol.* **2017**, *98*, 43–49. [CrossRef]
55. Fourie, I.; Williams, J.; Ismail, A.; Jansen van Vuren, P.; Stoltz, A.; Venter, M. Detection and genome characterization of Middelburg virus strains isolated from CSF and whole blood samples of humans with neurological manifestations in South Africa. *PLoS Negl. Trop. Dis.* **2022**, *16*, e0010020. [CrossRef]
56. Attoui, H.; Sailleau, C.; Mohd Jaafar, F.; Belhouchet, M.; Biagini, P.; Cantaloube, J.F.; de Micco, P.; Mertens, P.; Zientara, S. Complete nucleotide sequence of Middelburg virus, isolated from the spleen of a horse with severe clinical disease in Zimbabwe. *J. Gen. Virol.* **2007**, *88*, 3078–3088. [CrossRef]
57. Tscha, M.K.; Suzukawa, A.A.; Graf, T.; Piancini, L.D.S.; da Silva, A.M.; Faoro, H.; Riediger, I.N.; Medeiros, L.C.; Wowk, P.F.; Zanluca, C.; et al. Identification of a novel alphavirus related to the encephalitis complexes circulating in southern Brazil. *Emerg. Microbes Infect.* **2019**, *8*, 920–933. [CrossRef]
58. Du, Q.; Miao, Y.; He, W.; Zheng, H. ADP-Ribosylation in Antiviral Innate Immune Response. *Pathogens* **2023**, *12*, 303. [CrossRef]
59. Luscher, B.; Verheirstraeten, M.; Krieg, S.; Korn, P. Intracellular mono-ADP-ribosyltransferases at the host-virus interphase. *Cell. Mol. Life Sci.* **2022**, *79*, 288. [CrossRef]
60. Fehr, A.R.; Singh, S.A.; Kerr, C.M.; Mukai, S.; Higashi, H.; Aikawa, M. The impact of PARPs and ADP-ribosylation on inflammation and host-pathogen interactions. *Genes Dev.* **2020**, *34*, 341–359. [CrossRef]
61. Zhu, H.; Zheng, C. When PARPs Meet Antiviral Innate Immunity. *Trends Microbiol.* **2021**, *29*, 776–778. [CrossRef] [PubMed]
62. Todorova, T.; Bock, F.J.; Chang, P. Poly(ADP-ribose) polymerase-13 and RNA regulation in immunity and cancer. *Trends Mol. Med.* **2015**, *21*, 373–384. [CrossRef] [PubMed]
63. Yang, C.S.; Jividen, K.; Spencer, A.; Dworak, N.; Ni, L.; Oostdyk, L.T.; Chatterjee, M.; Kusmider, B.; Reon, B.; Parlak, M.; et al. Ubiquitin Modification by the E3 Ligase/ADP-Ribosyltransferase Dtx3L/Parp9. *Mol. Cell* **2017**, *66*, 503–516.e505. [CrossRef] [PubMed]

pathogens

Review

Mechanism and Modulation of SidE Family Proteins in the Pathogenesis of *Legionella pneumophila*

Yongchao Xie [1,2,†], Yi Zhang [1,†], Yong Wang [2] and Yue Feng [1,*]

1 Beijing Advanced Innovation Center for Soft Matter Science and Engineering, Beijing Key Laboratory of Bioprocess, State Key Laboratory of Chemical Resource Engineering, College of Life Science and Technology, Beijing University of Chemical Technology, Beijing 100029, China;
2 State Key Laboratory of Crop Biology, College of Life Sciences, Shandong Agricultural University, Tai'an 271002, China
* Correspondence: fengyue@mail.buct.edu.cn
† These authors contributed equally to this work.

Abstract: *Legionella pneumophila* is the causative agent of *Legionnaires'* disease, causing fever and lung infection, with a death rate up to 15% in severe cases. In the process of infection, *Legionella pneumophila* secretes over 330 effectors into host cell via the Dot/Icm type IV secretion system to modulate multiple host cellular physiological processes, thereby changing the environment of the host cell and promoting the growth and propagation of the bacterium. Among these effector proteins, SidE family proteins from *Legionella pneumophila* catalyze a non-canonical ubiquitination reaction, which combines mono-ADP-ribosylation and phosphodiesterase activities together to attach ubiquitin onto substrates. Meanwhile, the activity of SidE family proteins is also under multiple modulations by other effectors. Herein we summarize the key insights into recent studies in this area, emphasizing the tight link between the modular structure of SidE family proteins and the pathogen virulence as well as the fundamental mechanism and modulation network for further extensive research.

Keywords: *Legionella pneumophila*; SidE family; effector; PR-ubiquitination; host-pathogen interaction

check for updates

Citation: Xie, Y.; Zhang, Y.; Wang, Y.; Feng, Y. Mechanism and Modulation of SidE Family Proteins in the Pathogenesis of *Legionella pneumophila*. *Pathogens* **2023**, *12*, 629. https://doi.org/10.3390/pathogens12040629

Academic Editors: Anthony K L Leung, Anthony Fehr and Rachy Abraham

Received: 19 February 2023
Revised: 14 April 2023
Accepted: 18 April 2023
Published: 21 April 2023

1. Introduction

Gram-negative bacterium *L. pneumophila* was identified in 1976 at the annual convention of American legion, which caused a serious pneumophila, resulting in a lethality rate of 15.9% [1]. It has been reported that the pathogenic bacteria *L. pneumophila* has a versatile arsenal of effectors, keeping its virulence by expressing over 330 individual effectors through the Dot/Icm secretion system [2,3]. Moreover, further studies of pathogenic strategies revealed that after entering the cytoplasm of the host cell, the bacterium avoids its lysosomal-mediated degradation by escaping the endosomal-trafficking pathway and establishes *Legionella*-containing vacuoles (LCV) [4]. These specialized membrane-bound organelles are rich in nutrients and without lysosome hydrolases, providing *Legionella* with an ideal environment for its intracellular replication [5]. During the formation of LCVs, many post-translational modifications (PTMs) are involved, removing chemical moieties from protein residues, or attaching modifying groups to target protein, which mediates numerous physiological processes by their unique biochemical activities. Up to now, over 400 different types of PTMs have been identified in eukaryotic cells such as phosphorylation, glycosylation, acetylation, ADP-ribosylation and ubiquitination [6–8]. Among these, ubiquitination, a ubiquitous post-translational modification, which regulates a variety of physiological processes in eukaryotic cells, such as protein homeostasis, cell cycle, immune response, DNA repair and vesicle transport, has been studied for several decades [9].

The function and mechanism of canonical ubiquitination has been already well established. It occurs through a series of enzymatic reactions. First, the ubiquitin activating E1 consumes ATP and activates the C-terminal carboxyl group of ubiquitin and forms a thioester bond with cysteine at the active site of ubiquitin conjugating enzyme E2. Then, ubiquitin ligase E3 transfers ubiquitin from E2-Ub to a specific substrate. Finally, an isopeptide bond is formed between the carboxyl group of glycine at position 76 of ubiquitin and the ε-amino group of Lys or the α-amino group of Met1 of a substrate protein [10,11]. Intriguingly, owing to the key role of ubiquitination in the life of eukaryotic cells, many pathogens have derived a series of effector proteins targeting the host ubiquitination process during the long-term evolution with host cells, to construct a conducive internal environment for the reproduction of pathogens [12,13].

L. pneumophila as the pathogen causing pneumonia, also derived numerous effector proteins to modulate host ubiquitination and the most striking example of these to date is the SidE effector family [14–16]. The SidE family contains four highly conserved members SidE, SdeA, SdeB, and SdeC that mediate a noncanonical ubiquitination system to facilitate the optimal *Legionella* virulence. While the importance and the inherent understanding of canonical ubiquitination has been known for a long time, the atypical ubiquitination catalyzed by the SidE family shows an unprecedented aspect in this area, which has attracted a lot of attention. Here we will review the recent progress regarding the mechanism and modulation of SidE family effectors and the pathogenic strategies of *L. pneumophila* related to this ubiquitination process.

2. SidE Family Effectors Catalyze a Non-Canonical Ubiquitination Process

The non-canonical ubiquitination by SidE family proteins differs from the canonical ubiquitination system in several aspects, including structure characteristics of enzymes, energy consumption and the number of reaction enzymes or steps. Firstly, for structural features, SidE family members are large proteins (approximately 1500 residues), which contain a DUB (deubiquitinase) domain, a PDE (phosphodiesterase) domain, an mART (mono-ADP ribosyltransferase) domain and a C-terminal domain (CTD) (Figure 1a,b) [17,18]. Secondly, for the energy source, nicotinamide adenine dinucleotide (NAD^+) is required by the mART domain of SidE, which is a putative catalytic motif typically found in bacterial toxins [19,20]. Thirdly, this non-canonical ubiquitination is catalyzed only by one protein in an all-in-one mode rather than the three steps mode of canonical ubiquitination. Finally, Arg42 of ubiquitin and primarily a serine residue of substrate are linked by a phosphoribosyl moiety, so this type of ubiquitination is also named PR (phosphoribosyl)-ubiquitination. Recent studies also found that SdeC-mediated PR-ubiquitination also modifies tyrosine residues in host proteins [21].

2.1. The Structural Features of SidE Family

The SidE family protein contains four domains, including DUB, mART, PDE and CTD domains, and each one has its independent function or regulates another. The DUB domain, comprising ~200 residues in the N-terminus of SidE, was first characterized to have deubiquitinase activity, with a preference for Lys-63 Linked polyubiquitin chains [22]. The PDE domain spans residues approximately 200–600, which is formed by two lobes: a larger helical core lobe containing the catalytic pocket and a smaller cap lobe covering from the top [23] (Figure 1c). Structural comparison revealed that human SAMHD1, the dNTP hydrolase related to innate immune response, is the closest structural homologue in mammals of SdeA [24]. The SdeA mART domain contains a typical Rossmann fold and shows the conserved characteristics among all known mART toxins in bacteria [19]. Two-lobe structures constitute the SdeA mART domain, one with an N-terminal α-helical lobe and the other with a C-terminal β-sandwich lobe (Figure 1d). Even though the similarity between SdeA mART and other mART toxins exist, there are some weak differences in structural details. For example, the PN loop and ARTT loop in the mART domain are different from those of other mART proteins. Moreover, the plug loop, two consecutive helices connected by a loop, inserts into and interacts with the PDE domain, which is related to the activity of mART but not to the PDE domain [15].

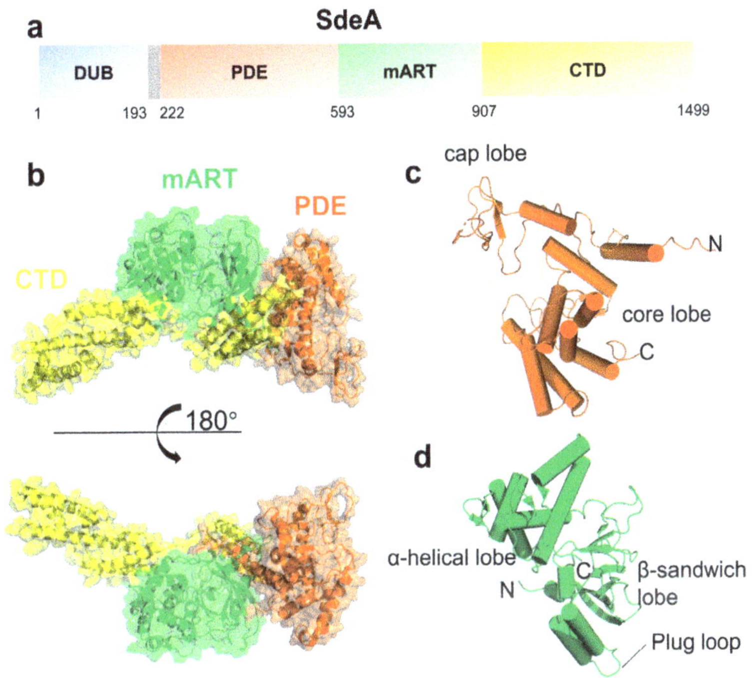

Figure 1. Overall structure of SdeA. (**a**) Domain diagram of SdeA (1–1499), SdeA contains four domains, DUB (blue), PDE (orange-yellow), mART (green) and CTD (golden-yellow), from N-terminus to C-terminus. (**b**) Two views of overall structure of SdeA (231–1190) colored as in a. (**c**) Structure of SdeA PDE domain. (**d**) Structure of SdeA mART domain, α-helical lobe and β-sandwich lobe are marked, and the "plug loop" was also labeled.

2.2. The Novel-Ubiquitination Machinery of the SidE Family

As mentioned above, ubiquitination as an important protein PTM, was well studied for decades [25]. However, in 2016, SdeA protein in *L. pneumophila* was identified to be capable of performing a non-canonical ubiquitination by itself [20]. In contrast to the conventional ATP-driving E1-E2-E3 cascade (Figure 2a), the ubiquitination catalyzed by SdeA effector requires NAD^+ as energy [26]. Overall, it is strikingly different between the three-enzyme systems and the all-in-one ubiquitination machinery SidE. While SidE family protein comprises four domains, only the enzymatic activities of the PDE and mART domain are involved in the ubiquitination process. The SidE ligase machinery was divided into two distinct parts, Ub activation and Ub-substrate ligation, which was catalyzed by the mART and PDE domain respectively (Figure 2b) [26].

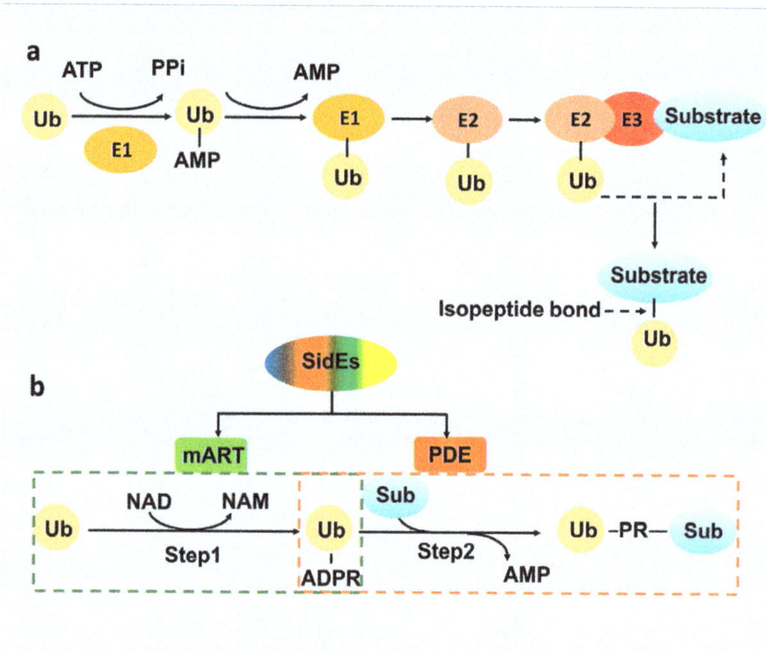

Figure 2. The mechanism diagram of E1-E2-E3 conventional ubiquitination and SidEs PR-ubiquitination. (**a**) Ubiquitin activating enzyme E1, ubiquitin conjugation enzyme E2 and ubiquitin ligase E3 are working together for the conventional type of ubiquitination. (**b**) SidE family proteins could catalyze the whole PR-ubiquitination reaction by itself, mART and PDE domain involved the first and second step respectively. NAM, nicotinamide.

First, the mART domain exhibits ADP-ribosyl transferase activity, using nucleotide cofactor NAD^+ as energy, leading to ADP-ribose group covalently added to Arg42 of Ub forming ADPR-Ub [26] (Figure 2b). ADP-ribosylation is also one of the most important types of protein PTMs, discovered in bacterial pathogen *Corynebacterium diphtheria* originally and in the eukaryotic cell subsequently, which regulate various cellular processes, including tumorigenesis and DNA repair [27,28]. Despite that ART protein Parp9 interacts with the E3 ligase Dtx3L to add mono-ADP-ribose group to the carboxyl terminus of ubiquitin molecule [29], ADP-ribosylation of ubiquitin catalyzed by SidE mART is also an example of a crosslink between ADP-ribosylation ubiquitination.

Second, the SidE PDE domain recognizes and binds the ADPR-Ub produced by the mART domain, exhibits phosphodiesterase activity to cleave the phosphoanhydride bond in ADPR-Ub and produce phosphoribosylated ubiquitin (PR-Ub) [23]. Meanwhile, in the presence of a substrate protein, the SdeA PDE domain utilizes a substrate binding cleft (constituted by N404, Q405, M408, L411 and S428), juxtaposed with the catalytic site, to position serines of the substrates for ubiquitination. During the reaction, a transient SdeA H277-PR-Ub intermediate was first formed and subsequently nucleophilic attacked by the OH group of the target serine of the substrate. Finally, PR-Ub was transferred to serine residues in target proteins, with the release of AMP [30] (Figure 2b). The PDE domain in the SidE family protein shares ~23% sequence identity with their closest similarity protein PA4781 from *Pseudomonas aeruginosa* and possesses the conserved catalytic residues, H277-H407-E340 catalytic triad. The reaction catalyzed by the PDE domain is similar to a phosphotransferase activity and akin in part to the activity of His kinases [31,32]. Notably, ADPR-Ub can be produced by the SdeA PDE mutant (H277A) and PR-Ub can still be transferred to a target protein, if the SdeA mART domain truncation was supplied with ADPR-Ub as a substrate, suggesting that these two reactions were separable [15,26,33].

3. Activity of the SidE Family Was Strictly Modulated by Many Effectors

Physiological processes in eukaryotic or prokaryotic cells are influenced and modulated extensively by other molecules, including chemical substances and proteins. Similarly, the activity of SidE family proteins is also strictly controlled by other proteins. Recently, *L. pneumophila* effectors, SidJ, SdjA, DupA and DupB have been proved to regulate the activity of SidE family proteins by some novel modes.

3.1. SidJ Interacts with Calmodulin to Modify SdeA

The ubiquitination activity of SdeA has a relatively strong toxic effect on host cells. However, this excessive toxic effect is not conducive to the proliferation of *L. pneumophila*. In 2015, the *L. pneumophila* effector protein SidJ was found to inhibit the toxicity of SdeA in the host [34]. In the subsequent study, it was proved that SidJ suppresses the ubiquitination activity of SdeA in vivo [35]. However, it was still unknown why SidJ can inhibit the activity of SidE family proteins only in the host cell at that time. In the process of exploring this question, calmodulin (CaM), the Ca^{2+} binding protein in eukaryotic cells, appears to participate in the regulation of SdeA by interacting with the *L. pneumophila* effector protein SidJ. Then, four independent studies revealed that SidJ and calmodulin form a stable complex, catalyzing a distinct PTM to the key catalytic residues of the SidE family protein, turning SidE into the "inactive state" (Figure 3). This unusual PTM was polyglutamylation and the exactly modified residue of SdeA was E860, a key catalytic residue in the mART domain [36–39]. The discovery of SidJ as a CaM-activated glutamylase explained that how SidJ-CaM complex inactivates the SidE family protein. However, there are still several intriguing questions to be further explored. First, for the mechanism details about CaM dependent activating mode, Sulpizio et.al., proposed that CaM-binding may stabilize the activation loop, which is vital for protein kinases, in an activated state via the CaM N-loop [39]. Second, for the substrate specificity of SidJ, it remains not fully understood whether SidJ only targets the SidE family protein. Bhogaraju et al., found that glutamylation signals still remained when the host cell was infected by *Legionella* strains lacking SidE family genes, indicating that the SidE family protein might not be the only substrates of SidJ-CaM. This finding was striking and interesting in that the pathogenic bacteria effectors along with the eukaryotic host protein might modify another effector together [37].

3.2. SdjA Reverses the Glutamylation Modification of SdeA

Remarkably, the modification mode of SidJ-CaM towards SdeA unveils an archetypal example that the pathogenic bacterial effector protein catalyzes glutamylation, modulating the PR-ubiquitination mediated by the SidE family protein in the host cell [40]. E860 is the key catalytic residue of the SdeA mART domain, which is polyglutamated by the SidJ-CaM complex, indicating that SidJ-CaM displays specificity towards this residue. Furthermore, Vincent et al., solved the cryo-EM structure of SdeA-SidJ-CaM intermediate complexes, proving that the kinase-like site of SidJ adenylates the active-site Glu in SidE in the presence of ATP and Mg^{2+}, forming a stable intermediate complex. At the same time, the insertion loop in the active site of the SidJ kinase domain accommodates the donor Glu near the acyl-adenylates site, facilitating the reaction of glutamylation [41].

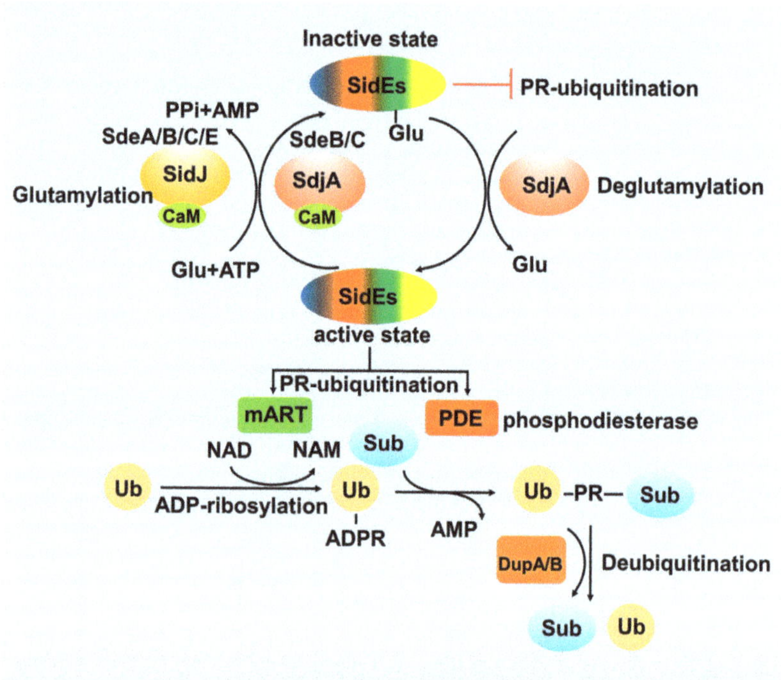

Figure 3. The modulation model of SidE family proteins. SidEs catalyze PR-ubiquitination by their mART and PDE domains. SidJ-CaM and SdjA-CaM mediate glutamylation to inactivate SidEs (SidJ-CaM inactivate SdeA/B/C/E, SdjA-CaM inactivate SdeB/C). The inactivated SidEs-Glu could be reactivated again by SdjA, named deglutamylation. The PR-ubiquitination substrates could be cleaved by DupA/B, releasing substrates and Ub again.

Furthermore, based on the fact that several *Legionella* or other bacteria effectors are working together to regulate one physiological process, we wondered whether other effectors also participate in the regulation of SidEs. Interestingly, SdjA, a paralog protein of SidJ which shows 57% sequence identity [42], shows glutamylation activity to SdeB and SdeC but not SdeA. Moreover, SdjA cannot complement the virulence defects displayed by a mutant lacking SidJ [43]. Due to the unusual characteristic of SdjA compared with SidJ, the function of effector SdjA still remains unknown. Coincidentally, two *Legionella* homologous effector proteins MavC and MvcA, have been proved to work together to stimulate and antagonize another unconventional ubiquitination, respectively [14,44,45]. In this distinct type of ubiquitination, MavC catalyzes the attachment of Ub to UBE2N by its transglutaminase activity (termed ubiquitination), while MvcA catalyzes the opposite process releasing ubiquitin from Ub-UBE2N by its deamidase activity (Deubiquitination) [14,16,45]. Interestingly, we found that SdjA contains deglutamylase activity, changing SdeA-Glu into SdeA in the absence of CaM, thereby antagonizing the activity of SidJ. Actually, SdjA was a bifunctional enzyme that exhibits distinct activities towards SidE family proteins and the specificity was dependent on its N-terminal region (Figure 3) [43].

3.3. DupA and DupB Function as Deubiquitinases for PR-Ubiquitination

Conventional ubiquitination is a reversible process, the substrate of ubiquitination can be re-cleaved into ubiquitin and substrate by deubiquitinating enzymes [46]. The novel ubiquitination mediated by the SidE-effector proteins involves the formation of thioester bonds between substrates serine hydroxyl and ubiquitin [26]. In this process, the

PDE domain of SidE can cleave ADPR-Ub to generate AMP and PR-Ub in the absence of substrates [26]. Interestingly, the *L. pneumophila* effectors DupA and DupB, two homologous proteins of the PDE domain with 70% sequence similarity to SdeA PDE, have been proved to exhibit activity to process ADPR-Ub to PR-Ub [47]. So that, the balance of PR-ubiquitination of multiple substrates in the host cell was controlled by these two specific deubiquitinases upon bacterial infection stringently. While SidEs catalyze PR-ubiquitination with its PDE domain in the second step, DupA and DupB catalyze deubiquitination also via their PDE domains [48]. This is reminiscent of the characteristics of SidJ/SdjA, or MavC/MvcA, which were mentioned above (Figure 3).

4. Multiple Host Proteins Targeted by SidE Family Effectors

Previous studies indicated that the host substrates of the SidE family are related with the endoplasmic reticulum (ER) and Golgi in the host cell, resulting in disturbances of their transport pathways, which modulates the internal host environment and promotes the formation of LCV. However, along with the deepening of studies into the biological significance of SidE-mediated ubiquitination, especially the use of DupA/B deletion bacterial strain, other physiological systems, such as endo-lysosomal system, mitochondria, proteasomal subunits, cytoskeleton and nuclear membrane related proteins, have also been reported to be regulated by this ubiquitination [48]. It is necessary to determine the exact relationship between the ubiquitination catalyzed by SidE and these cellular processes to cast light on how *L. pneumophila* exploits these effectors for survival and proliferation.

4.1. The Effects of SidE Family Proteins on Endoplasmic Reticulum

L. pneumophila is an intravacuolar pathogen, utilizing a type IVB secretion system (T4SS) to translocate effector proteins into the host cytosol to establish LCV, an endoplasmic reticulum (ER)-associated organelle [49,50]. However, these bacterial effector proteins are unable to form an LCV themselves which means that they need to make use of the substances or protein substrates in the host cell for this process. Endoplasmic reticulum is a continuous omental system, a cystic, vesicular, and tubular structure organelle formed by a single membrane, which is in charge of the production and movement of proteins and other molecules [51]. Endoplasmic reticulum could be classified into the perinuclear, ribosome-associated ER sheet and tubular ER and the tubular ER is a vast network of cylinders that are enriched with some structural ER membrane proteins, such as reticulon family proteins [52]. Previous studies identified that several ER-associated GTPases and reticulon 4 (Rtn4) are PR-ubiquitinated by SidE family proteins. During its infection, *L. pneumophila* exploits effectors to regulate the dynamics of membranes to create LCV. PR-ubiquitination was utilized by *Legionella* to modify ER-related proteins, such as RTN3, RTN4, TEX264, FAM134A, FAM134B and FAM134C, giving rise to ER membrane fragmentation and dynamic defect [48,53,54]. Among these ER-related proteins, RTN4 is required to induce the formation and stabilization of endoplasmic reticulum tubules, regulating membrane morphogenesis in the ER [55], and previously regarded as a critical substrate for the formation of LCV. FAM134 family proteins (FAM134A, FAM134B and FAM134C) are ER-anchored autophagy receptors, which mediate ER transports into lysosome, promoting membrane remodeling and ER dissociation. Furthermore, FAM134B targets the ER fragments into autophagosomes via interaction with ATG8 family proteins [54]. Taken together, these suggest that SidE family proteins mediated PR-ubiquitination of host substrates to affect the normal function of endoplasmic reticulum (Table 1).

Table 1. List of SidEs and their major related proteins in this review.

Gene ID	Species	Aliases	Function	Reference
lpg2157	*L. pneumophila*	SdeA	PR-ubiquitination	[20]
lpg2156	*L. pneumophila*	SdeB	PR-ubiquitination	
lpg2153	*L. pneumophila*	SdeC	PR-ubiquitination	
lpg0234	*L. pneumophila*	SidE	PR-ubiquitination	
lpg2155	*L. pneumophila*	SidJ	Inhibits SdeA, SdeB, SdeC and SidE ubiquitinating activity by Glutamylation.	[35,36,38]
lpg2508	*L. pneumophila*	SdjA	Reverses the SidJ-CaM modification of SdeA.	[43]
lpg2154	*L. pneumophila*	DupA	Regulates Phosphoribosyl-Linked Serine Ubiquitination by Deubiquitination.	[47,48]
lpg2509	*L. pneumophila*	DupB		
10313	Homo sapiens	RTN3	Induces the formation of ER	[56]
57142	Homo sapiens	RTN4	Stabilization of endoplasmic reticulum (ER)	[55–57]
162427	Homo sapiens	FAM134C	Endoplasmic reticulum remodeling, ER-phagy, and Collagen quality control.	[58]
51368	Homo sapiens	TEX264	ATG8-Interacting Protein Critical for ER Remodeling.ER-phage.	[59,60]
83452	Homo sapiens	Rab33b	ER-associated Rab small GTPases.Regulators of Golgi homeostasis and trafficking.	[61,62]
26003	Homo sapiens	GRASP55	Function in the connection of Golgi stack and the maintenance of Golgi structure	[63,64]
64689	Homo sapiens	GRASP65		
64746	Homo sapiens	GCP60	Affecting protein transport between the endoplasmic reticulum and Golgi.	[65]
10897	Homo sapiens	YIF1A	Role in transport between endoplasmic reticulum and Golgi.	[66]
27131	Homo sapiens	SNX5	Mediates retrograde transport of cargo proteins from endosomes to the trans-Golgi network.	[67]

Notes: this table only contains a small number of ER and Golgi related substrates.

4.2. The Effects of SidE Family Proteins on the Golgi Complex

In the early stage of infection, *L. pneumophila* exploits effectors, such as SidE family proteins, to manipulate Rab1 and other ER-related proteins to intercept the versicles to the LCV [68]. Actually, the downstream process after ER vesicles fusing to the Golgi complex is also disturbed. Recently, the relationship between PR-ubiquitination and the Golgi complex has received increasing attention. The Golgi complex, the cystic structure apparatus formed by the elementary membrane, is the component of the eukaryotic endomembrane system, which functions as the PTM factory for protein modification, classification and translocation [69]. The vesicles from the endoplasmic reticulum could fuse with the Golgi membrane, delivering the inclusions into the Golgi lumen, where the newly synthesized peptide chains continue to complete their modification and packing [70].

Most obviously, compared with the relative comprehensive understanding that *L. pneumophila* markedly disrupts the ER trafficking pathway, it is elusive how the SidE family proteins affect the function of the Golgi apparatus and which Golgi-related proteins in the host cell are taking part in the PR-ubiquitination. Notably, Shin et al., showed that two deubiquitinases (DupA and DupB) specifically cleave PR-ubiquitin from serine on substrates and take advantage of the catalytically inactive DupA and its affinity for PR-ubiquitinated protein to capture and identify nearly 180 host proteins targeted by SidE family proteins [48]. Among these substrates, some Golgi-related proteins were also identified, such as GRASP55, TMED8, GCP60, YIF1A, RAB33B and SNX5. Notably, GRASP55 and GCP60 had the highest ratio among these Golgi protein substrates (Figure 4, Table 1).

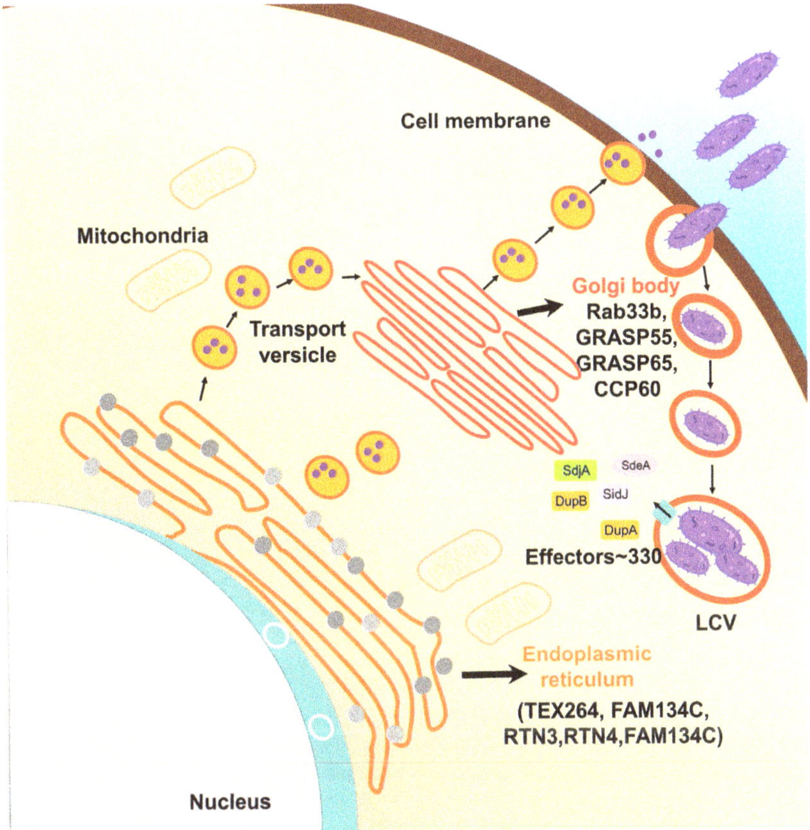

Figure 4. The spatial relationship between LCV, ER and Golgi and the localization of the various SidE target proteins.

GRASP55 plays a vital role in the maintenance of the Golgi integrality [64,65]. GRASP65 and GRASP55 are homologous proteins, both belong to the same protein family named GRASP, which function in the connection of the Golgi stack and the maintenance of the Golgi structure through self-interaction and interactions with other Golgi proteins [71]. They are localized to Golgi cisternae and required for the ER-to-Golgi transport of specific cargo, which contains C-terminal valine motif [72]. It has been known that the activities of mammalian GRASPs are regulated by serine phosphorylation, one of the most canonical PTMs, resulting in Golgi fragmentation [73]. Recently, Liu et al., confirmed that the C-terminus of SdeA is not only critical for its Golgi localization, but also for its ability to PR-ubiquitinate Golgi protein in the host cell. Taken together, the PR-ubiquitination of GRASP65 and GRASP55 by SidE family proteins, causes disruption of Golgi integrity, thus preventing their ability to aggregate and form oligomeric states. In fact, the presence or absence of PR-ubiquitination of GRASPs can have an important impact on the host secretory pathway, while is not linked directly to the recruitment of Golgi membranes to the growing LCV [74].

5. Conclusions

In this review, we summarize the mechanisms, modulation and protein substrates related to endoplasmic reticulum and Golgi apparatus of non-canonical ubiquitination by SidE family proteins during the pathogenesis of *L. pneumophila*. Harboring three enzymatic activities, SidE family proteins also undergo extensive modulations. In terms of activity modulation, the important role of calmodulin and the need to study the structure and function of SdjA are emphasized. Regarding substrates, we summarized mainly the substrates related to endoplasmic reticulum and Golgi apparatus, which have been studied more extensively at present, and pointed out the perspectives for subsequent research on substrates of other physiological processes.

Specifically, SidJ inactivates SdeA and SdjA renders SdeA to regain its activity of PR-ubiquitination [43]. This SdeA regulation mode is associated with calmodulin, the calcium binding protein in eukaryotic cells, which participates in many physiology processes and especially plays a vital role in the calcium signal transduction pathway [75]. However, it still remains unknown whether calmodulin is only used by *L. pneumophila* to control its virulence of SidE family effectors or is simultaneously influencing other physiology processes involved in signal transduction. This will be interesting to be investigated by future studies. Moreover, pathogens need to strictly control their virulence to proliferate normally. From the aspect of host-pathogen interaction, when the host cell was infected by pathogens, they also need to evolve approaches to counteract the influence of pathogens. Therefore, it is an interesting question whether the need of calmodulin binding for the activity of SidJ in *L. pneumophila* is a beneficial approach for the pathogen to modulate the activity of its effectors or a strategy exploited by the host to inactivate the toxic effectors of pathogens. Based on the recent study, SdjA seems to eb a critical member in the regulation network of the SidE family and its deglutamylation activity was not dependent on calmodulin binding. However, the key domains and residues for deglutamylation activity in SdjA still need further investigation [43].

With a growing number of PR-ubiquitination substrates identified, more and more related physiological processes have been found. This means that the functions of SidE family effectors are more complicated and significant for the pathogen than what we have ever known. Whereas some advances of PR-ubiquitination in endoplasmic reticulum and the Golgi complex have been achieved, the effects on other related processes, such as the Endo-lysosomal system, mitochondria, proteasomal subunits, cytoskeleton and nuclear membrane related proteins, still need to be further explored. Moreover, temporal and spatial regulation of the activity of SidEs by the modulation effectors in these physiological processes will also be interesting subjects in future studies.

Author Contributions: Y.X., Y.Z. and Y.W. wrote the original manuscript. Y.F. revised and finalized the manuscript. All authors have read and agreed to the published version of the manuscript.

Funding: This work was supported by the National Natural Science Foundation of China (32000901) and National key research and development program of China (2022YFC3401500 and 2022YFC2104800).

Institutional Review Board Statement: Not applicable.

Informed Consent Statement: Not applicable.

Data Availability Statement: No new data were created or analyzed in this study. Data sharing is not applicable to this article.

Conflicts of Interest: The authors have nothing to disclose. No private companies had any role in the study design.

References

1. Fraser, D.W.; Tsai, T.R.; Orenstein, W.; Parkin, W.E.; Beecham, H.J.; Sharrar, R.G.; Harris, J.; Mallison, G.F.; Martin, S.M.; Mcdade, J.E.; et al. Legionnaires' Disease. *N. Engl. J. Med.* **1977**, *297*, 1189–1197. [CrossRef] [PubMed]
2. Luo, Z.Q.; Isberg, R.R. Multiple substrates of the Legionella pneumophila Dot/Icm system identified by interbacterial protein transfer. *Proc. Natl. Acad. Sci. USA* **2004**, *101*, 841–846. [CrossRef] [PubMed]
3. Ensminger, A.W. Legionella pneumophila, armed to the hilt: Justifying the largest arsenal of effectors in the bacterial world. *Curr. Opin. Microbiol.* **2016**, *29*, 74–80. [CrossRef] [PubMed]
4. Isberg, R.R.; O'connor, T.J.; Heidtman, M. The Legionella pneumophila replication vacuole: Making a cosy niche inside host cells. *Nat. Rev. Microbiol.* **2009**, *7*, 13–24. [CrossRef] [PubMed]
5. Robinson, C.G.; Roy, C.R. Attachment and fusion of endoplasmic reticulum with vacuoles containing Legionella pneumophila. *Cell. Microbiol.* **2006**, *8*, 793–805. [CrossRef]
6. Ramazi, S.; Zahiri, J. Posttranslational modifications in proteins: Resources, tools and prediction methods. *Database* **2021**, baab012. [CrossRef]
7. Mann, M.; Jensen, O.N.J.N.B. Proteomic analysis of post-translational modifications. *Nat. Biotechnol.* **2003**, *21*, 255–261. [CrossRef]
8. Olsen, J.V.; Mann, M.J.M.; Mcp, C.P. Status of Large-scale Analysis of Post-translational Modifications by Mass Spectrometry. *Mol. Cell. Proteom.* **2013**, *12*, 3444–3452. [CrossRef]
9. Sun, S.C. Deubiquitylation and regulation of the immune response. *Nat. Rev. Immunol.* **2008**, *8*, 501–511. [CrossRef]
10. Hershko, A.; Ciechanover, A. The ubiquitin system. *Annu. Rev. Biochem.* **1998**, *67*, 425–479. [CrossRef]
11. Zheng, N.; Shabek, N. Ubiquitin Ligases: Structure, Function, and Regulation. *Annu. Rev. Biochem.* **2017**, *86*, 129–157. [CrossRef] [PubMed]
12. Yan, F.; Huang, C.; Wang, X.; Tan, J.; Cheng, S.; Wan, M.; Wang, Z.; Wang, S.; Luo, S.; Li, A.; et al. Threonine ADP-Ribosylation of Ubiquitin by a Bacterial Effector Family Blocks Host Ubiquitination. *Mol. Cell* **2020**, *78*, 641–652.e649. [CrossRef] [PubMed]
13. Lin, Y.; Hu, Q.; Zhou, J.; Yin, W.; Yao, D.; Shao, Y.; Zhao, Y.; Guo, B.; Xia, Y.; Chen, Q.; et al. Phytophthora sojae effector Avr1d functions as an E2 competitor and inhibits ubiquitination activity of GmPUB13 to facilitate infection. *Proc. Natl. Acad. Sci. USA* **2021**, *118*, e2018312118. [CrossRef] [PubMed]
14. Gan, N.; Nakayasu, E.S.; Hollenbeck, P.J.; Luo, Z.Q. Legionella pneumophila inhibits immune signalling via MavC-mediated transglutaminase-induced ubiquitination of UBE2N. *Nat. Microbiol.* **2019**, *4*, 134–143. [CrossRef] [PubMed]
15. Dong, Y.; Mu, Y.; Xie, Y.; Zhang, Y.; Han, Y.; Zhou, Y.; Wang, W.; Liu, Z.; Wu, M.; Wang, H.; et al. Structural basis of ubiquitin modification by the Legionella effector SdeA. *Nature* **2018**, *557*, 674–678. [CrossRef] [PubMed]
16. Mu, Y.; Wang, Y.; Huang, Y.; Li, D.; Han, Y.; Chang, M.; Fu, J.; Xie, Y.; Ren, J.; Wang, H.; et al. Structural insights into the mechanism and inhibition of transglutaminase-induced ubiquitination by the Legionella effector MavC. *Nat. Commun.* **2020**, *11*, 1774. [CrossRef] [PubMed]
17. Bardill, J.P.; Miller, J.L.; Vogel, J.P. IcmS-dependent translocation of SdeA into macrophages by the Legionella pneumophila type IV secretion system. *Mol. Microbiol.* **2005**, *56*, 90–103. [CrossRef]
18. Krissinel, E.; Henrick, K. Inference of macromolecular assemblies from crystalline state. *J. Mol. Biol.* **2007**, *372*, 774–797. [CrossRef] [PubMed]
19. Han, S.; Arvai, A.S.; Clancy, S.B.; Tainer, J.A. Crystal structure and novel recognition motif of Rho ADP-ribosylating C3 exoenzyme from Clostridium botulinum: Structural insights for recognition specificity and catalysis1 1Edited by D. Rees. *J. Mol. Biol.* **2001**, *305*, 95–107. [CrossRef]
20. Qiu, J.; Sheedlo, M.J.; Yu, K.; Tan, Y.; Nakayasu, E.S.; Das, C.; Liu, X.; Luo, Z.Q. Ubiquitination independent of E1 and E2 enzymes by bacterial effectors. *Nature* **2016**, *533*, 120–124. [CrossRef]
21. Zhang, M.; Mcewen, J.M.; Sjoblom, N.M.; Kotewicz, K.M.; Isberg, R.R.; Scheck, R.A. Members of the Legionella pneumophila Sde family target tyrosine residues for phosphoribosyl-linked ubiquitination. *RSC Chem. Biol.* **2021**, *2*, 1509–1519. [CrossRef] [PubMed]
22. Sheedlo, M.J.; Qiu, J.; Tan, Y.; Paul, L.N.; Luo, Z.Q.; Das, C. Structural basis of substrate recognition by a bacterial deubiquitinase important for dynamics of phagosome ubiquitination. *Proc. Natl. Acad. Sci. USA* **2015**, *112*, 15090–15095. [CrossRef] [PubMed]
23. Wang, Y.; Shi, M.; Feng, H.; Zhu, Y.; Liu, S.; Gao, A.; Gao, P. Structural Insights into Non-canonical Ubiquitination Catalyzed by SidE. *Cell* **2018**, *173*, 1231–1243.e1216. [CrossRef] [PubMed]
24. Ji, X.; Wu, Y.; Yan, J.; Mehrens, J.; Yang, H.; Delucia, M.; Hao, C.; Gronenborn, A.M.; Skowronski, J.; Ahn, J.; et al. Mechanism of allosteric activation of SAMHD1 by dGTP. *Nat. Struct. Mol. Biol.* **2013**, *20*, 1304–1309. [CrossRef] [PubMed]
25. Varshavsky, A. The ubiquitin system. *Trends Biochem. Sci.* **1997**, *22*, 383–387. [CrossRef] [PubMed]
26. Bhogaraju, S.; Kalayil, S.; Liu, Y.; Bonn, F.; Colby, T.; Matic, I.; Dikic, I. Phosphoribosylation of Ubiquitin Promotes Serine Ubiquitination and Impairs Conventional Ubiquitination. *Cell* **2016**, *167*, 1636–1649.e1613. [CrossRef] [PubMed]
27. Honjo, T.; Nishizuka, Y.; Hayaishi, O.; Kato, I. Diphtheria Toxin-dependent Adenosine Diphosphate Ribosylation of Aminoacyl Transferase II and Inhibition of Protein Synthesis. *J. Biol. Chem.* **1968**, *243*, 3553–3555. [CrossRef]
28. Corda, D.; Di Girolamo, M. Functional aspects of protein mono-ADP-ribosylation. *EMBO J.* **2003**, *22*, 1953–1958. [CrossRef]
29. Yang, C.S.; Jividen, K.; Spencer, A.; Dworak, N.; Ni, L.; Oostdyk, L.T.; Chatterjee, M.; Kuśmider, B.; Reon, B.; Parlak, M.; et al. Ubiquitin Modification by the E3 Ligase/ADP-Ribosyltransferase Dtx3L/Parp9. *Mol. Cell* **2017**, *66*, 503–516.e505. [CrossRef]

30. Kotewicz, K.M.; Ramabhadran, V.; Sjoblom, N.; Vogel, J.P.; Haenssler, E.; Zhang, M.; Behringer, J.; Scheck, R.A.; Isberg, R.R. A Single Legionella Effector Catalyzes a Multistep Ubiquitination Pathway to Rearrange Tubular Endoplasmic Reticulum for Replication. *Cell Host Microbe* **2017**, *21*, 169–181. [CrossRef]
31. Kalayil, S.; Bhogaraju, S.; Bonn, F.; Shin, D.; Liu, Y.; Gan, N.; Basquin, J.; Grumati, P.; Luo, Z.-Q.; Dikic, I. Insights into catalysis and function of phosphoribosyl-linked serine ubiquitination. *Nature* **2018**, *557*, 734–738. [CrossRef] [PubMed]
32. Klumpp, S.; Krieglstein, J. Phosphorylation and dephosphorylation of histidine residues in proteins. *Eur. J. Biochem.* **2002**, *269*, 1067–1071. [CrossRef] [PubMed]
33. Akturk, A.; Wasilko, D.J.; Wu, X.; Liu, Y.; Zhang, Y.; Qiu, J.; Luo, Z.-Q.; Reiter, K.H.; Brzovic, P.S.; Klevit, R.E.; et al. Mechanism of phosphoribosyl-ubiquitination mediated by a single Legionella effector. *Nature* **2018**, *557*, 729–733. [CrossRef] [PubMed]
34. Havey, J.C.; Roy, C.R. Toxicity and SidJ-Mediated Suppression of Toxicity Require Distinct Regions in the SidE Family of Legionella pneumophila Effectors. *Infect. Immun.* **2015**, *83*, 3506–3514. [CrossRef] [PubMed]
35. Qiu, J.; Yu, K.; Fei, X.; Liu, Y.; Nakayasu, E.S.; Piehowski, P.D.; Shaw, J.B.; Puvar, K.; Das, C.; Liu, X.; et al. A unique deubiquitinase that deconjugates phosphoribosyl-linked protein ubiquitination. *Cell Res.* **2017**, *27*, 865–881. [CrossRef]
36. Black, M.H.; Osinski, A.; Gradowski, M.; Servage, K.A.; Pawłowski, K.; Tomchick, D.R.; Tagliabracci, V.S. Bacterial pseudokinase catalyzes protein polyglutamylation to inhibit the SidE-family ubiquitin ligases. *Science* **2019**, *364*, 787–792. [CrossRef]
37. Bhogaraju, S.; Bonn, F.; Mukherjee, R.; Adams, M.; Pfleiderer, M.M.; Galej, W.P.; Matkovic, V.; Lopez-Mosqueda, J.; Kalayil, S.; Shin, D.; et al. Inhibition of bacterial ubiquitin ligases by SidJ-calmodulin catalysed glutamylation. *Nature* **2019**, *572*, 382–386. [CrossRef]
38. Gan, N.; Zhen, X.; Liu, Y.; Xu, X.; He, C.; Qiu, J.; Liu, Y.; Fujimoto, G.M.; Nakayasu, E.S.; Zhou, B.; et al. Regulation of phosphoribosyl ubiquitination by a calmodulin-dependent glutamylase. *Nature* **2019**, *572*, 387–391. [CrossRef]
39. Sulpizio, A.; Minelli, M.E.; Wan, M.; Burrowes, P.D.; Wu, X.; Sanford, E.J.; Shin, J.H.; Williams, B.C.; Goldberg, M.L.; Smolka, M.B.; et al. Protein polyglutamylation catalyzed by the bacterial calmodulin-dependent pseudokinase SidJ. *eLife* **2019**, *8*, e51162. [CrossRef]
40. Sulpizio, A.G.; Minelli, M.E.; Mao, Y. Glutamylation of Bacterial Ubiquitin Ligases by a Legionella Pseudokinase. *Trends Microbiol.* **2019**, *27*, 967–969. [CrossRef]
41. Osinski, A.; Black, M.H.; Pawłowski, K.; Chen, Z.; Li, Y.; Tagliabracci, V.S. Structural and mechanistic basis for protein glutamylation by the kinase fold. *Mol. Cell* **2021**, *81*, 4527–4539.e4528. [CrossRef]
42. Liu, Y.; Luo, Z.Q. The Legionella pneumophila effector SidJ is required for efficient recruitment of endoplasmic reticulum proteins to the bacterial phagosome. *Infect. Immun.* **2007**, *75*, 592–603. [CrossRef] [PubMed]
43. Song, L.; Xie, Y.; Li, C.; Wang, L.; He, C.; Zhang, Y.; Yuan, J.; Luo, J.; Liu, X.; Xiu, Y.; et al. The Legionella Effector SdjA Is a Bifunctional Enzyme That Distinctly Regulates Phosphoribosyl Ubiquitination. *mBio* **2021**, *12*, e0231621. [CrossRef] [PubMed]
44. Valleau, D.; Quaile, A.T.; Cui, H.; Xu, X.; Evdokimova, E.; Chang, C.; Cuff, M.E.; Urbanus, M.L.; Houliston, S.; Arrowsmith, C.H.; et al. Discovery of Ubiquitin Deamidases in the Pathogenic Arsenal of Legionella pneumophila. *Cell Rep.* **2018**, *23*, 568–583. [CrossRef] [PubMed]
45. Gan, N.; Guan, H.; Huang, Y.; Yu, T.; Fu, J.; Nakayasu, E.S.; Puvar, K.; Das, C.; Wang, D.; Ouyang, S.; et al. Legionella pneumophila regulates the activity of UBE2N by deamidase-mediated deubiquitination. *EMBO J.* **2020**, *39*, e102806. [CrossRef] [PubMed]
46. Komander, D.; Clague, M.J.; Urbé, S. Breaking the chains: Structure and function of the deubiquitinases. *Nat. Rev. Mol. Cell Biol.* **2009**, *10*, 550–563. [CrossRef] [PubMed]
47. Wan, M.; Sulpizio, A.G.; Akturk, A.; Beck WH, J.; Lanz, M.; Faça, V.M.; Smolka, M.B.; Vogel, J.P.; Mao, Y. Deubiquitination of phosphoribosyl-ubiquitin conjugates by phosphodiesterase-domain-containing Legionella effectors. *Proc. Natl. Acad. Sci. USA* **2019**, *116*, 23518–23526. [CrossRef]
48. Shin, D.; Mukherjee, R.; Liu, Y.; Gonzalez, A.; Bonn, F.; Liu, Y.; Rogov, V.V.; Heinz, M.; Stolz, A.; Hummer, G.; et al. Regulation of Phosphoribosyl-Linked Serine Ubiquitination by Deubiquitinases DupA and DupB. *Mol. Cell* **2020**, *77*, 164–179.e166. [CrossRef]
49. Rowbotham, T.J. Preliminary report on the pathogenicity of Legionella pneumophila for freshwater and soil amoebae. *J. Clin. Pathol.* **1980**, *33*, 1179–1183. [CrossRef]
50. Swanson, M.S.; Isberg, R.R. Association of Legionella pneumophila with the macrophage endoplasmic reticulum. *Infect. Immun.* **1995**, *63*, 3609–3620. [CrossRef]
51. Schwarz, D.S.; Blower, M.D. The endoplasmic reticulum: Structure, function and response to cellular signaling. *Cell. Mol. Life Sci. CMLS* **2016**, *73*, 79–94. [CrossRef] [PubMed]
52. Nixon-Abell, J.; Obara, C.J.; Weigel, A.V.; Li, D.; Legant, W.R.; Xu, C.S.; Pasolli, H.A.; Harvey, K.; Hess, H.F.; Betzig, E.; et al. Increased spatiotemporal resolution reveals highly dynamic dense tubular matrices in the peripheral ER. *Science* **2016**, *354*, aaf3928. [CrossRef] [PubMed]
53. Grumati, P.; Morozzi, G.; Hölper, S.; Mari, M.; Harwardt, M.I.; Yan, R.; Müller, S.; Reggiori, F.; Heilemann, M.; Dikic, I. Full length RTN3 regulates turnover of tubular endoplasmic reticulum via selective autophagy. *eLife* **2017**, *6*, e25555. [CrossRef]
54. Khaminets, A.; Heinrich, T.; Mari, M.; Grumati, P.; Huebner, A.K.; Akutsu, M.; Liebmann, L.; Stolz, A.; Nietzsche, S.; Koch, N.; et al. Regulation of endoplasmic reticulum turnover by selective autophagy. *Nature* **2015**, *522*, 354–358. [CrossRef] [PubMed]
55. Wang, S.; Tukachinsky, H.; Romano, F.B.; Rapoport, T.A. Cooperation of the ER-shaping proteins atlastin, lunapark, and reticulons to generate a tubular membrane network. *eLife* **2016**, *5*, e18605. [CrossRef] [PubMed]

56. Urade, T.; Yamamoto, Y.; Zhang, X.; Ku, Y.; Sakisaka, T. Identification and characterization of TMEM33 as a reticulon-binding protein. *Kobe J. Med. Sci.* **2014**, *60*, E57–E65. [PubMed]
57. Yamamoto, Y.; Yoshida, A.; Miyazaki, N.; Iwasaki, K.; Sakisaka, T. Arl6IP1 has the ability to shape the mammalian ER membrane in a reticulon-like fashion. *Biochem. J.* **2014**, *458*, 69–79. [CrossRef] [PubMed]
58. Reggio, A.; Buonomo, V.; Berkane, R.; Bhaskara, R.M.; Tellechea, M.; Peluso, I.; Polishchuk, E.; Di Lorenzo, G.; Cirillo, C.; Esposito, M.; et al. Role of FAM134 paralogues in endoplasmic reticulum remodeling, ER-phagy, and Collagen quality control. *EMBO Rep.* **2021**, *22*, e52289. [CrossRef]
59. An, H.; Ordureau, A.; Paulo, J.A.; Shoemaker, C.J.; Denic, V.; Harper, J.W. TEX264 Is an Endoplasmic Reticulum-Resident ATG8-Interacting Protein Critical for ER Remodeling during Nutrient Stress. *Mol. Cell* **2019**, *74*, 891–908.e810. [CrossRef]
60. Chino, H.; Hatta, T.; Natsume, T.; Mizushima, N. Intrinsically Disordered Protein TEX264 Mediates ER-phagy. *Mol. Cell* **2019**, *74*, 909–921.e906. [CrossRef]
61. Starr, T.; Sun, Y.; Wilkins, N.; Storrie, B. Rab33b and Rab6 are functionally overlapping regulators of Golgi homeostasis and trafficking. *Traffic* **2010**, *11*, 626–636. [CrossRef] [PubMed]
62. Nottingham, R.M.; Ganley, I.G.; Barr, F.A.; Lambright, D.G.; Pfeffer, S.R. RUTBC1 protein, a Rab9A effector that activates GTP hydrolysis by Rab32 and Rab33B proteins. *J. Biol. Chem.* **2011**, *286*, 33213–33222. [CrossRef] [PubMed]
63. Zhang, Y.; Seemann, J. Rapid degradation of GRASP55 and GRASP65 reveals their immediate impact on the Golgi structure. *J. Cell Biol.* **2021**, *220*, e202007052. [CrossRef] [PubMed]
64. Grond, R.; Veenendaal, T.; Duran, J.M.; Raote, I.; Van Es, J.H.; Corstjens, S.; Delfgou, L.; El Haddouti, B.; Malhotra, V.; Rabouille, C. The function of GORASPs in Golgi apparatus organization in vivo. *J. Cell Biol.* **2020**, *219*, e202004191. [CrossRef] [PubMed]
65. Sohda, M.; Misumi, Y.; Yamamoto, A.; Yano, A.; Nakamura, N.; Ikehara, Y. Identification and Characterization of a Novel Golgi Protein, GCP60, That Interacts with the Integral Membrane Protein Giantin*. *J. Biol. Chem.* **2001**, *276*, 45298–45306. [CrossRef]
66. Jin, C.; Zhang, Y.; Zhu, H.; Ahmed, K.; Fu, C.; Yao, X. Human Yip1A specifies the localization of Yif1 to the Golgi apparatus. *Biochem. Biophys. Res. Commun.* **2005**, *334*, 16–22. [CrossRef]
67. Van Weering, J.R.; Sessions, R.B.; Traer, C.J.; Kloer, D.P.; Bhatia, V.K.; Stamou, D.; Carlsson, S.R.; Hurley, J.H.; Cullen, P.J. Molecular basis for SNX-BAR-mediated assembly of distinct endosomal sorting tubules. *EMBO J.* **2012**, *31*, 4466–4480. [CrossRef]
68. Kawabata, M.; Matsuo, H.; Koito, T.; Murata, M.; Kubori, T.; Nagai, H.; Tagaya, M.; Arasaki, K. Legionella hijacks the host Golgi-to-ER retrograde pathway for the association of Legionella-containing vacuole with the ER. *PLoS Pathog.* **2021**, *17*, e1009437. [CrossRef]
69. Kulkarni-Gosavi, P.; Makhoul, C.; Gleeson, P.A. Form and function of the Golgi apparatus: Scaffolds, cytoskeleton and signalling. *FEBS Lett.* **2019**, *593*, 2289–2305. [CrossRef]
70. Li, J.; Ahat, E.; Wang, Y. Golgi Structure and Function in Health, Stress, and Diseases. *Results Probl. Cell Differ.* **2019**, *67*, 441–485.
71. Rabouille, C.; Linstedt, A.D. GRASP: A Multitasking Tether. *Front. Cell Dev.* **2016**, *4*, 1. [CrossRef]
72. Shorter, J.; Watson, R.; Giannakou, M.E.; Clarke, M.; Warren, G.; Barr, F.A. GRASP55, a second mammalian GRASP protein involved in the stacking of Golgi cisternae in a cell-free system. *EMBO J.* **1999**, *18*, 4949–4960. [CrossRef] [PubMed]
73. Feinstein, T.N.; Linstedt, A.D. GRASP55 regulates Golgi ribbon formation. *Mol. Biol. Cell* **2008**, *19*, 2696–2707. [CrossRef]
74. Liu, Y.; Mukherjee, R.; Bonn, F.; Colby, T.; Matic, I.; Glogger, M.; Heilemann, M.; Dikic, I. Serine-ubiquitination regulates Golgi morphology and the secretory pathway upon Legionella infection. *Cell Death Differ.* **2021**, *28*, 2957–2969. [CrossRef] [PubMed]
75. Zhang, M.; Abrams, C.; Wang, L.; Gizzi, A.; He, L.; Lin, R.; Chen, Y.; Loll Patrick, J.; Pascal John, M.; Zhang, J.-F. Structural Basis for Calmodulin as a Dynamic Calcium Sensor. *Structure* **2012**, *20*, 911–923. [CrossRef] [PubMed]

pathogens

MDPI

Review

IFN-Induced PARPs—Sensors of Foreign Nucleic Acids?

Katharina Biaesch †, Sarah Knapp † and Patricia Korn *

Institute of Biochemistry and Molecular Biology, Medical Faculty, RWTH Aachen University, Pauwelsstraße 30, 52074 Aachen, Germany

* Correspondence: pkorn@ukaachen.de

† These authors contributed equally to this work.

Abstract: Cells have developed different strategies to cope with viral infections. Key to initiating a defense response against viruses is the ability to distinguish foreign molecules from their own. One central mechanism is the perception of foreign nucleic acids by host proteins which, in turn, initiate an efficient immune response. Nucleic acid sensing pattern recognition receptors have evolved, each targeting specific features to discriminate viral from host RNA. These are complemented by several RNA-binding proteins that assist in sensing of foreign RNAs. There is increasing evidence that the interferon-inducible ADP-ribosyltransferases (ARTs; PARP9—PARP15) contribute to immune defense and attenuation of viruses. However, their activation, subsequent targets, and precise mechanisms of interference with viruses and their propagation are still largely unknown. Best known for its antiviral activities and its role as RNA sensor is PARP13. In addition, PARP9 has been recently described as sensor for viral RNA. Here we will discuss recent findings suggesting that some PARPs function in antiviral innate immunity. We expand on these findings and integrate this information into a concept that outlines how the different PARPs might function as sensors of foreign RNA. We speculate about possible consequences of RNA binding with regard to the catalytic activities of PARPs, substrate specificity and signaling, which together result in antiviral activities.

Keywords: ADP-ribosylation; MARylation; hydrolase; interferon; macrodomain; PARP; RNA-virus

Citation: Biaesch, K.; Knapp, S.; Korn, P. IFN-Induced PARPs—Sensors of Foreign Nucleic Acids? *Pathogens* **2023**, *12*, 457. https://doi.org/10.3390/pathogens12030457

Academic Editors: Anthony K L Leung, Anthony Fehr and Rachy Abraham

Received: 5 February 2023
Revised: 10 March 2023
Accepted: 12 March 2023
Published: 14 March 2023

1. Introduction

In order to establish an innate immune response to invading viruses, cells need to be able to distinguish self from foreign. This is enabled in part by a repertoire of proteins that specifically sense foreign nucleic acids. These proteins belong to the so-called pattern recognition receptors (PRRs) that recognize and bind pathogen-associated molecular pattern (PAMPs), including different pathogen-associated nucleic acids [1–3]. In general, upon PAMP binding these PRRs are activated to trigger downstream signaling events via activation of transcription factors, such as interferon regulatory factors 3 and 7 (IRF3, IRF7) and nuclear factor kappa B (NF-κB). This results in the activation of a gene expression program, which includes the induction of interferon (IFN) as well as other cytokine genes. IFNs act in a paracrine and autocrine manner to induce the expression of interferon-stimulated genes (ISGs) by which an antipathogenic state is promoted [1,3].

The nucleic acid-sensing PRRs can be subdivided into two groups, the compartmentalized, endosomal and the cytosolic nucleic acid sensors. A subset of Toll-like receptors (TLRs) belongs to the first subgroup, whereas the second group includes retinoic acid-inducible gene I (RIG-I)-like receptors (RLRs), Protein kinase R (PKR), 2′–5′ oligoadenylate synthetase proteins (OAS1-3), nucleotide-binding oligomerization domain (NOD)-like receptors (NLRs), absent in melanoma 2 (AIM2)-like receptors (ALRs) and cyclic GMP-AMP synthase (cGAS) [2,4–10].

In addition to these classical PRRs, a growing list of nucleic acid sensor proteins or accessory proteins have been described. These include helicases, ubiquitin ligases and ADP-ribosyltransferases, that can sense certain nucleic acids, assist in, or mediate the recognition of foreign nucleic acids and accelerate downstream signaling thereby contributing to and modulating an antiviral immune response [11–15].

Best known for its viral ribonucleic acid (RNA)-binding activities is PARP13 [11]. In addition, PARP9 has recently been identified as sensor of foreign RNA [15]. For PARP13, RNA binding is facilitated by zinc finger domains, whereas for PARP9 the macrodomain has been identified as viral RNA-binding module. PARP9 and PARP13 belong to the adenosine diphosphate (ADP)-ribosyltransferase diphtheria toxin-like (ARTD) family, of which a small subset of proteins has been linked to innate immunity due to their responsiveness to IFNs (for further reading on *PARPs* as ISGs we refer to a recent excellent review [16]). These proteins share a conserved ADP-ribosyltransferase (ART) domain, which, with exception of PARP13, possesses mono-ADP-ribosylation (MARylation) activity. All these PARPs are characterized with a range of additional protein domains, among them macrodomains, RNA-recognition domains or zinc fingers. Although the functions of these associated domains are largely unknown, many of these have been associated with RNA-binding. Thus, they provide these proteins with the potential ability to function as RNA sensors similar to what has been proposed for PARP9 and PARP13 [11,15]. Together, we hypothesize that IFN-inducible PARPs function as RNA sensing PRRs and expand the RNA-binding modalities of the known classical PRRs with regard to sequence and/or structure specificities. Moreover, also RNA binding might regulate their modes of action and functionality.

2. The Classical Pathogen Recognition Receptors

2.1. Compartmentalized PRRs

Toll-like receptors involved in sensing pathogenic nucleic acids are TLR3 and TLR7-9 [4,17–19]. These TLRs are integrated into the membranes of endosomes with their N-terminal nucleic acid-binding ectodomain facing the inside of these vesicles [4,17,18]. Nucleic acid binding provokes dimerization of two TLRs, upon which diverse signaling processes are initiated [4]. Due to their localization, these TLRs are capable to respond to endocytosed or phagocytosed pathogens that may be disassembled in this compartment through the action of endosomal proteases and nucleases. As a result, pathogen-derived RNA or deoxyribonucleic acid (DNA) are processed and exposed, providing PAMPs that can interact with the endosomal TLRs [18]. This initiates a first wave of antiviral signaling [4,18,19].

To cover the recognition of a broad range of different pathogens, these TLRs have evolved different preferences for nucleic acids [4,18,19]. TLR3 recognizes and binds double-stranded RNA species based on electrostatic interactions between positively charged amino acids as part of the leucine-rich repeats in the ectodomain and the negatively charged ribose-phosphate backbone of the RNA. Binding occurs independently of specific RNA sequences [19]. Recently, its activation by cellular R-loop derived RNA-DNA hybrids has been demonstrated, which provokes subsequent immune signaling resulting in activation of IRF3 has been demonstrated [20]. However, how R-loop processing is regulated and how these hybrids, originally generated in the nucleus, reach the cytosol or even are capable of activating this endosomal receptor remains unclear. Of note is, that R-Loop forming sequence have also been identified among viruses, but whether these indeed form R-Loop structures and are able to trigger TLR3 activation needs to be investigated [21].

TLR7 and TLR8, which are closely related, sense single-stranded RNA and RNA breakdown products. Both, TLR7 and TLR8 harbor two RNA binding motifs, of which the first recognizes a single guanosine or uridine, respectively, whereas the second has been demonstrated to mediate some sequence specificity. TLR7 preferentially binds polyU 3-mers, while TLR8 senses UG/UUG oligoribonucleotides [22,23]. In contrast, TLR9 has been shown to bind to single-stranded CpG motif-containing DNA [4,18].

2.2. Cytosolic PRRs

Key sensors of viral nucleic acids in the cytosol, present upon virus infection, are the RLRs [2,7,24]. The eponymous member of these cytosolic receptors is RIG-I. Additional members include melanoma differentiation association gene 5 (MDA5) and laboratory of genetics and physiology 2 (LGP2). All three proteins share a similar domain organization with a central RNA-helicase domain that in concert with their C-terminal domain (CTD) mediates RNA binding [2,7,24]. In contrast to LGP2, RIG-I and MDA5 share two caspase-activation and recruitment domains (CARDs) at their N-terminus that triggers downstream signaling events [2,7]. In case of RIG-I these CARDs are intramolecularly bound by the helicase domain and CTD, provoking a closed conformation of the protein and thereby preventing downstream signaling in absent of a ligand [7,25]. Nucleic acid recognition entails the hydrolysis of ATP by RIG-I and provokes the change to an open conformation and its oligomerization. This allows a closer interaction of the RNA-binding part with nucleic acids while the CARDs are released to interact with mitochondrial interactor of virus signaling (MAVS) for downstream signal transduction [7,24]. This autoinhibitory state shown for RIG-I does not occur for MDA5. Instead, MDA5 rather shows an open and flexible and thus uninhibited conformation. This entails downstream signaling upon overexpression of MDA5 in the absence of an RNA ligand [26–28]. Due to the lack of CARDs LGP2 cannot directly initiate downstream signaling via MAVS. But it seems to function as modulator of MDA5 signaling. At low protein levels, LGP2 accelerates and stabilizes MDA5-RNA interaction, whereas high levels of LGP2 lead to MDA5 inhibition [27,29,30].

For all three family members, the recognition of nucleic acids is facilitated by the central helicase domain and the CTD [2,7,24]. These protein domains facilitate the scanning of biochemical features located at the 5′ end of RNA molecules. Despite sharing comparable helicase domains and CTDs, RIG-I and MDA5 sense slightly different features within RNAs [31]. RIG-I prefers shorter double-stranded (ds)RNAs or ssRNAs and is activated by 5′-PPP-dsRNA or 5′-pp-dsRNA, whereas 5′ monophosphorylated RNA stays undetected by RIG-I [32]. Further, RNAs enriched in poly-U/UC or AU regions as well as circular viral RNAs are recognized by RIG-I [33–35]. Binding to circular RNAs is proposed to be mediated by RNA structural features or through accessory RNA-binding proteins, which need to be identified [33]. MDA5 preferentially binds to long dsRNAs and AU-rich regions [28,36,37]. LGP2 has been shown to detect a wide range of diverse RNAs. Neither the phosphorylation status of the 5′-end nor the length of the RNA seem to influence recognition and binding by LGP2 [38,39].

RNA sensing by PKR or OAS family proteins 1–3 is also known to contribute to an antiviral defense response [9,10]. PKR recognizes dsRNA molecules longer than 30 bp in a cap-independent fashion [40], but also ssRNA and structured 5′-PPP-RNA binding has been described [41,42]. Binding is facilitated by two tandem RNA-binding domains located in its N-terminal half, which upon RNA binding initiate dimerization of PKR and subsequent kinase activation [43]. OAS1-3 bind to dsRNA [10,44–46]. Upon dsRNA binding OAS1-3 synthesizes 2′–5′ phosphodiester-linked oligoadenylates, which serve as second messenger to trigger dimerization and in turn activation of Ribonuclease (RNase) L and thus cleavage of RNA [10,47]. Cleaved RNA fragments serve to amplify antiviral signaling as they are sensed by PRRs [9].

An additional line of immune defense is displayed by NLRs and ALRs [1,6,48]. Upon activation, some NLRs and ALRs have been shown to initiate the assembly of so called inflammasomes, multiprotein enzymatic complexes in which they oligomerize and bind to apoptosis-associated speck-like protein containing CARD (ASC) domains to mediate the proteolytic activation of caspase-1. This in turn enables the maturation of cytokines such as Interleukin 1β (IL-1β) and IL-18, thereby contributing to an antiviral immune response.

Among the NLRs, NLRP3 has been shown to be activated by a broad range of diverse RNAs [8,49]. However, direct interaction with RNAs has not been demonstrated. Instead, NLRP3 assembles with accessory proteins, among them DExD/H-box RNA helicases or TRIM ubiquitin ligases, which have been shown to enable RNA-sensing and subsequently the activation of the inflammasome [8,49]. In contrast to NLRP3, AIM2 as representative of the ALRs is activated by DNA [6,48,50].

In addition to AIM2, cGAS functions as cytosolic sensor of DNA [5]. Full activation of cGAS occurs upon binding to longer DNA molecules. These allow for dimerization of cGAS, a prerequisite for full activation. However, cGAS has been shown to recognize a variety of DNA molecules, among them dsDNA, ssDNA providing secondary structures that result in dsDNA, or RNA-DNA hybrids (as e.g., derived from R-loops). Upon binding, signaling is propagated through cGAMP-mediated activation of stimulator of interferon genes (STING), resulting in the activation of IRF3 [5]. Thus, cGAS can be activated by pathogenic DNA but also by cellular DNA, for example in response to cytosolic DNA as a result of missegregation of chromosomes, micronuclei and DNA shattering [51].

Besides these classical PRRs, several additional host factors have been identified serving as sensors for foreign nucleic acids, among them DExD/H box helicases, trispartite motif family (TRIM) ubiquitin ligases and a growing number of various additional RNA-binding proteins [12–14,52,53]. Also, the very heterogeneous family of scavenger receptors has been implicated in innate immunity and some members have been shown to bind to foreign nucleic acids [54]. For some of these RNA-binding proteins scaffolding function has been implicated [3,5,12]. They sense and bind foreign RNA, and present it to RLRs, thereby contributing to and amplifying antiviral signaling [3,5,12].

3. PARP13—A Sensor of Viral RNA

One of these scaffolding proteins referred to above, which is involved in the innate immune response, is the zinc finger antiviral protein (ZAP), also known as PARP13 (Figure 1). Even though it does not possess catalytic activity, it is known for its efficient antiviral activity [11]. PARP13 exists in four different isoforms, arising from alternative splicing and polyadenylation [11,16]. The two best studied isoforms are PARP13.1 (ZAPL) and PARP13.2 (ZAPS), the latter lacks the PARP-like domain [11]. While PARP13.1 seems to be constitutively expressed, PARP13.2 is induced upon interferon signaling [55]. An interaction with the 3′ untranslated region (3′UTR) of the interferon messenger RNA (mRNA) has been described for PARP13.2, which is therefore considered to be involved in a negative feedback response to IFN signaling [55]. Interestingly, PARP13.2 was found to colocalize with RIG-I when stimulated with 5′-PPP-double stranded RNA (dsRNA) and appears to play a role in promoting interferon production [56].

All isoforms of PARP13 have an RNA-binding domain (RBD) consisting of four CCCH zinc finger (ZnF) motifs, the second of which is known for its hydrophobic binding pocket with a high affinity for CpG-dinucleotides [11,57]. The other ZnFs display weak affinity for RNA of unknown sequence [11]. PARP13 is able to dimerize, and even multimerization of PARP13 on target RNA has been suggested for efficient defense against pathogens [11].

Recently, a severe acute respiratory syndrome coronavirus type 2 (SARS-CoV-2) RNA interactome screen identified PARP13, as well as its cofactor TRIM25, to bind directly to the viral RNA [58]. Following ectopic expression of PARP13.1 and PARP13.2, PARP13.2 but not PARP13.1 appeared to have an antiviral effect, as evidenced by a significant reduction in SARS-CoV-2 non-structural protein 12 (nsP12) RNA levels, encoding the viral RNA-dependent RNA polymerase [58]. In contrast, Nchioua and colleagues reported a reduction in the accumulation of SARS-CoV-2 full length RNA only in PARP13.1 overexpression experiments [59]. However, for both isoforms a reduction in the RNA levels of SARS-CoV-2 structural spike- and nucleocapsid protein was observed [59]. Differences in cellular localization might account for these findings, as PARP13.2 has a diffuse cytoplasmic distribution, while PARP13.1 can be S-farnesylated, which localizes it to endolysosomes or the endoplasmic reticulum (ER) [11]. SARS-CoV-2 forms ER-derived double-membrane vesicles for replication [60]. Indeed, it was later demonstrated that S-farnesylation of PARP13.1 is needed for SARS-CoV-2 attenuation [61]. Antiviral activity has also been described against influenza A virus (IAV). While PARP13.1 seems to modulate viral protein expression, PARP13.2 has been described to directly target IAV RNA [11,62]. Liu and colleagues reported PARP13.1 to promote the poly-ADP-ribosylation (PARylation) of IAV polymerase proteins, which leads to their subsequent ubiquitination and degradation [62].

However, as PARP13 has no reported catalytic activity, another ADP-ribosylating protein needs to be involved in this process. The shorter isoform, PARP13.2, is able to bind to IAV basic RNA polymerase 2 (PB2) mRNA and leads to its degradation as well as preventing its translation [63]. This process is counteracted by the non-structural protein 1 (NS1) protein of IAV, which was found to prevent viral RNA binding by PARP13.2 [63]. Interestingly, also the NS1 mRNA seems to be unaffected by PARP13.2 [63]. Potentially, this could be attributed to secondary structures within the NS1 RNA, which has been demonstrated to form hairpins resulting in large parts of this RNA being double stranded [64]. Another genus of viruses restricted by PARP13 are alphaviruses like Sindbis virus, which is targeted by PARP13.1 in stress granules (SGs) [55].

Recently, different groups found in crystallization experiments that the WWE2 pocket of PARP13 is able to bind to an ADP-ribose (ADPr)-moiety of poly-ADP-ribose (PAR) chains [65,66]. Xue and colleagues also confirmed these results in vitro and revealed an essential role of two amino acids in the WWE2 domain, W611 and Q668, for this binding. Further, they demonstrated that the ZnF5, WWE1 and WWE2 of PARP13 combine to form a domain they termed central domain (CD), and that this CD binds to PAR in cells. The long isoform of PARP13, PARP13.1, was also shown to bind PAR in cells, although not as efficiently as the isolated CD [66]. This binding plays an important role in stress granules (SGs), where PAR binding is a prerequisite for PARP13-CD and PARP13.1 re-localization [66]. In addition, mutational impairment of PARP13.1 binding to PAR was found to attenuate its antiviral activity [66]. Localization to stress granules has also been described for PARP13.2, which is increasingly PARylated upon stress [67]. Thus, stress granules (SGs) allow the accumulation of RNA, PAR and PARP13 [66,68]. Whether clustering contributes to the antiviral activity of PARP13, namely promoting RNA degradation or inhibiting translation will need to be addressed. Worth mentioning is, that similar to PARP13 additional PARP proteins have been shown to associate to SGs, suggesting a concerted action or a synergistic role of PARPs in SG formation and/or function. Pointing to a similar direction is the finding, that PARP13, although lacking catalytic activity itself, is ADP-ribosylated and therefore must closely interact with other PARP enzymes [67]. This ADP-ribosylation may control PARP13 function as e.g., shown for PARP7, which MARylates cysteine residues in ZnFs, thereby interfering with the ability of PARP13 to interact with RNA [16]. We expect may additional interaction between PARP proteins as well as other PRRs and downstream effectors. Thus, how PARPs synergize for efficient recognition of nucleic acids and defense against pathogens are exciting questions in the field of innate immune defense.

4. The IFN-Regulated Subclass of PARPs

4.1. The PARP Family

Based on domain organization and structural analysis PARP13 is assigned to the family of ADP-ribosyltransferases diphtheria toxin-like (ARTDs), which in total encompasses 17 members [69–71]. They all share a highly conserved ART domain, which with exception of PARP13 enables these proteins to catalyze ADP-ribosylation. ADP-ribosylation is a reversible posttranslational modification (PTM), which is characterized by the addition of one or several ADP-ribose moieties onto a substrate [70]. Based in part on the amino acid composition of the catalytical triade individual enzymes can either catalyze PARylation (PARP1, PARP2, TNKS1 and TNKS2) or MARylation (PARP3, PARP4, PARP7-PARP12, PARP14-PARP16) [70,72]. To do so, they consume nicotinamide adenine dinucleotide (NAD^+) as a cofactor and transfer ADP-ribose, either a single moiety (MARylation) or in an iterative process (PARylation) multiple units with release of nicotinamide [70]. PARP13 is the only family member lacking ADP-ribosylation activity due to its inability to properly bind NAD^+ [73].

In the following we will concentrate on the interferon-responsive PARPs (PARP9-15; Figure 1) [16], MARylation and the (potential) nucleic acid sensing capabilities of this subset of PARPs.

4.2. Regulation and Propagation of MARylation

Like other PTMs MARylation needs to be read and the signal propagated. The macrodomains 2 and 3 of PARP14 have been identified as readers of MARylation [70,74,75]. Further, MARylation displays a fully reversible PTM enabled by the hydrolytic activity some macrodomains possess [70]. Cellular erasers of MARylation include MacroD1, MacroD2 and TARG1. De-MARylation is enabled by their active macrodomain [70]. The macrodomain fold is highly conserved among all species of life and is embedded in non-structural proteins of several positive sense single-stranded ((+)ss) RNA viruses as well [16,76,77].

The induction of MARylating PARPs by the innate IFN system in combination with the ability of several viral macrodomains to revert MARylation indicates an antiviral role of IFN-inducible PARPs. Further, it has been shown that PARPs have evolved under strong positive selection, additionally pointing to a function in innate immunity [78,79].

However, insights into mechanisms and the exact function of IFN-inducible PARPs remain elusive. One possibility how IFN-inducible PARPs might contribute to an antiviral response is by recognition of foreign nucleic acids. As outlined before, adaptor proteins like DExD/H box helicases or PARP13 can serve as scaffolds bringing nucleic acids and effector proteins in close proximity. Similarly, the IFN-responsive PARPs could function as scaffolds thereby assisting in RNA-recognition by one of the classical PRRs. On top of that, their MARylation activity could add another level of regulation to fine tune the innate immune response. There are indications that the presence of viral RNA might trigger MARylation activity of these enzymes [80,81]. Postulating that RNA binding determines catalytic activity, it might also allow for redirecting catalytic activity to distinct substrates. These could be both viral and host factors. Moreover, the altered specificity might also affect protein stability, for example by reducing automodification, thereby conferring stability of certain PARP enzymes [82]. Additionally, viral RNA might represent a substrate for MARylation, as RNA has been identified to be MARylated both in vitro and in cells [83,84].

4.3. Domain Organization of IFN-Regulated PARPs

Of note is, that the IFN-responsive PARPs all display domain and motifs potentially implicated in nucleic acid binding (Figure 1).

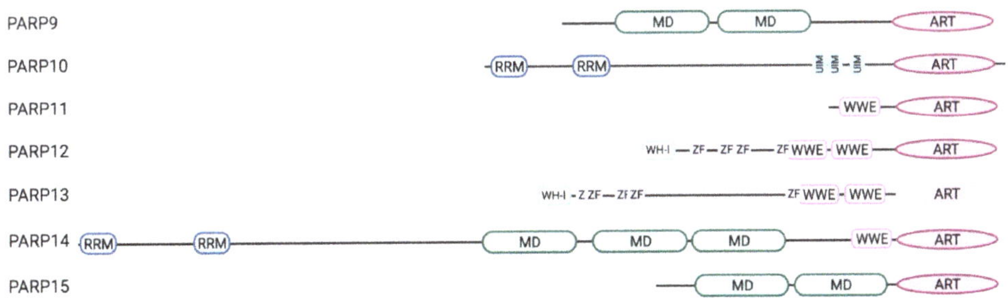

Figure 1. Domain architecture of the IFN-responsive PARPs. All IFN-responsive PARP family members contain the conserved ADP-ribosyltransferase (ART) domain at their C-terminus. Except for PARP13 the ART domain of the other PARPs possesses MARylation activity [72,73]. PARP9, PARP14 and PARP15 contain macrodomain (MD) repeats, either 2 as in case of PARP9 and PARP15 or three as seen for PARP14. In addition to the three macrodomains PARP14 is also equipped with two RNA-recognition motifs (RRM) at its N-terminus, known to mediate RNA-binding. Similarly, PARP10 carries two RRMs at its N-terminus. PARP11-PARP14 harbor one (PARP11, PARP14) or two WWE (PARP12, PARP13) modules, known to facilitate poly-ADP-ribose binding. N-terminally PARP12 and PARP13 both contain Winged-Helix-like (WH-l) DNA-binding domains followed by five zinc finger motifs (ZF), known to mediate binding to nucleic acids. PARP10 is unique, as it is the only family member equipped with ubiquitin-interaction motifs (UIMs), of which it carries three in its C-terminal half (Created with BioRender.com).

PARP12 resembles the overall domain structure of PARP13 and similarly is equipped with several ZnFs. These domains are well described as nucleic acid-binding modules, among other functions, and as such broadly involved in host-pathogen interactions [85]. This provokes questions as to which function(s) can be assigned to the ZnFs of PARP12 and whether these are implicated in RNA sensing.

There is accumulating evidence that macrodomains represent an additional nucleic acid-binding module. Recently, PARP9 has been shown to bind to viral RNA mediated by its first macrodomain [15]. The capability of binding to RNA has been demonstrated for TARG1 as well [86]. The macrodomain as binding module for nucleic acids has also been established from findings with some viral macrodomains (vMDs). The vMD of Chikungunya virus (CHIKV) or Venezuelan encephalitis virus (VEEV) have been shown to bind ssRNA [87], whereas the second and third vMD (SARS unique domains, SUD) of SARS-Coronavirus have been demonstrated to bind G-quadruplexes [88,89]. Besides PARP9, PARP14 and PARP15 belong to the macrodomain-containing IFN-stimulated PARPs. Whereas PARP14 macro2 and macro3 as well as PARP15 macro2 have been identified to bind to MAR [75,90], the function of the first macrodomain within both proteins remains elusive. However, based on sequence comparisons they are phylogenetically closer related to the hydrolytic macrodomains encoded by ssRNA viruses, maybe allowing to postulate an ability in RNA binding as well (Figure 2).

In addition to its macrodomains, PARP14 displays two RNA-recognition motifs (RRMs) near its N-terminus, which are separated by an intrinsically disordered region (IDR, according to the amino acid sequence analysis using PONDR) from its other functional domains. This is also the case for PARP10 (analysis by PONDR) (Figure 1). RRMs but also IDRs individually or cooperatively can mediate RNA binding [91–93]. Generally, multiple RRMs work in tandem thereby facilitating proper RNA binding and conferring RNA specificity [94]. It will be of interest to evaluate nucleic acid binding modes of these subset of PARPs. Do these domains indeed sense foreign nucleic acids to contribute to a robust antiviral response?

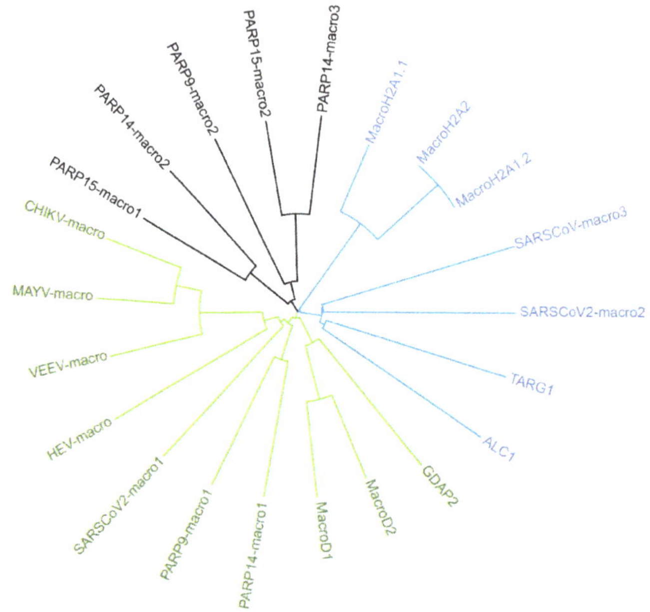

Figure 2. Phylogenetic tree of human and some selected viral macrodomains. Amino acid sequences (>sp | O75367 | 184-370_MacroH2A1.1; >sp | Q9P0M6 | 184-370_MacroH2A1.2; >sp | Q9P0M6 | 184

-370_MacroH2A2; >sp|Q86WJ1|704-897_ALC1; >sp|Q9Y530|2-152_TARG; >sp|Q8IXQ6|107-296_PARP9-macro1; >sp|Q8IXQ6|306-487_PARP9-macro2; >sp|Q460N5|791-978_PARP14-macro1; >sp|Q460N5|1003-1190_PARP14-macro2; >sp|Q460N5|1216-1387_PARP14-macro3; >sp|Q460N3|78-267_PARP15-macro1; >sp|Q460N3|293-464_PARP15-macro2; >sp|Q9BQ69|141-322_MacroD1; >sp|A1Z1Q3|59-240_MacroD2; >sp|Q9NXN4|43-223_GDAP2; >sp|P36328|1330-1489_VEEV-macro; >sp|Q8JUX6|1334-1493_CHIKV-macro; >sp|Q8QZ73|1335-1493_MAYV-macro; >sp|P0DTD1|1025-1194_SARSCoV2-macro1; >sp|P0DTD1|1231-1359_SARSCoV2-macro2; >sp|P0DTD1|1367-1494_SARSCoV-macro3; >sp|Q9WC28|775-921_HEV-macro; >sp|K9N7C7|1110-1276_MERS-macro1; >sp|K9N7C7|1278-1404_MERS-macro2) were analyzed by CLUSTAL 2.1 and the phylogenetic tree file uploaded to iTOL 6.6 to generate this phylogenetic tree [95].

4.4. IFN-Regulated PARPs as Host Restriction Factors

As already stated, PARP12 possesses a similar domain organization as PARP13 but its ART domain displays enzymatic activity [16] (Figure 1). While PARP13 is already known for its role as a PRR in the innate immune response, a similar function might be postulated for PARP12 [11,96]. However, for PARP12 RNA binding has not been confirmed so far experimentally, but there is evidence coming from PARP12 being recruited to SGs [67,97,98]. SGs are condensates enriched in mRNA due to the stress-dependent stalled translation complexes and PAR [67,99]. Localization of PARP12 to these condensates is dependent on its ZnFs and WWE domains, suggesting that the ability to potentially bind both RNA and PAR provokes PARP12 to localize to SGs [97,98]. Of note is, that like RNA binding, PAR-binding by the WWE domain of PARP12 has not been experimentally validated. A functional role of PARP12 in SG biology granules has yet to be found, but as SGs are discussed as first line response to viral infections, the regulation and/or modulation of these condensates might be one mode of antiviral action of PARP12 [100]. It is worth pointing out is that in addition to PARP13 and PARP12, PARP14 and PARP15 have been identified as SG proteins as well, at least when overexpressed [67]. It will be interesting to analyze whether PARP12, analogous to PARP13, regulates RNA turnover and/or translation and whether this is restricted to viral RNAs or might also be relevant for host mRNAs in infected and thus stressed cells. An additional line of evidence for PARP12 as RNA-binding protein is deduced from recent SARS-CoV-2 research. The identification of host factors interacting with the SARS-CoV-2 RNA genome revealed PARP12 and PARP13 as interacting proteins [58,101].

Indeed, PARP12 has been identified as restriction factor for some viruses [81,102,103]. One potential mechanism being discussed is limiting alphavirus replication by modulation of cellular translation [102]. Upon VEEV infection, PARP12 seems to complex with ribosomes and several proteins known to play a role in translation [102]. This might also provide a link to SG biology and/or the modulation of these condensates as they are enriched in stalled translation complexes [100]. In addition, PARP12 limits Zika virus (ZIKV) replication in fact upon interaction with PARP11 via their WWE domains [104,105]. Here, the restrictive effect is mediated by promoting PARylation of the viral non-structural proteins NS1 and NS3 targeting them for proteasomal degradation [104,105]. This resembles the mode of action shown for PARP13.1 with regard to IAV proteins being primed by PAR for proteasomal degradation [62]. Again, presumably other PARP enzymes are also involved in this process, as PARylation is neither catalyzed by PARP12 nor PARP11 [72].

PARP11 has been identified as regulator of IFN signaling. It has been shown to catalyze MARylation of β-TrcP, a ubiquitin E3 ligase. This results in the subsequent ubiquitination and turnover of the IFNα/β receptor 1 (IFNAR1) indicating a feedback control of IFN signaling by PARP11 [106].

PARP9, along with PARP14 and PARP15, is one of the macrodomain-containing PARPs [16] (Figure 1). However, to date it has not been fully elucidated for PARP9, whether or not it has ADP-ribosylating activity [16]. The PARP9 macrodomains have been identified to bind PAR enabling PARP9 colocalization with the PARylating enzyme PARP1 upon DNA damage [107,108]. Furthermore, an antiviral role for PARP9 has been discussed. In dendritic cells, influenza A, a minus-strand RNA virus, induces the expression of PARP9 [15]. Further, Xing and colleagues reported a protective effect of PARP9 against minus-sense RNA virus vesicular stomatitis virus and dsRNA reovirus infection in mice, whereas this effect does not occur with the DNA-virus Herpes simplex virus type 1 (HSV-1) [15]. They found the first macrodomain of PARP9 to be essential for binding of viral dsRNA ranging from 1100 base pairs (bp) to 1400 bp (Table 1). Furthermore, PARP9 contributes substantially to the type-I IFN production by activating the phosphoinositide-3-kinase/protein kinase B (PI3K/AKT) signaling pathway [15].

For many processes, however, PARP9 forms a heterodimer with the E3 ubiquitin ligase deltex E3 ubiquitin ligase L (DTX3L). Together they play a role in DNA damage repair and antiviral defense [15,108]. The DTX3L/PARP9 heterodimer is capable of selectively MARylating ubiquitin [108]. The authors suggest that this modification depends on the catalytic activity of PARP9 [108]. Russo and colleagues found that the DTX3L/PARP9 heterodimer plays a central role in ADP-ribosylation induced upon induction of ISGs. This seems to be independent of PARP9 activity itself, suggesting a potential crosstalk with other MARylating PARPs or a concerted action of these proteins. The increase in overall MARylation is reversed by the hydrolase activity of the SARS-CoV-2 nsP3 macrodomain1 [109,110].

In 2016, Iwata and colleagues found signal transducer and activator of transcription 1 (STAT1) and STAT6 to be ADP-ribosylated in vitro by PARP14, a process suppressed by PARP9. They further claimed STAT1α phosphorylation to be inhibited by PARP14 mediated STAT1α ADP-ribosylation [111]. Additionally, an anti-inflammatory role of PARP14 in macrophages, promoting the interleukin (IL)-4 response and suppressing IFN-γ induced responses, was observed [111]. Although this work has received strong criticism [112], at least the PARP9-PARP14 interaction has been confirmed in co-immunoprecipitation experiments by other groups [113]. Grunewald and colleagues suggest that PARP14 can regulate the IFN response both, dependent on ADP-ribosylation, but also independent of its catalytic activity [114]. Further, they observed increased viral replication of mouse hepatitis virus (MHV) in Parp14 inhibition and knockdown experiments, suggesting antiviral capacities of PARP14 [114]. In viral crosslinking and solid-phase purification (VIR-CLASP) experiments for Chikungunya virus (CHIKV), PARP14 and PARP9 were identified as CHIKV-RNA interactors [115]. A screen for interactors of the IAV-genome in contrast, did not reveal interaction of any of the mono-ARTDs [115].

PARP14 has three macrodomains and macro2 and macro3 have been reported to bind to MARylated PARP10 but seem to lack hydrolase activity and therefore are considered as readers of MARylation [75]. Interestingly, the PARP14 macrodomain1 has been described to resemble, at least at the sequence level, the SARS-CoV-2 macrodomain (Figure 2) [116,117]. PARP14 is the largest of the PARP enzymes and has an RNA recognition motif (RRM) at its N-terminus followed by a long intrinsically disordered region, the function of which are as yet unknown [118].

PARP14 binds to the 3'UTR of tissue factor mRNA in synergy with tristetraprolin (TTP) upon Lipopolysaccharide (LPS) stimulation (Table 1) [119]. However, which domains of PARP14 are involved in this interaction or if PARP14 mediated ADP-ribosylation contributes to this interaction remains to be determined [119]. Nucleic acid binding of PARP14 has also been reported by Riley and colleagues, who found two putative DNA motifs recognized by PARP14 (Table 1). These motifs are present in the promoter region of *interleukin-4 (Il-4)* and *Il-5* and PARP14 seems to have a role in the expression of T helper type 2 (Th2) cytokines [120]. This is further supported by findings of a role of PARP14 in allergic reactions in mice [121].

PARP14 was found to be localized mainly in the cytosol and translocates to the nucleus upon LPS treatment [113]. It also seems to be involved in the translocation of other proteins to the nucleus, especially those that are IFN inducible [113].

PARP10 is highly expressed in hematopoietic cells, supporting a functional role in innate immunity [122]. Like PARP12, PARP10 has been shown to be restrictive for viral replication [81,102,103]. Atasheva and colleagues showed that expression of PARP10 from a second subgenomic promotor within the VEEV genome results in translation inhibition [102]. However, how PARP10 interferes with translation remains open. Similarly, whether this possible modulation of translation confers to its antiviral activity is unclear.

Table 1. Overview on RNA-binding modalities of the classical PRRs and the IFN-regulated PARPs.

Protein	RNA	Reference
TLR3	double-stranded RNA; sequence independent	[4,17–20]
TLR7	single-stranded RNA and RNA breakdown products; preferentially binds polyU 3-mers	[4,17–19,23]
TLR8	single-stranded RNA and RNA breakdown products; recognizes UG/UUG oligoribonucleotides	[4,17–19,22]
RIG-I	5′-PPP-dsRNA or 5′-pp-dsRNA;RNAs enriched in poly-U/UC or AU regions; circular viral RNAs	[2,7,24,31–35]
MDA5	long dsRNAs; AU-rich regions	[2,7,24,28,31,36,37]
LGP2	range of diverse RNAs	[38,39]
PKR	dsRNA > 30 bp; ssRNA; 5′-PPP-RNA	[9,40–42]
OAS1-3	dsRNA	[9,10,44–46]
DExD/H box helicases	Adapter proteins; enables RNA sensing and activating of PRRs	[13,53]
TRIM ubiquitin ligases	Adapter protein; enables RNA sensing and activating of PRRs; preferentially binds to positive strand RNAs	[14]
PARP13	ssRNA (CpG bound by ZnF2); weak binding of RNA (of unknown sequence) by ZnF 1+3+4	[11,57]
PARP9	Macrodomain: viral dsRNA binding ranging from 1100 base pairs (bp) to 1400 bp	[15]
PARP10	*RRMs potentially mediate RNA-binding*	
PARP11	Unknown	
PARP12	*ZnFs potentially mediate interaction with host and viral RNA*	
PARP14	Binds some host mRNAs via 3′UTR;two putative DNA-motifs bound by PARP14 (Motif 1: CACTGAGTGGAG; Motif 2: TCCAAGGATC)*RRMs and macrodomains potentially mediate interaction with host and viral RNA*	[119,120]
PARP15	*Macrodomains potentially facilitate RNA binding*	

Recently, non-structural protein (nsP) 2 of CHIKV has been identified as PARP10 substrate. MARylation impairs proteolytic activity of nsP2, which is essential for replication [81]. CHIKV nsPs are translated as polyprotein in need to be processed into the individual nsPs, which subsequently form the functional replication complex [123]. Thus, the antiviral activity of PARP10 might be mediated at least in part by modification and regulation of viral proteins.

Interestingly, MARylation of CHIKV-nsP2 was only observed when mimicking a viral infection by transfection of an in vitro transcribed RNA replicon. Plasmid-based co-expression of GFP-nsP2 and PARP10 was not sufficient to induce MARylation [81]. Similar results were observed studying ADP-ribosylation in context of an infection with the murine hepatitis virus (MHV), a coronavirus. The nucleocapsid (N) protein of MHV was only found to be ADP-ribosylated upon MHV infection and failed modification when expressed exogenously in cells [80]. These findings foster speculation. Is the presence of viral RNA necessary for full activation of PARP10 as well as other PARPs?

N-terminally PARP10 possesses two RRMs near the N-terminus. This is followed by an intrinsically disordered, glycine-rich domain (Figure 1). Whether these enable nucleic acid binding needs to be investigated to address the question whether PARP10 might function as PRR.

As pointed out above, RNA has been identified as substrate for MARylation [83,84,124,125]. The isolated catalytic domains of PARP10 as well as PARP15 are capable to MARylate the terminal 5′ phosphate of ssRNA in vitro. However, the full-length variants of these proteins failed to do so in vitro [83,84]. The ADP-ribosyltransferase identified to MARylate RNA as full-length protein in vitro and in cells, is TRPT-1. MARylation of 5′-P-RNA has been shown to prevent translation [84].

4.5. Perspective on IFN-Regulated PARPs as Sensors of Viral RNA

What can be drawn from these findings? Quite clearly, PARPs are involved in antiviral defense. There is increasing evidence linking this subset of IFN-responsive PARP enzymes to innate immunity, as summarized in recent reviews [16,109,118]. But as this is quite an emerging and rapidly developing research field, there are ample open questions to be addressed and answered.

Besides induction by IFNs, we hardly understand how the expression of these PARP genes and the function of the encoded proteins are regulated. How is their catalytic activity regulated? Is a precise regulation of MAR activity needed? How is turnover of these proteins achieved? What are the functions of the diverse protein domains these PARP proteins are equipped with? Is there crosstalk between these different domains and, extending on this, do they provide functionality separated from MAR activity? Further, how do the individual enzymes synergize to contribute to the establishment of a robust antiviral response? What are substrate molecules (protein or nucleic acids) to fine tune an immune response to one or the other pathogen? How is specificity achieved? In this last section we like to speculate on possible answers to these questions.

Based on their domain organization (Figure 1), we speculate that this subset of PARPs interacts with foreign but possibly also cellular nucleic acids. Viral RNAs exhibit a lot of secondary structures, which along with sequence and/or modification of the RNA might allow for recognition and binding [126,127]. These complex secondary structures, mostly located in the 5′UTR and 3′UTR of viral RNAs shield them from recognition by many ssRNA sensors [128]. In addition, viruses have evolved different strategies, such as cap-snatching (stealing the cap from host mRNA) or cap-imitation to evade recognition by the classical PRRs [129]. Thus, it is conceivable that PARPs, such as PARP9, come into play (Figure 3). By binding RNA, they might assist in activation of the classical PRRs, as has been shown for PARP13 in concert with RIG-I [11].

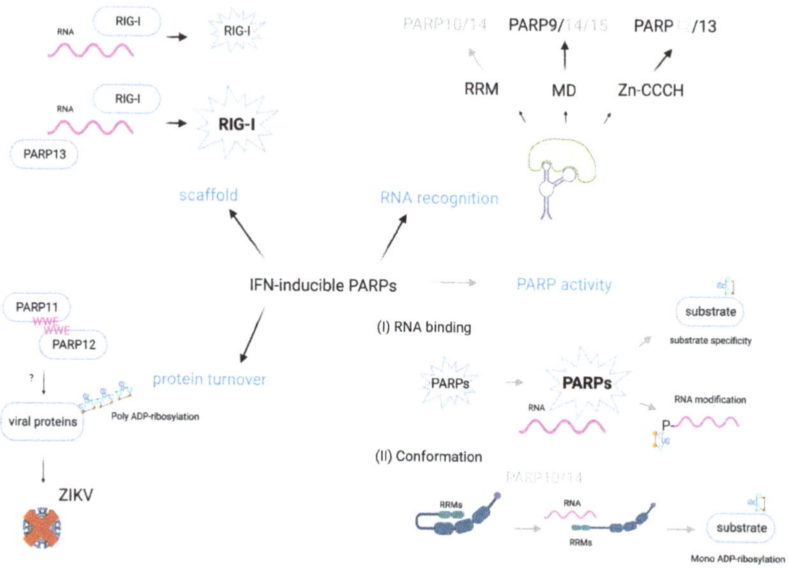

Figure 3. IFN-regulated PARPs as sensors of foreign RNA and possible consequences of this interaction. IFN-inducible PARPs might play a role in foreign nucleic acid recognition through specific protein

domains, such as RNA-recognition motifs (RRMs), macrodomains, or CCCH zinc fingers. Similar to PARP13, these PARPs might act as scaffolds, presenting RNA to classical PRRs thereby accelerating downstream signaling. RNA binding might trigger MARylation activity, maybe by a conformational switch, allowing for specific substrate modification. A concerted action of PARPs and interaction between them might contribute to their antiviral activities, as already shown for PARP11 and PARP12, which contribute to degradation of viral proteins. (Black: known mechanisms; grey: speculative but possible mechanisms; created with BioRender.com).

A direct link to inflammasome activation has not been elucidated yet, but NLRP3, which is activated upon a broad range of RNAs does fully rely on accessory proteins, as it has no intrinsic RNA-binding capability [49]. In such scenarios PARPs could come into play to sense nucleic acids and as a consequence bind to and mediate PRR activation. This might be controlled by MARylation. Indeed, ADP-ribosylation of NLRP3 has already been shown. PARylation by PARP1 contributes to its activation and subsequent inflammasome assembly [130]. Further MARylation of NLRP3 by bacterial toxins has been demonstrated to contribute to inflammasome activation [131]. It will be interesting to test whether IFN-regulated PARPs might bridge RNA-sensing and inflammasome activation and whether this is independent on MARylation.

RNA-binding might trigger the activation of PARP enzymes and contribute to specificity, suggested by findings with CHIKV-nsP2 or the N-protein of MHV (Figure 3) [80,81]. In these studies, modification of the viral proteins could only be observed after infection and thus presence of viral RNA. The concept of nucleic acid dependent enzyme activation has long been known for PARP1, which is fully activated only upon presence of nicked DNA due to the crosstalk between ZnF III and the ART domain [132]. Such domain crosstalk is well imaginable for the IFN-regulated PARPs as well. Another mode of activation, although highly speculative at present, might be comparable to how RIG-I is activated [25]. The RRMs and the long intrinsically disordered glycine-rich region present in PARP10 and PARP14 might contribute to an inactive conformation, which opens when the proteins interact with RNA (Figure 3). Such a more open conformation might then allow catalytic activity and/or recognition of substrates. Thus, it will be interesting to clarify whether such intramolecular interactions occur and how these are regulated.

Further, promiscuity of PARP enzymes has been discussed recently [133]. Promiscuity might be overcome by co-factors. For example, HPF-1 directs PARP1 activity towards modification of serine and DTX3L has been discussed to confer to PARP9 catalytic activity [108,134]. An interesting idea is, that in addition to proteins acting as co-factors, RNA might also convey specificity thereby shifting a potential repertoire of substrates (Figure 3). Thinking this further, RNA binding might also result in specific substrate modification instead of automodification. Moreover, inhibition of catalytic activity of some PARPs was shown to increase their stability, indicating that automodification provokes their proteasomal degradation [82]. Thus, RNA-binding of these PARPs might reduce automodification due to the changes in substrate specificity, thereby promoting stability of IFN-responsive PARPs. This increase in protein might be important to enhance the cellular capacity to recognize pathogen nucleic acids. Moreover, once the infection stress is resolved and foreign RNA is eliminated from the cells, PARPs would switch back to automodification, promoting their degradation. Thus, such a scenario would initially enhance and subsequently participate in the timely turn-off of the innate immune response, thus preventing toxic effects due to the fact of overshooting immunity.

The RNA-binding capacities of PARP-enzymes might interfere with viral translation. Alphaviruses, for example, contain high CpG-content and are therefore recognized and targeted by PARP13 [129]. PARP13 in turn was shown to interact with eukaryotic translation initiation factor 4G (eIF-4G) and eIF-4A [129]. Macrodomain-associated PARPs might interfere with SARS-CoV-2 RNA translation. The SARS-CoV-2 nsP3 localizes to ER-derived double membrane vesicles [60]. The SUD of nsP3, consisting of two viral macrodomains and the domain preceding ubiquitin-like 2 (Ubl2) and papain-like protease 2 ($PL2^{pro}$)

(DPUP), has been shown to interact with ribosomes and polyadenylate-binding protein-interacting protein 1 (PAIP1) [128]. This interaction is thought to be crucial for viral translation. Furthermore, the macrodomains in the SUD are known to be capable of binding G-quadruplexes and, in the case of macro3, probably poly-A [128]. The binding of viral RNA to the nsP3 SUD macrodomains, could also shield them from recognition by human macrodomains. However, this assumption is still rather vague, as it has not yet been shown that viral RNA binds to the CoV-2 SUD MDs, nor that the human MDs would be able to engage with the viral RNA here. Alternatively, the viral macrodomains might bind to host mRNAs and thus hinder their translation together with nsP1 [128]. However, in both cases it is also interesting to note the proximity of the viral macro1 adjacent N-terminal to the SUD. Macro1 has hydrolase activity, suggesting that PARPs or ADP-ribosylation are involved in attenuating viruses by interfering with their translation [135].

Viral RNA itself could be a substrate (Figure 3). In vitro studies failed to show MARylation by full length PARP10 or PARP15 [84]. However, given the artificial nature of the 5′-P-RNA-stretch used in these in vitro experiments, modification of RNA by PARP enzymes cannot be excluded. Again, structural and/or sequence motifs of RNA might be important for binding and for altering activity and/or specificity of MARylation, aspects to be elucidated by future research.

Interaction and collaboration between PARPs could also be mediated by RNA. Several of these PARPs, at least when overexpressed appear to form condensates in cells [136]. RNA plays an important role as scaffold in many condensates. MARylation might in addition to RNA allow for recruiting these PARPs to such condensates. Based on studies with TARG1, RNA binding and APD-ribose binding appear to be not exclusive, suggesting that macrodomains might well be capable of recognizing MAR signals as well as RNA at the same time [86].

After this rather speculative ideas on PARPs as sensors of foreign RNA and potential consequence of this RNA-interaction, there is still one obvious question to be answered. Do the IFN-regulated PARPs need tight regulation? As they are involved in innate immunity, does deregulation or activating mutations link PARPs to autoimmune disorders? Of note is also that PARP enzymes might act as double-edged sword, not only playing an antiviral role but also being exploited by some viruses. One such candidate might be PARP11, as counterbalance for IFN signaling [106].

5. Conclusions

The last years have created several lines of evidence indicating a subset of the MARylating ARTDs plays a role in innate immunity. With proteins and recently also RNA being substrates of MARylation potential mechanisms how they confer to a robust antiviral response are being discussed and evaluated. In addition to the ART domain and modulation by catalytic activity, potential roles of various other domains, with which the IFN-regulated PARPs are equipped, are coming in focus for their possible contributions to antiviral activities. Certainly, we are far away from understanding the necessary details of the functions of these PARP proteins to draw a comprehensive picture of their involvement in innate immunity. Nevertheless, in this review we present possibilities how the additional domains besides the ART domain might contribute to innate immune signaling. Obtaining a more complete understanding of their functions and interplay with viral and host factors, both protein and RNA, will most certainly define novel starting points for pharmacological intervention.

Author Contributions: All authors contributed to the text; P.K. performed most editing work; K.B. and P.K. created the figures, the table was created by S.K. and P.K. All authors have read and agreed to the published version of the manuscript.

Funding: The authors gratefully acknowledge funding from the Volkswagen-Foundation (9A870) to P.K. and the START program of the Medical School of the RWTH Aachen University (117/15) to P.K.

Acknowledgments: We thank Bernhard Lüscher for thoroughly reading the final version of the manuscript.

Conflicts of Interest: The authors declare that they have no competing interests.

References

1. Carty, M.; Guy, C.; Bowie, A.G. Detection of Viral Infections by Innate Immunity. *Biochem. Pharmacol.* **2021**, *183*, 114316. [CrossRef] [PubMed]
2. Chow, K.T.; Gale, M., Jr.; Loo, Y.M. RIG-I and Other RNA Sensors in Antiviral Immunity. *Annu. Rev. Immunol.* **2018**, *36*, 667–694. [CrossRef] [PubMed]
3. Said, E.A.; Tremblay, N.; Al-Balushi, M.S.; Al-Jabri, A.A.; Lamarre, D. Viruses Seen by Our Cells: The Role of Viral RNA Sensors. *J. Immunol. Res.* **2018**, *2018*, 9480497. [CrossRef] [PubMed]
4. Fitzgerald, K.A.; Kagan, J.C. Toll-like Receptors and the Control of Immunity. *Cell* **2020**, *180*, 1044–1066. [CrossRef]
5. Hopfner, K.P.; Hornung, V. Molecular mechanisms and cellular functions of cGAS-STING signalling. *Nat. Rev. Mol. Cell Biol.* **2020**, *21*, 501–521. [CrossRef]
6. Lugrin, J.; Martinon, F. The AIM2 inflammasome: Sensor of pathogens and cellular perturbations. *Immunol. Rev.* **2018**, *281*, 99–114. [CrossRef]
7. Rehwinkel, J.; Gack, M.U. RIG-I-like receptors: Their regulation and roles in RNA sensing. *Nat. Rev. Immunol.* **2020**, *20*, 537–551. [CrossRef]
8. Zhao, C.; Zhao, W. NLRP3 Inflammasome-A Key Player in Antiviral Responses. *Front. Immunol.* **2020**, *11*, 211. [CrossRef]
9. Schlee, M.; Hartmann, G. Discriminating self from non-self in nucleic acid sensing. *Nat. Rev. Immunol.* **2016**, *16*, 566–580. [CrossRef]
10. Schwartz, S.L.; Conn, G.L. RNA regulation of the antiviral protein 2′-5′-oligoadenylate synthetase. *Wiley Interdiscip. Rev. RNA* **2019**, *10*, e1534. [CrossRef]
11. Ficarelli, M.; Neil, S.J.D.; Swanson, C.M. Targeted Restriction of Viral Gene Expression and Replication by the ZAP Antiviral System. *Annu. Rev. Virol.* **2021**, *8*, 265–283. [CrossRef]
12. Oshiumi, H.; Kouwaki, T.; Seya, T. Accessory Factors of Cytoplasmic Viral RNA Sensors Required for Antiviral Innate Immune Response. *Front. Immunol.* **2016**, *7*, 200. [CrossRef]
13. Su, C.; Tang, Y.D.; Zheng, C. DExD/H-box helicases: Multifunctional regulators in antiviral innate immunity. *Cell. Mol. Life Sci.* **2021**, *79*, 2. [CrossRef]
14. Williams, F.P.; Haubrich, K.; Perez-Borrajero, C.; Hennig, J. Emerging RNA-binding roles in the TRIM family of ubiquitin ligases. *Biol. Chem.* **2019**, *400*, 1443–1464. [CrossRef]
15. Xing, J.; Zhang, A.; Du, Y.; Fang, M.; Minze, L.J.; Liu, Y.J.; Li, X.C.; Zhang, Z. Identification of poly(ADP-ribose) polymerase 9 (PARP9) as a noncanonical sensor for RNA virus in dendritic cells. *Nat. Commun.* **2021**, *12*, 2681. [CrossRef]
16. Luscher, B.; Verheirstraeten, M.; Krieg, S.; Korn, P. Intracellular mono-ADP-ribosyltransferases at the host-virus interphase. *Cell. Mol. Life Sci.* **2022**, *79*, 288. [CrossRef]
17. Duan, T.; Du, Y.; Xing, C.; Wang, H.Y.; Wang, R.F. Toll-Like Receptor Signaling and Its Role in Cell-Mediated Immunity. *Front. Immunol.* **2022**, *13*, 812774. [CrossRef]
18. Lind, N.A.; Rael, V.E.; Pestal, K.; Liu, B.; Barton, G.M. Regulation of the nucleic acid-sensing Toll-like receptors. *Nat. Rev. Immunol.* **2022**, *22*, 224–235. [CrossRef]
19. Vierbuchen, T.; Stein, K.; Heine, H. RNA is taking its Toll: Impact of RNA-specific Toll-like receptors on health and disease. *Allergy* **2019**, *74*, 223–235. [CrossRef]
20. Crossley, M.P.; Song, C.; Bocek, M.J.; Choi, J.H.; Kousorous, J.; Sathirachinda, A.; Lin, C.; Brickner, J.R.; Bai, G.; Lans, H.; et al. R-loop-derived cytoplasmic RNA-DNA hybrids activate an immune response. *Nature* **2023**, *613*, 187–194. [CrossRef]
21. Wongsurawat, T.; Gupta, A.; Jenjaroenpun, P.; Owens, S.; Forrest, J.C.; Nookaew, I. R-loop-forming Sequences Analysis in Thousands of Viral Genomes Identify a New Common Element in Herpesviruses. *Sci. Rep.* **2020**, *10*, 6389. [CrossRef] [PubMed]
22. Tanji, H.; Ohto, U.; Shibata, T.; Taoka, M.; Yamauchi, Y.; Isobe, T.; Miyake, K.; Shimizu, T. Toll-like receptor 8 senses degradation products of single-stranded RNA. *Nat. Struct. Mol. Biol.* **2015**, *22*, 109–115. [CrossRef] [PubMed]
23. Zhang, Z.; Ohto, U.; Shibata, T.; Krayukhina, E.; Taoka, M.; Yamauchi, Y.; Tanji, H.; Isobe, T.; Uchiyama, S.; Miyake, K.; et al. Structural Analysis Reveals that Toll-like Receptor 7 Is a Dual Receptor for Guanosine and Single-Stranded RNA. *Immunity* **2016**, *45*, 737–748. [CrossRef] [PubMed]
24. Thoresen, D.; Wang, W.; Galls, D.; Guo, R.; Xu, L.; Pyle, A.M. The molecular mechanism of RIG-I activation and signaling. *Immunol. Rev.* **2021**, *304*, 154–168. [CrossRef] [PubMed]
25. Wang, W.; Pyle, A.M. The RIG-I receptor adopts two different conformations for distinguishing host from viral RNA ligands. *Mol. Cell* **2022**, *82*, 4131–4144.e6. [CrossRef]
26. Berke, I.C.; Li, Y.; Modis, Y. Structural basis of innate immune recognition of viral RNA. *Cell. Microbiol.* **2013**, *15*, 386–394. [CrossRef]
27. Bruns, A.M.; Horvath, C.M. LGP2 synergy with MDA5 in RLR-mediated RNA recognition and antiviral signaling. *Cytokine* **2015**, *74*, 198–206. [CrossRef]
28. Wu, B.; Peisley, A.; Richards, C.; Yao, H.; Zeng, X.; Lin, C.; Chu, F.; Walz, T.; Hur, S. Structural basis for dsRNA recognition, filament formation, and antiviral signal activation by MDA5. *Cell* **2013**, *152*, 276–289. [CrossRef]

29. Satoh, T.; Kato, H.; Kumagai, Y.; Yoneyama, M.; Sato, S.; Matsushita, K.; Tsujimura, T.; Fujita, T.; Akira, S.; Takeuchi, O. LGP2 is a positive regulator of RIG-I- and MDA5-mediated antiviral responses. *Proc. Natl. Acad. Sci. USA* **2010**, *107*, 1512–1517. [CrossRef]
30. Zhu, Z.; Zhang, X.; Wang, G.; Zheng, H. The laboratory of genetics and physiology 2: Emerging insights into the controversial functions of this RIG-I-like receptor. *BioMed Res. Int.* **2014**, *2014*, 960190. [CrossRef]
31. Sanchez David, R.Y.; Combredet, C.; Sismeiro, O.; Dillies, M.A.; Jagla, B.; Coppee, J.Y.; Mura, M.; Guerbois Galla, M.; Despres, P.; Tangy, F.; et al. Comparative analysis of viral RNA signatures on different RIG-I-like receptors. *Elife* **2016**, *5*, e11275. [CrossRef]
32. Ren, X.; Linehan, M.M.; Iwasaki, A.; Pyle, A.M. RIG-I Selectively Discriminates against 5′-Monophosphate RNA. *Cell Rep.* **2019**, *26*, 2019–2027.e2014. [CrossRef]
33. Li, X.; Liu, C.X.; Xue, W.; Zhang, Y.; Jiang, S.; Yin, Q.F.; Wei, J.; Yao, R.W.; Yang, L.; Chen, L.L. Coordinated circRNA Biogenesis and Function with NF90/NF110 in Viral Infection. *Mol. Cell* **2017**, *67*, 214–227.e217. [CrossRef]
34. Saito, T.; Owen, D.M.; Jiang, F.; Marcotrigiano, J.; Gale, M., Jr. Innate immunity induced by composition-dependent RIG-I recognition of hepatitis C virus RNA. *Nature* **2008**, *454*, 523–527. [CrossRef]
35. Schnell, G.; Loo, Y.M.; Marcotrigiano, J.; Gale, M., Jr. Uridine composition of the poly-U/UC tract of HCV RNA defines non-self recognition by RIG-I. *PLoS Pathog.* **2012**, *8*, e1002839. [CrossRef]
36. Peisley, A.; Jo, M.H.; Lin, C.; Wu, B.; Orme-Johnson, M.; Walz, T.; Hohng, S.; Hur, S. Kinetic mechanism for viral dsRNA length discrimination by MDA5 filaments. *Proc. Natl. Acad. Sci. USA* **2012**, *109*, E3340–E3349. [CrossRef]
37. Peisley, A.; Lin, C.; Wu, B.; Orme-Johnson, M.; Liu, M.; Walz, T.; Hur, S. Cooperative assembly and dynamic disassembly of MDA5 filaments for viral dsRNA recognition. *Proc. Natl. Acad. Sci. USA* **2011**, *108*, 21010–21015. [CrossRef]
38. Pippig, D.A.; Hellmuth, J.C.; Cui, S.; Kirchhofer, A.; Lammens, K.; Lammens, A.; Schmidt, A.; Rothenfusser, S.; Hopfner, K.P. The regulatory domain of the RIG-I family ATPase LGP2 senses double-stranded RNA. *Nucleic Acids Res.* **2009**, *37*, 2014–2025. [CrossRef]
39. Uchikawa, E.; Lethier, M.; Malet, H.; Brunel, J.; Gerlier, D.; Cusack, S. Structural Analysis of dsRNA Binding to Anti-viral Pattern Recognition Receptors LGP2 and MDA5. *Mol. Cell* **2016**, *62*, 586–602. [CrossRef]
40. Lemaire, P.A.; Anderson, E.; Lary, J.; Cole, J.L. Mechanism of PKR Activation by dsRNA. *J. Mol. Biol.* **2008**, *381*, 351–360. [CrossRef]
41. Zheng, X.; Bevilacqua, P.C. Activation of the protein kinase PKR by short double-stranded RNAs with single-stranded tails. *RNA* **2004**, *10*, 1934–1945. [CrossRef] [PubMed]
42. Nallagatla, S.R.; Hwang, J.; Toroney, R.; Zheng, X.; Cameron, C.E.; Bevilacqua, P.C. 5′-triphosphate-dependent activation of PKR by RNAs with short stem-loops. *Science* **2007**, *318*, 1455–1458. [CrossRef] [PubMed]
43. Cole, J.L. Activation of PKR: An open and shut case? *Trends Biochem. Sci.* **2007**, *32*, 57–62. [CrossRef] [PubMed]
44. Donovan, J.; Dufner, M.; Korennykh, A. Structural basis for cytosolic double-stranded RNA surveillance by human oligoadenylate synthetase 1. *Proc. Natl. Acad. Sci. USA* **2013**, *110*, 1652–1657. [CrossRef]
45. Ibsen, M.S.; Gad, H.H.; Thavachelvam, K.; Boesen, T.; Despres, P.; Hartmann, R. The 2′-5′-oligoadenylate synthetase 3 enzyme potently synthesizes the 2′-5′-oligoadenylates required for RNase L activation. *J. Virol.* **2014**, *88*, 14222–14231. [CrossRef]
46. Koul, A.; Deo, S.; Booy, E.P.; Orriss, G.L.; Genung, M.; McKenna, S.A. Impact of double-stranded RNA characteristics on the activation of human 2′-5′-oligoadenylate synthetase 2 (OAS2). *Biochem. Cell. Biol.* **2020**, *98*, 70–82. [CrossRef]
47. Huang, H.; Zeqiraj, E.; Dong, B.; Jha, B.K.; Duffy, N.M.; Orlicky, S.; Thevakumaran, N.; Talukdar, M.; Pillon, M.C.; Ceccarelli, D.F.; et al. Dimeric structure of pseudokinase RNase L bound to 2-5A reveals a basis for interferon-induced antiviral activity. *Mol. Cell* **2014**, *53*, 221–234. [CrossRef]
48. Man, S.M.; Karki, R.; Kanneganti, T.D. AIM2 inflammasome in infection, cancer, and autoimmunity: Role in DNA sensing, inflammation, and innate immunity. *Eur. J. Immunol.* **2016**, *46*, 269–280. [CrossRef]
49. Xiao, T.S. The nucleic acid-sensing inflammasomes. *Immunol. Rev.* **2015**, *265*, 103–111. [CrossRef]
50. Kumar, V. The Trinity of cGAS, TLR9, and ALRs Guardians of the Cellular Galaxy Against Host-Derived Self-DNA. *Front. Immunol.* **2020**, *11*, 624597. [CrossRef]
51. Krupina, K.; Goginashvili, A.; Cleveland, D.W. Causes and consequences of micronuclei. *Curr. Opin. Cell Biol.* **2021**, *70*, 91–99. [CrossRef]
52. Bohn, J.A.; DaSilva, J.; Kharytonchyk, S.; Mercedes, M.; Vosters, J.; Telesnitsky, A.; Hatziioannou, T.; Smith, J.L. Flexibility in Nucleic Acid Binding Is Central to APOBEC3H Antiviral Activity. *J. Virol.* **2019**, *93*, e01275-19. [CrossRef]
53. Fullam, A.; Schroder, M. DExD/H-box RNA helicases as mediators of anti-viral innate immunity and essential host factors for viral replication. *Biochim. Biophys. Acta* **2013**, *1829*, 854–865. [CrossRef]
54. Canton, J.; Neculai, D.; Grinstein, S. Scavenger receptors in homeostasis and immunity. *Nat. Rev. Immunol.* **2013**, *13*, 621–634. [CrossRef]
55. Schwerk, J.; Soveg, F.W.; Ryan, A.P.; Thomas, K.R.; Hatfield, L.D.; Ozarkar, S.; Forero, A.; Kell, A.M.; Roby, J.A.; So, L.; et al. RNA-binding protein isoforms ZAP-S and ZAP-L have distinct antiviral and immune resolution functions. *Nat. Immunol.* **2019**, *20*, 1610–1620. [CrossRef]
56. Hayakawa, S.; Shiratori, S.; Yamato, H.; Kameyama, T.; Kitatsuji, C.; Kashigi, F.; Goto, S.; Kameoka, S.; Fujikura, D.; Yamada, T.; et al. ZAPS is a potent stimulator of signaling mediated by the RNA helicase RIG-I during antiviral responses. *Nat. Immunol.* **2011**, *12*, 37–44. [CrossRef]

57. Meagher, J.L.; Takata, M.; Goncalves-Carneiro, D.; Keane, S.C.; Rebendenne, A.; Ong, H.; Orr, V.K.; MacDonald, M.R.; Stuckey, J.A.; Bieniasz, P.D.; et al. Structure of the zinc-finger antiviral protein in complex with RNA reveals a mechanism for selective targeting of CG-rich viral sequences. *Proc. Natl. Acad. Sci. USA* **2019**, *116*, 24303–24309. [CrossRef]
58. Lee, S.; Lee, Y.S.; Choi, Y.; Son, A.; Park, Y.; Lee, K.M.; Kim, J.; Kim, J.S.; Kim, V.N. The SARS-CoV-2 RNA interactome. *Mol. Cell* **2021**, *81*, 2838–2850.e6. [CrossRef]
59. Nchioua, R.; Kmiec, D.; Muller, J.A.; Conzelmann, C.; Gross, R.; Swanson, C.M.; Neil, S.J.D.; Stenger, S.; Sauter, D.; Munch, J.; et al. SARS-CoV-2 Is Restricted by Zinc Finger Antiviral Protein despite Preadaptation to the Low-CpG Environment in Humans. *mBio* **2020**, *11*, e01930-20. [CrossRef]
60. Tabata, K.; Prasad, V.; Paul, D.; Lee, J.Y.; Pham, M.T.; Twu, W.I.; Neufeldt, C.J.; Cortese, M.; Cerikan, B.; Stahl, Y.; et al. Convergent use of phosphatidic acid for hepatitis C virus and SARS-CoV-2 replication organelle formation. *Nat. Commun.* **2021**, *12*, 7276. [CrossRef]
61. Kmiec, D.; Lista, M.J.; Ficarelli, M.; Swanson, C.M.; Neil, S.J.D. S-farnesylation is essential for antiviral activity of the long ZAP isoform against RNA viruses with diverse replication strategies. *PLoS Pathog.* **2021**, *17*, e1009726. [CrossRef] [PubMed]
62. Liu, C.H.; Zhou, L.; Chen, G.; Krug, R.M. Battle between influenza A virus and a newly identified antiviral activity of the PARP-containing ZAPL protein. *Proc. Natl. Acad. Sci. USA* **2015**, *112*, 14048–14053. [CrossRef] [PubMed]
63. Tang, Q.; Wang, X.; Gao, G. The Short Form of the Zinc Finger Antiviral Protein Inhibits Influenza A Virus Protein Expression and Is Antagonized by the Virus-Encoded NS1. *J. Virol.* **2017**, *91*, e01909-16. [CrossRef] [PubMed]
64. Baranovskaya, S.; Shevtsov, S.; Maksimova, S.; Kuzmin, A.; Schwartz, E. The mutations and VNTRs in the phenylalanine hydroxylase gene of phenylketonuria in St Petersburg. *J. Inherit. Metab. Dis.* **1996**, *19*, 705. [CrossRef]
65. Kuttiyatveetil, J.R.A.; Soufari, H.; Dasovich, M.; Uribe, I.R.; Mirhasan, M.; Cheng, S.J.; Leung, A.K.L.; Pascal, J.M. Crystal structures and functional analysis of the ZnF5-WWE1-WWE2 region of PARP13/ZAP define a distinctive mode of engaging poly(ADP-ribose). *Cell Rep.* **2022**, *41*, 111529. [CrossRef]
66. Xue, G.; Braczyk, K.; Goncalves-Carneiro, D.; Dawidziak, D.M.; Sanchez, K.; Ong, H.; Wan, Y.; Zadrozny, K.K.; Ganser-Pornillos, B.K.; Bieniasz, P.D.; et al. Poly(ADP-ribose) potentiates ZAP antiviral activity. *PLoS Pathog.* **2022**, *18*, e1009202. [CrossRef]
67. Leung, A.K.; Vyas, S.; Rood, J.E.; Bhutkar, A.; Sharp, P.A.; Chang, P. Poly(ADP-ribose) regulates stress responses and microRNA activity in the cytoplasm. *Mol. Cell* **2011**, *42*, 489–499. [CrossRef]
68. Mateju, D.; Chao, J.A. Stress granules: Regulators or by-products? *FEBS J.* **2022**, *289*, 363–373. [CrossRef]
69. Hottiger, M.O.; Hassa, P.O.; Luscher, B.; Schuler, H.; Koch-Nolte, F. Toward a unified nomenclature for mammalian ADP-ribosyltransferases. *Trends Biochem. Sci.* **2010**, *35*, 208–219. [CrossRef]
70. Luscher, B.; Butepage, M.; Eckei, L.; Krieg, S.; Verheugd, P.; Shilton, B.H. ADP-Ribosylation, a Multifaceted Posttranslational Modification Involved in the Control of Cell Physiology in Health and Disease. *Chem. Rev.* **2018**, *118*, 1092–1136. [CrossRef]
71. Luscher, B.; Ahel, I.; Altmeyer, M.; Ashworth, A.; Bai, P.; Chang, P.; Cohen, M.; Corda, D.; Dantzer, F.; Daugherty, M.D.; et al. ADP-ribosyltransferases, an update on function and nomenclature. *FEBS J.* **2021**, *289*, 7399–7410. [CrossRef]
72. Vyas, S.; Matic, I.; Uchima, L.; Rood, J.; Zaja, R.; Hay, R.T.; Ahel, I.; Chang, P. Family-wide analysis of poly(ADP-ribose) polymerase activity. *Nat. Commun.* **2014**, *5*, 4426. [CrossRef]
73. Karlberg, T.; Klepsch, M.; Thorsell, A.G.; Andersson, C.D.; Linusson, A.; Schuler, H. Structural basis for lack of ADP-ribosyltransferase activity in poly(ADP-ribose) polymerase-13/zinc finger antiviral protein. *J. Biol. Chem.* **2015**, *290*, 7336–7344. [CrossRef]
74. Butepage, M.; Krieg, S.; Eckei, L.; Li, J.; Rossetti, G.; Verheugd, P.; Luscher, B. Assessment of Intracellular Auto-Modification Levels of ARTD10 Using Mono-ADP-Ribose-Specific Macrodomains 2 and 3 of Murine Artd8. *Methods Mol. Biol.* **2018**, *1813*, 41–63. [CrossRef]
75. Forst, A.H.; Karlberg, T.; Herzog, N.; Thorsell, A.G.; Gross, A.; Feijs, K.L.; Verheugd, P.; Kursula, P.; Nijmeijer, B.; Kremmer, E.; et al. Recognition of mono-ADP-ribosylated ARTD10 substrates by ARTD8 macrodomains. *Structure* **2013**, *21*, 462–475. [CrossRef]
76. Eckei, L.; Krieg, S.; Butepage, M.; Lehmann, A.; Gross, A.; Lippok, B.; Grimm, A.R.; Kummerer, B.M.; Rossetti, G.; Luscher, B.; et al. The conserved macrodomains of the non-structural proteins of Chikungunya virus and other pathogenic positive strand RNA viruses function as mono-ADP-ribosylhydrolases. *Sci. Rep.* **2017**, *7*, 41746. [CrossRef]
77. Rack, J.G.; Perina, D.; Ahel, I. Macrodomains: Structure, Function, Evolution, and Catalytic Activities. *Annu. Rev. Biochem.* **2016**, *85*, 431–454. [CrossRef]
78. Daugherty, M.D.; Young, J.M.; Kerns, J.A.; Malik, H.S. Rapid evolution of PARP genes suggests a broad role for ADP-ribosylation in host-virus conflicts. *PLoS Genet.* **2014**, *10*, e1004403. [CrossRef]
79. Kerns, J.A.; Emerman, M.; Malik, H.S. Positive selection and increased antiviral activity associated with the PARP-containing isoform of human zinc-finger antiviral protein. *PLoS Genet.* **2008**, *4*, e21. [CrossRef]
80. Grunewald, M.E.; Fehr, A.R.; Athmer, J.; Perlman, S. The coronavirus nucleocapsid protein is ADP-ribosylated. *Virology* **2018**, *517*, 62–68. [CrossRef]
81. Krieg, S.; Pott, F.; Potthoff, L.; Verheirstraeten, M.; Butepage, M.; Golzmann, A.; Lippok, B.; Goffinet, C.; Luscher, B.; Korn, P. Mono-ADP-ribosylation by PARP10 inhibits Chikungunya virus nsP2 proteolytic activity and viral replication. *Cell. Mol. Life Sci.* **2023**, *80*, 72. [CrossRef] [PubMed]

82. Sanderson, D.J.; Cohen, M.S. Mechanisms governing PARP expression, localization, and activity in cells. *Crit. Rev. Biochem. Mol. Biol.* **2020**, *55*, 541–554. [CrossRef] [PubMed]
83. Munnur, D.; Bartlett, E.; Mikolcevic, P.; Kirby, I.T.; Rack, J.G.M.; Mikoc, A.; Cohen, M.S.; Ahel, I. Reversible ADP-ribosylation of RNA. *Nucleic Acids Res.* **2019**, *47*, 5658–5669. [CrossRef] [PubMed]
84. Weixler, L.; Feijs, K.L.H.; Zaja, R. ADP-ribosylation of RNA in mammalian cells is mediated by TRPT1 and multiple PARPs. *Nucleic Acids Res.* **2022**, *50*, 9426–9441. [CrossRef] [PubMed]
85. Wang, G.; Zheng, C. Zinc finger proteins in the host-virus interplay: Multifaceted functions based on their nucleic acid-binding property. *FEMS Microbiol. Rev.* **2021**, *45*, fuaa059. [CrossRef]
86. Butepage, M.; Preisinger, C.; von Kriegsheim, A.; Scheufen, A.; Lausberg, E.; Li, J.; Kappes, F.; Feederle, R.; Ernst, S.; Eckei, L.; et al. Nucleolar-nucleoplasmic shuttling of TARG1 and its control by DNA damage-induced poly-ADP-ribosylation and by nucleolar transcription. *Sci. Rep.* **2018**, *8*, 6748. [CrossRef]
87. Malet, H.; Coutard, B.; Jamal, S.; Dutartre, H.; Papageorgiou, N.; Neuvonen, M.; Ahola, T.; Forrester, N.; Gould, E.A.; Lafitte, D.; et al. The crystal structures of Chikungunya and Venezuelan equine encephalitis virus nsP3 macro domains define a conserved adenosine binding pocket. *J. Virol.* **2009**, *83*, 6534–6545. [CrossRef]
88. Tan, J.; Kusov, Y.; Mutschall, D.; Tech, S.; Nagarajan, K.; Hilgenfeld, R.; Schmidt, C.L. The "SARS-unique domain" (SUD) of SARS coronavirus is an oligo(G)-binding protein. *Biochem. Biophys. Res. Commun.* **2007**, *364*, 877–882. [CrossRef]
89. Tan, J.; Vonrhein, C.; Smart, O.S.; Bricogne, G.; Bollati, M.; Kusov, Y.; Hansen, G.; Mesters, J.R.; Schmidt, C.L.; Hilgenfeld, R. The SARS-unique domain (SUD) of SARS coronavirus contains two macrodomains that bind G-quadruplexes. *PLoS Pathog.* **2009**, *5*, e1000428. [CrossRef]
90. Ekblad, T.; Verheugd, P.; Lindgren, A.E.; Nyman, T.; Elofsson, M.; Schuler, H. Identification of Poly(ADP-Ribose) Polymerase Macrodomain Inhibitors Using an AlphaScreen Protocol. *SLAS Discov.* **2018**, *23*, 353–362. [CrossRef]
91. Corley, M.; Burns, M.C.; Yeo, G.W. How RNA-Binding Proteins Interact with RNA: Molecules and Mechanisms. *Mol. Cell* **2020**, *78*, 9–29. [CrossRef]
92. Hentze, M.W.; Castello, A.; Schwarzl, T.; Preiss, T. A brave new world of RNA-binding proteins. *Nat. Rev. Mol. Cell Biol.* **2018**, *19*, 327–341. [CrossRef]
93. Liu, S.; Li, B.; Liang, Q.; Liu, A.; Qu, L.; Yang, J. Classification and function of RNA-protein interactions. *Wiley Interdiscip. Rev. RNA* **2020**, *11*, e1601. [CrossRef]
94. Conte, M.R.; Grune, T.; Ghuman, J.; Kelly, G.; Ladas, A.; Matthews, S.; Curry, S. Structure of tandem RNA recognition motifs from polypyrimidine tract binding protein reveals novel features of the RRM fold. *EMBO J.* **2000**, *19*, 3132–3141. [CrossRef]
95. Letunic, I.; Bork, P. Interactive Tree of Life (iTOL) v5: An online tool for phylogenetic tree display and annotation. *Nucleic Acids Res.* **2021**, *49*, W293–W296. [CrossRef]
96. Yang, E.; Nguyen, L.P.; Wisherop, C.A.; Kan, R.L.; Li, M.M.H. The Role of ZAP and TRIM25 RNA Binding in Restricting Viral Translation. *Front. Cell. Infect. Microbiol.* **2022**, *12*, 886929. [CrossRef]
97. Catara, G.; Grimaldi, G.; Schembri, L.; Spano, D.; Turacchio, G.; Lo Monte, M.; Beccari, A.R.; Valente, C.; Corda, D. PARP1-produced poly-ADP-ribose causes the PARP12 translocation to stress granules and impairment of Golgi complex functions. *Sci. Rep.* **2017**, *7*, 14035. [CrossRef]
98. Welsby, I.; Hutin, D.; Gueydan, C.; Kruys, V.; Rongvaux, A.; Leo, O. PARP12, an interferon-stimulated gene involved in the control of protein translation and inflammation. *J. Biol. Chem.* **2014**, *289*, 26642–26657. [CrossRef]
99. Campos-Melo, D.; Hawley, Z.C.E.; Droppelmann, C.A.; Strong, M.J. The Integral Role of RNA in Stress Granule Formation and Function. *Front. Cell Dev. Biol.* **2021**, *9*, 621779. [CrossRef]
100. McCormick, C.; Khaperskyy, D.A. Translation inhibition and stress granules in the antiviral immune response. *Nat. Rev. Immunol.* **2017**, *17*, 647–660. [CrossRef]
101. Zhang, S.; Huang, W.; Ren, L.; Ju, X.; Gong, M.; Rao, J.; Sun, L.; Li, P.; Ding, Q.; Wang, J.; et al. Comparison of viral RNA-host protein interactomes across pathogenic RNA viruses informs rapid antiviral drug discovery for SARS-CoV-2. *Cell Res.* **2022**, *32*, 9–23. [CrossRef] [PubMed]
102. Atasheva, S.; Frolova, E.I.; Frolov, I. Interferon-stimulated PARPs are potent inhibitors of cellular translation and virus replication. *J. Virol.* **2014**, *88*, 2116–2130. [CrossRef] [PubMed]
103. Atasheva, S.; Akhrymuk, M.; Frolova, E.I.; Frolov, I. New PARP gene with an anti-alphavirus function. *J. Virol.* **2012**, *86*, 8147–8160. [CrossRef] [PubMed]
104. Li, L.; Shi, Y.; Li, S.; Liu, J.; Zu, S.; Xu, X.; Gao, M.; Sun, N.; Pan, C.; Peng, L.; et al. ADP-ribosyltransferase PARP11 suppresses Zika virus in synergy with PARP12. *Cell Biosci.* **2021**, *11*, 116. [CrossRef]
105. Li, L.; Zhao, H.; Liu, P.; Li, C.; Quanquin, N.; Ji, X.; Sun, N.; Du, P.; Qin, C.F.; Lu, N.; et al. PARP12 suppresses Zika virus infection through PARP-dependent degradation of NS1 and NS3 viral proteins. *Sci. Signal.* **2018**, *11*, eaas9332. [CrossRef]
106. Guo, T.; Zuo, Y.; Qian, L.; Liu, J.; Yuan, Y.; Xu, K.; Miao, Y.; Feng, Q.; Chen, X.; Jin, L.; et al. ADP-ribosyltransferase PARP11 modulates the interferon antiviral response by mono-ADP-ribosylating the ubiquitin E3 ligase beta-TrCP. *Nat. Microbiol.* **2019**, *4*, 1872–1884. [CrossRef]
107. Yan, Q.; Xu, R.; Zhu, L.; Cheng, X.; Wang, Z.; Manis, J.; Shipp, M.A. BAL1 and its partner E3 ligase, BBAP, link Poly(ADP-ribose) activation, ubiquitylation, and double-strand DNA repair independent of ATM, MDC1, and RNF8. *Mol. Cell. Biol.* **2013**, *33*, 845–857. [CrossRef]

108. Yang, C.S.; Jividen, K.; Spencer, A.; Dworak, N.; Ni, L.; Oostdyk, L.T.; Chatterjee, M.; Kusmider, B.; Reon, B.; Parlak, M.; et al. Ubiquitin Modification by the E3 Ligase/ADP-Ribosyltransferase Dtx3L/Parp9. *Mol. Cell* **2017**, *66*, 503–516.e5. [CrossRef]
109. Alhammad, Y.M.O.; Kashipathy, M.M.; Roy, A.; Gagne, J.P.; McDonald, P.; Gao, P.; Nonfoux, L.; Battaile, K.P.; Johnson, D.K.; Holmstrom, E.D.; et al. The SARS-CoV-2 Conserved Macrodomain Is a Mono-ADP-Ribosylhydrolase. *J. Virol.* **2021**, *95*, e01969-20. [CrossRef]
110. Russo, L.C.; Tomasin, R.; Matos, I.A.; Manucci, A.C.; Sowa, S.T.; Dale, K.; Caldecott, K.W.; Lehtio, L.; Schechtman, D.; Meotti, F.C.; et al. The SARS-CoV-2 Nsp3 macrodomain reverses PARP9/DTX3L-dependent ADP-ribosylation induced by interferon signaling. *J. Biol. Chem.* **2021**, *297*, 101041. [CrossRef]
111. Iwata, H.; Goettsch, C.; Sharma, A.; Ricchiuto, P.; Goh, W.W.; Halu, A.; Yamada, I.; Yoshida, H.; Hara, T.; Wei, M.; et al. PARP9 and PARP14 cross-regulate macrophage activation via STAT1 ADP-ribosylation. *Nat. Commun.* **2016**, *7*, 12849. [CrossRef]
112. Begitt, A.; Cavey, J.; Droescher, M.; Vinkemeier, U. On the role of STAT1 and STAT6 ADP-ribosylation in the regulation of macrophage activation. *Nat. Commun.* **2018**, *9*, 2144. [CrossRef]
113. Caprara, G.; Prosperini, E.; Piccolo, V.; Sigismondo, G.; Melacarne, A.; Cuomo, A.; Boothby, M.; Rescigno, M.; Bonaldi, T.; Natoli, G. PARP14 Controls the Nuclear Accumulation of a Subset of Type I IFN-Inducible Proteins. *J. Immunol.* **2018**, *200*, 2439–2454. [CrossRef]
114. Grunewald, M.E.; Chen, Y.; Kuny, C.; Maejima, T.; Lease, R.; Ferraris, D.; Aikawa, M.; Sullivan, C.S.; Perlman, S.; Fehr, A.R. The coronavirus macrodomain is required to prevent PARP-mediated inhibition of virus replication and enhancement of IFN expression. *PLoS Pathog.* **2019**, *15*, e1007756. [CrossRef]
115. Kim, B.; Arcos, S.; Rothamel, K.; Jian, J.; Rose, K.L.; McDonald, W.H.; Bian, Y.; Reasoner, S.; Barrows, N.J.; Bradrick, S.; et al. Discovery of Widespread Host Protein Interactions with the Pre-replicated Genome of CHIKV Using VIR-CLASP. *Mol. Cell* **2020**, *78*, 624–640.e7. [CrossRef]
116. Shang, J.; Smith, M.R.; Anmangandla, A.; Lin, H. NAD+-consuming enzymes in immune defense against viral infection. *Biochem. J.* **2021**, *478*, 4071–4092. [CrossRef]
117. Webb, T.E.; Saad, R. Sequence homology between human PARP14 and the SARS-CoV-2 ADP ribose 1′-phosphatase. *Immunol. Lett.* **2020**, *224*, 38–39. [CrossRef]
118. Parthasarathy, S.; Fehr, A.R. PARP14: A key ADP-ribosylating protein in host-virus interactions? *PLoS Pathog.* **2022**, *18*, e1010535. [CrossRef]
119. Iqbal, M.B.; Johns, M.; Cao, J.; Liu, Y.; Yu, S.C.; Hyde, G.D.; Laffan, M.A.; Marchese, F.P.; Cho, S.H.; Clark, A.R.; et al. PARP-14 combines with tristetraprolin in the selective posttranscriptional control of macrophage tissue factor expression. *Blood* **2014**, *124*, 3646–3655. [CrossRef]
120. Riley, J.P.; Kulkarni, A.; Mehrotra, P.; Koh, B.; Perumal, N.B.; Kaplan, M.H.; Goenka, S. PARP-14 binds specific DNA sequences to promote Th2 cell gene expression. *PLoS ONE* **2013**, *8*, e83127. [CrossRef]
121. Eddie, A.M.; Chen, K.W.; Schenkel, L.B.; Swinger, K.K.; Molina, J.R.; Kunii, K.; Raybuck, A.L.; Keilhack, H.; Gibson-Corley, K.N.; Niepel, M.; et al. Selective Pharmaceutical Inhibition of PARP14 Mitigates Allergen-Induced IgE and Mucus Overproduction in a Mouse Model of Pulmonary Allergic Response. *Immunohorizons* **2022**, *6*, 432–446. [CrossRef] [PubMed]
122. Yu, M.; Schreek, S.; Cerni, C.; Schamberger, C.; Lesniewicz, K.; Poreba, E.; Vervoorts, J.; Walsemann, G.; Grotzinger, J.; Kremmer, E.; et al. PARP-10, a novel Myc-interacting protein with poly(ADP-ribose) polymerase activity, inhibits transformation. *Oncogene* **2005**, *24*, 1982–1993. [CrossRef] [PubMed]
123. Kril, V.; Aiqui-Reboul-Paviet, O.; Briant, L.; Amara, A. New Insights into Chikungunya Virus Infection and Pathogenesis. *Annu. Rev. Virol.* **2021**, *8*, 327–347. [CrossRef] [PubMed]
124. Groslambert, J.; Prokhorova, E.; Ahel, I. ADP-ribosylation of DNA and RNA. *DNA Repair* **2021**, *105*, 103144. [CrossRef]
125. Weixler, L.; Scharinger, K.; Momoh, J.; Luscher, B.; Feijs, K.L.H.; Zaja, R. ADP-ribosylation of RNA and DNA: From in vitro characterization to in vivo function. *Nucleic Acids Res.* **2021**, *49*, 3634–3650. [CrossRef]
126. Kiening, M.; Ochsenreiter, R.; Hellinger, H.J.; Rattei, T.; Hofacker, I.; Frishman, D. Conserved Secondary Structures in Viral mRNAs. *Viruses* **2019**, *11*, 401. [CrossRef]
127. Smyth, R.P.; Negroni, M.; Lever, A.M.; Mak, J.; Kenyon, J.C. RNA Structure-A Neglected Puppet Master for the Evolution of Virus and Host Immunity. *Front. Immunol.* **2018**, *9*, 2097. [CrossRef]
128. Lei, J.; Ma-Lauer, Y.; Han, Y.; Thoms, M.; Buschauer, R.; Jores, J.; Thiel, V.; Beckmann, R.; Deng, W.; Leonhardt, H.; et al. The SARS-unique domain (SUD) of SARS-CoV and SARS-CoV-2 interacts with human Paip1 to enhance viral RNA translation. *EMBO J.* **2021**, *40*, e102277. [CrossRef]
129. Markiewicz, L.; Drazkowska, K.; Sikorski, P.J. Tricks and threats of RNA viruses-towards understanding the fate of viral RNA. *RNA Biol.* **2021**, *18*, 669–687. [CrossRef]
130. Chiu, L.Y.; Huang, D.Y.; Lin, W.W. PARP-1 regulates inflammasome activity by poly-ADP-ribosylation of NLRP3 and interaction with TXNIP in primary macrophages. *Cell. Mol. Life Sci.* **2022**, *79*, 108. [CrossRef]
131. Bose, S.; Segovia, J.A.; Somarajan, S.R.; Chang, T.H.; Kannan, T.R.; Baseman, J.B. ADP-ribosylation of NLRP3 by Mycoplasma pneumoniae CARDS toxin regulates inflammasome activity. *mBio* **2014**, *5*, e02186-14. [CrossRef]
132. Langelier, M.F.; Planck, J.L.; Roy, S.; Pascal, J.M. Structural basis for DNA damage-dependent poly(ADP-ribosyl)ation by human PARP-1. *Science* **2012**, *336*, 728–732. [CrossRef]
133. Feijs, K.L.H.; Zaja, R. Are PARPs promiscuous? *Biosci. Rep.* **2022**, *42*, BSR20212489. [CrossRef]

134. Bonfiglio, J.J.; Fontana, P.; Zhang, Q.; Colby, T.; Gibbs-Seymour, I.; Atanassov, I.; Bartlett, E.; Zaja, R.; Ahel, I.; Matic, I. Serine ADP-Ribosylation Depends on HPF1. *Mol. Cell* **2017**, *65*, 932–940.e6. [CrossRef]
135. Kim, D.S.; Challa, S.; Jones, A.; Kraus, W.L. PARPs and ADP-ribosylation in RNA biology: From RNA expression and processing to protein translation and proteostasis. *Genes Dev.* **2020**, *34*, 302–320. [CrossRef]
136. Vyas, S.; Chesarone-Cataldo, M.; Todorova, T.; Huang, Y.H.; Chang, P. A systematic analysis of the PARP protein family identifies new functions critical for cell physiology. *Nat. Commun.* **2013**, *4*, 2240. [CrossRef]

pathogens

MDPI

Article

Discovery and Development Strategies for SARS-CoV-2 NSP3 Macrodomain Inhibitors

Marion Schuller [1,*], Tryfon Zarganes-Tzitzikas [2], James Bennett [2], Stephane De Cesco [2], Daren Fearon [3,4], Frank von Delft [2,3,4,5,6], Oleg Fedorov [2], Paul E. Brennan [2] and Ivan Ahel [1,*]

1 Sir William Dunn School of Pathology, University of Oxford, Oxford OX1 3RE, UK
2 Centre for Medicines Discovery, University of Oxford, Headington OX3 7DQ, UK
3 Diamond Light Source Ltd., Harwell Science and Innovation Campus, Didcot OX11 0DE, UK
4 Research Complex at Harwell, Harwell Science and Innovation Campus, Didcot OX11 0FA, UK
5 Structural Genomics Consortium, University of Oxford, Headington OX3 7DQ, UK
6 Department of Biochemistry, University of Johannesburg, Auckland Park, Johannesburg 2006, South Africa
* Correspondence: marion.schuller@path.ox.ac.uk (M.S.); ivan.ahel@path.ox.ac.uk (I.A.)

Abstract: The worldwide public health and socioeconomic consequences caused by the COVID-19 pandemic highlight the importance of increasing preparedness for viral disease outbreaks by providing rapid disease prevention and treatment strategies. The NSP3 macrodomain of coronaviruses including SARS-CoV-2 is among the viral protein repertoire that was identified as a potential target for the development of antiviral agents, due to its critical role in viral replication and consequent pathogenicity in the host. By combining virtual and biophysical screening efforts, we discovered several experimental small molecules and FDA-approved drugs as inhibitors of the NSP3 macrodomain. Analogue characterisation of the hit matter and crystallographic studies confirming binding modes, including that of the antibiotic compound aztreonam, to the active site of the macrodomain provide valuable structure–activity relationship information that support current approaches and open up new avenues for NSP3 macrodomain inhibitor development.

Keywords: ADP-ribosylation; macrodomain; SARS-CoV-2; COVID-19; non-structural protein 3 (NSP3); drug discovery and development; virtual screening

Citation: Schuller, M.; Zarganes-Tzitzikas, T.; Bennett, J.; De Cesco, S.; Fearon, D.; von Delft, F.; Fedorov, O.; Brennan, P.E.; Ahel, I. Discovery and Development Strategies for SARS-CoV-2 NSP3 Macrodomain Inhibitors. *Pathogens* **2023**, *12*, 324. https://doi.org/10.3390/pathogens12020324

Academic Editors: Anthony K L Leung and Rachy Abraham

Received: 12 January 2023
Revised: 3 February 2023
Accepted: 10 February 2023
Published: 15 February 2023

1. Introduction

The outbreak of coronavirus disease 2019 (COVID-19) caused by severe acute respiratory syndrome coronavirus 2 (SARS-CoV-2) has become a major public health challenge over the last two years, claiming over 6 million lives so far while being accompanied by severe socioeconomic consequences worldwide [1–3]. The impact of this recent pandemic together with previous coronaviral outbreaks within the past two decades, including SARS-CoV in 2002–2003 and MERS-CoV in 2012–2015 [4], underlines the importance of developing strategies for effectively gaining control of such and general viral disease outbreaks. Apart from non-pharmacologic interventions and prevention measures achievable by vaccines, the development of antiviral agents presents an alternative to increase preparedness by providing rapid disease treatment possibilities.

SARS-CoV-2 is characterised as an enveloped single-stranded positive sense RNA β-coronavirus whose genome encodes for 29 proteins essential for the viral life cycle and its modulation of host immune responses [5–7]. Proteins involved in the viral replication machinery are thereby in particular focus as drug targets. Thus, the RNA polymerase is targeted with nucleoside analogues (e.g., remdesivir or molnupiravir) to inhibit the genome replication and gene transcription of SARS-CoV-2 [8,9]. Moreover, the viral proteases, the main protease (M^{pro}), and the papain-like protease 2 ($PL2^{pro}$), are inhibited by peptide analogues (e.g., nirmatrelvir) and small molecules to prevent the processing of two polypeptides into constituent viral non-structural proteins (NSP) required for viral replication [9–11]. NSP3 is thereby the largest multi-domain protein produced by coronaviruses and is itself an essential component of the replication and transcription complex [12]. SARS-CoV-2 NSP3 features eight (out of 15) domains that exist in all known coronaviruses, including ubiquitin-like domains, PL2^{pro}, transmembrane regions, and a macrodomain (Mac1) [12]. Macrodomains are highly conserved domains found in all kingdoms of life [13] and recognise ADP-ribosylation modifications on proteins and nucleic acids catalysed by poly(ADP-ribosyl)polymerases (PARPs) [14,15]. The interferon (IFN) response triggered through viral infections thereby induces the gene expression of several PARP family members, i.e., PARP7 and PARPs 9–14, whose ADP-ribosylation signalling activity establishes an antiviral environment [16,17]. While, for instance, the antiviral effect of PARP12 was shown to be achieved at least partially through the inhibition of protein translation and by promoting the ADP-ribosylation-dependent degradation of viral proteins [16,18], PARP9 provides a possibility of viral infection control in complex with DTX3L by targeting EMCV 3C protease for ubiquitination and degradation [19]. Furthermore, PARP14 was shown to promote anti-inflammatory interleukin-4-mediated signalling pathways by activating STAT6-dependent gene expression and inhibiting STAT-1-dependent gene expression [20,21]. However, PARP14 expression is also induced by interferon (IFN), and it enhances host IFN responses to lipopolysaccharide (LPS), poly(I:C), and viral infection, indicating a role for PARP14 in restricting viral and bacterial infections [22–24]. However, viral macrodomains such as the SARS-CoV-2 macrodomain evolved with ADP-ribosyl hydrolase activity to reverse PARP-catalysed ADP-ribosylation, thus providing the virus with a strategy to counteract these host defence mechanisms [25,26]. Studies in mice confirmed that mutations of the SARS-CoV macrodomain impairing its catalytic activity led to virus attenuation, a reduction in viral loads, and a stronger immune response following infection compared to the wild-type virus, thereby rendering the virus nonlethal [23,27]. Thus, with the NSP3 macrodomain being critical for replication and pathogenicity in the host for coronaviruses, and similarly for alphaviruses and Hepatitis E virus [28], the macrodomain was established as a therapeutic target for SARS-CoV-2 infection [26,29].

Although the molecular physiological substrates and exact mechanisms of the enzymatic 'arms race' between antiviral PARP and coronaviral macrodomain are still unclear, the macrodomain itself has been in intense focus to pave the way for the development of a new antiviral drug. Its well-defined binding pocket along with its high amenability for structural and biochemical characterisation fostered rapid assay development for in vitro compound screening and discovery [30–33], allowed the elucidation of its catalytic mechanism [34], and gave insights into druggability and plasticity by crystallographic [35,36], NMR [37], and computational molecular dynamics [38,39] approaches. Furthermore, due to the general conservation of the macrodomain fold, the screening of focused chemical libraries curated from inhibitor development programmes of the PARG macrodomain could be performed to further support SARS-CoV-2 NSP3 macrodomain drug development [40].

In this study, we present our approaches to contribute to the initial drug discovery phase for the SARS-CoV-2 NSP3 macrodomain, also referred to as 'Mac1'. We performed computational docking studies which provided insights into the chemical matter to be considered for targeting the active site of the macrodomain. Using an established HTRF-based screening assay for NSP3 Mac1, we furthermore screened medium-sized libraries comprising either experimental small molecules or FDA-approved drugs. The former library screening approach enabled the identification of four molecular scaffolds with inhibitory effects on NSP3 Mac1, whereby initial structure–activity relations were obtained by analogue characterisation. Moreover, we discovered with our FDA-approved library screening that the active site of the SARS-CoV-2 macrodomain can be inhibited by antibiotic agents including aztreonam, whose binding we confirmed by crystallographic studies. Altogether, our studies provide valuable chemical starting points for future inhibitor development for the NSP3 macrodomain.

2. Methods

2.1. Materials, Reagents, and Chemicals

Crystallisation screens were procured from Hampton Research. The ADPr-peptide with sequence ARTK(Bio)QTARK(Aoa-RADP)S used for HTRF assays was purchased from Cambridge Peptides. All remaining chemicals were purchased from Sigma unless stated otherwise. The BioAscent library of 125,000 compounds was purchased from BioAscent (https://www.bioascent.com/integrated-drug-discovery/in-house-diversity-and-fragment-libraries, accessed on 24 April 2020). The MIDAS library was a generous gift from Allan Jordan of Cancer Research UK.

2.2. Constructs

SARS-CoV-2 NSP3 Mac1 (residues 206–379) cloned into a pDEST17 vector with N-terminal His_6-tag was used for performing HTRF assays [26]. SARS-CoV-2 NSP3 Mac1 (residues 207–373) cloned into a pNIC28-Bsa4 expression vector with N-terminal His_6-TEV cleavage site [36] was used for protein crystallisation.

2.3. Protein Expression and Purification for Crystallisation

E. coli Rosetta strain BL21(DE3) was transformed with the constructs encoding SARS-CoV-2 NSP3 macrodomains and grown at 37°C in Terrific Broth (Merck Millipore, Burlington, MA, US), which was supplemented with 50 µg/mL of kanamycin and 35 µg/mL of chloramphenicol. After reaching an OD_{600nm} of 1.0–1.2, the temperature was lowered to 18°C prior to the induction of protein expression overnight (O/N) by adding 0.5 mM of IPTG. The harvested cells were resuspended in lysis buffer (50 mM of HEPES (pH 7.4), 500 mM of NaCl, 5% glycerol, 20 mM of imidazole, 0.5 mM of TCEP, cOmplete EDTA-free protease inhibitors (Roche, Basel, Switzerland) and stored at −20°C until purification.

For protein purification, pellets were gently thawed and lysed by high-pressure homogenisation. DNA was digested using Benzonase Nuclease (Merck Life Science, Darmstadt, Germany). Proteins were purified by immobilised metal affinity chromatography (IMAC) using Ni-Sepharose resin (GE Healthcare, Chicago, IL, US) and eluted stepwise in binding buffer containing 40–500 mM imidazole. Typically, a high salt wash with 1 M of NaCl was combined with the first elution step including 40 mM of imidazole. Protein purified for performing HTRF assays was further purified by size exclusion chromatography (SEC) (Superdex 75, GE Healthcare) in a buffer consisting of 25 mM of HEPES (pH 7.5), 300 mM of NaCl, 5% glycerol, and 0.5 mM of TCEP. For protein purified for the crystallisation experiments, the removal of the hexahistidine tag was carried out after the first Ni-IMAC step by the addition of recombinant TEV protease during O/N dialysis into buffer without imidazole, followed by purification on a second IMAC column. Finally, protein was purified by SEC (Superdex 75, GE Healthcare) in a buffer consisting of 20 mM of HEPES (pH 8.0), 250 mM of NaCl and 2 mM of DTT. The proteins were characterised by SDS-PAGE, then flash-frozen in liquid nitrogen and stored at −80°C until required.

2.4. HTRF Assay

The inhibition of SARS-CoV-2 NSP3 Mac1 was assessed by the displacement of an ADP-ribose-conjugated biotin peptide from His_6-tagged protein using an HTRF-technology-based screening assay, which was performed as previously described [36]. Compound library screens (including the MIDAS and FDA-approved screening set and the curated BioAscent hit compound library) were performed at a compound concentration of 25 µM in duplicate measurements, while for hit confirmation, IC_{50} curves were acquired with a top compound concentration of 125 µM (MIDAS and FDA-approved hit compounds) or 187 µM (BioAscent hit compounds), followed by an 8-point 1:1 dilution series in duplicate measurements. The compounds were dispensed into ProxiPlate-384 Plus (PerkinElmer, Waltham, MA, US) assay plates using an Echo 525 liquid handler (Labcyte, San Jose, CA, US). Binding assays were conducted in a final volume of 16 µL with 12.5 nM of SARS-CoV-2 NSP3 Mac1, 400 nM of peptide ARTK(Bio)QTARK(Aoa-RADP)S, 1:20,000 Anti-His_6-

Eu^{3+} cryptate (HTRF donor, PerkinElmer), and 1:125 Streptavidin-XL665 (HTRF acceptor, PerkinElmer) in assay buffer (25 mM of HEPES pH 7.0, 20 mM of NaCl, 0.05% bovine serum albumin and 0.05% Tween-20). Assay reagents were dispensed into plates using a Multidrop combi (Thermo Scientific, Waltham, MA, US). Macrodomain protein and peptide were first dispensed and incubated for 30 min at room temperature. This was followed by the addition of the HTRF reagents and incubation at room temperature for 1 h. Fluorescence was measured using a PHERAstar microplate reader (BMG) using the HTRF module with dual emission protocol (A = excitation of 320 nm, emission of 665 nm, and B = excitation of 320 nm, emission of 620 nm). Raw data were processed to give an HTRF ratio (channel A/B × 10,000), which was used to generate IC_{50} curves. The IC_{50} values were determined by nonlinear regression using GraphPad Prism v.9 (GraphPad Software, San Diego, CA, USA). Of note is that we judged—based on our experience in medicinal chemistry and FRET-based assays, as well as references in the literature (e.g., Baell and Walters, 2014 [41])—the screening compounds for the presence of chemical features known to cause assay interference and promiscuous binding behaviour. Compounds were excluded from the hit validation processes without further biophysical testing or computational predictions when certain motifs were identified. However, where stated as "showed assay effects at higher concentrations", the compounds did not show structural features suspicious for assay interference per se and were tested in the HTRF-based assay. At higher compound concentrations of the titration, we observed a decrease or were even unable to determine Mac1 inhibition values, indicating that these compounds had unspecific assay effects unrelated to true Mac1 inhibition.

2.5. Crystallisation, Crystal Soaking, and Data Processing

The purified SARS-CoV-2 NSP3 Mac1 protein was concentrated to 47 mg/mL, and crystallisation drops were set-up in MRC two-well crystallization microplates (Swissci, Buckinghamshire, UK) using the Mosquito Crystal robot (TTP Labtech, Cambridgeshire, UK) with protein to reservoir ratios of 1:1 and 1:2, in a 150 nl total volume equilibrated against 75 µL of reservoir solution containing 100 mM CHES pH 9.5 and 30% PEG3000. To ease crystallisation for soaking experiments, ~5 crystals were harvested and prepared as seed stock using a Seed Bead Kit (Hampton Research, Aliso Viejo, CA, US) in 100 nl of reservoir solution. An amount of 20 nl of a 1:500 dilution of the resulting seed stock was added to the crystallisation experiments. The compounds were soaked into crystals by adding 0.5 µL of dissolved compounds directly to the crystallisation drops. After incubation for 1–3 h, the crystals were harvested using reservoir solution supplemented with 20% ethylene glycol (*v*/*v*) as a cryo-protectant prior to flash freezing in liquid nitrogen. X-ray data were collected at beamline I03 at Diamond Light Source (Rutherford Appleton Laboratory, Harwell, UK) and data collection statistics are given in Supplementary Table S1.

The X-ray data were processed using the XIA2-DIALS platform [42], and phase information was obtained using the molecular replacement method with PHASER [43] using 7KQP as template. Atomic models were improved following consecutive cycles of manual building in COOT [44] and structure refinement in REFMAC5 [45]. The structures were refined to good Ramachandran statistics, and MolProbity [46] was used to validate the models prior to deposition in the PDB. The processing and refinement statistics are given in Supplementary Table S1. Structural alignments and analyses, as well as figure preparation, were carried out using PyMol (Molecular Graphics System, Version 2.3.3 Schrödinger, LLC., New York, NY, USA).

2.6. Virtual Screening

The structure of the NSP3 macrodomain ('Mac1') was downloaded from the Protein Data Bank (rcsb.org) as a PDB file (6W02). The macrodomain displayed a closed conformation. All water molecules were removed, except four in the binding site that formed water-mediated hydrogen bonds between ADP-ribose and the protein (wb32, wb60, wb71, and wb107). The protein was prepared for docking in Schrodinger. The BioAscent library was prepared for docking using Schrodinger ligprep with racemic compounds being expanded to include discreet enantiomers. The compounds were docked into the protein using Glide SP with default parameters and the top scoring enantiomer kept for evaluation. The top 2000 highest scoring compounds were selected for IC_{50} determination, of which 1786 compounds could be supplied by the company for biophysical characterisation in the HTRF assay. ChemDraw 21.0.0 was used for the visualization and drawing of compound structures.

3. Results

3.1. NSP3 Macrodomain Virtual Ligand Screen

Initial ligand discovery efforts focused on the virtual screening of NSP3 Mac1. ADP-ribose was removed from the 3D structure of the bound macrodomain (PDB ID 6W02). Redocking returned the ligand bound structure with high overlap compared to the X-ray structure (RMSD 1.05 Å), confirming the docking validity (Supplementary Figure S1). A 125,000-compound virtual copy of the BioAscent library was then screened using Schrodinger Glide SP and the top 2000 compound selected for profiling. A total of 1786 compounds could be supplied by the company and were screened against NSP3 Mac1 using the HTRF assay as described below. The most potent inhibitors, IAL-MD0305 and IAL-MD0306, showed 28 μM and 18 μM IC_{50}s, respectively (Figure 1A,B), although attempts at co-crystallisation and soaking did not yield experimental ligand-bound protein structures with NSP3 Mac1. Yet, the models of the macrodomain-hit compound complexes suggest a binding mode of both compounds in the open ribose-phosphate binding site of the ADP-ribose (Figure 1C). While IAL-MD0305 may be stabilised only through hydrogen bonding to the Ser128 backbone amine and hydrophobic interactions, IAL-MD0306 interacts with its carboxyl group to the backbone amines of S128, Phe132 and I131, thus rationalising its slightly higher inhibitory activity in the HTRF assay.

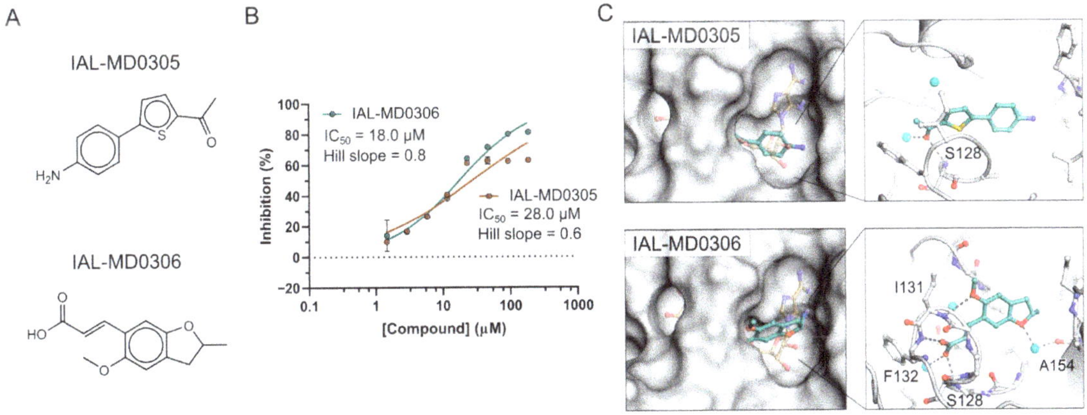

Figure 1. NSP3 macrodomain hit compounds identified by virtual ligand screening. (**A**) The molecular structure of the most potent hit compounds. (**B**) NSP3 Mac1 IC_{50} curves and parameters of the virtual hit compounds obtained in confirmatory HTRF assays. (**C**) Docking models of NSP3 Mac1 in complex with the hit compounds (cyan stick model). (**Right**) Surface representation showing as reference the binding mode of ADP-ribose (brown stick model in low transparency; generated by structure overlay with PDB ID 7KQP). (**Left**) Molecular interactions of the hit compounds with NSP3 Mac1. Water molecules are shown as blue spheres.

3.2. NSP3 Macrodomain Hit Discovery by MIDAS Compound Library Screen

To discover and characterise additional new hit matter for SARS-CoV-2 NSP3 Mac1, we performed in vitro primary and confirmatory screening using an established HTRF technology-based screening assay previously set-up for the characterisation of fragment hits for this target [36]. The assay involves an ADP-ribose mimic-conjugated peptide [47] that binds via its biotinylated lysine to a streptavidin-labelled XL665 HTRF acceptor fluorophore, while the macrodomain protein is complexed by its hexahistidine tag with an anti-His_6-antibody, which itself is conjugated to the Europium HTRF donor fluorophore (Figure 2A). Binding of the macrodomain to the ADP-ribose imitating part of the peptide produces a FRET-based HTRF signal, which is disrupted by inhibitors targeting the active site of the macrodomain. ADP-ribose, which is recognised by NSP3 Mac1 with a K_D of 13 µM [48], is used as positive control, showing an IC_{50} of 1.1 µM in the HTRF assay (Supplementary Figure S2). The measured IC_{50} for ADP-ribose thus matches the ADP-ribose IC_{50} of 1.5 µM obtained in a similar set-up using the peptide in an AlphaScreen-based assay [33].

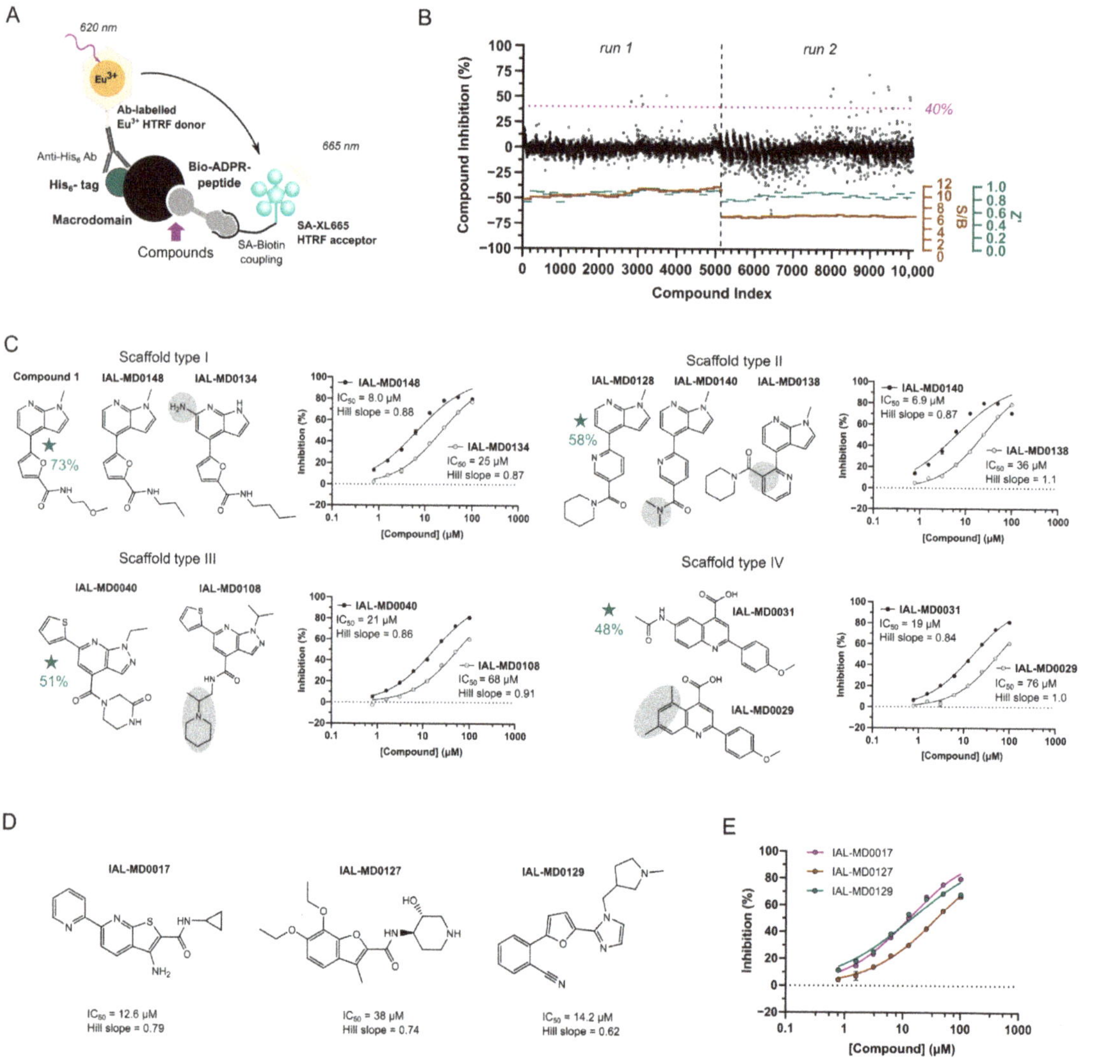

Figure 2. NSP3 macrodomain hit discovery by MIDAS compound library screen. (**A**) HTRF assay principle for performing the compound library screening on SARS-CoV-2 NSP3 Mac1. (**B**) Compound

inhibition profile and assay quality monitoring of the MIDAS compound library screen. The cut-off for hit compounds was set to a macrodomain inhibition of ≥40%. S/B: Signal-to-Background; Z′: assay quality parameter. (**C**) Four scaffold types identified from the MIDAS screening for classifying hit compounds. Representative examples of scaffold analogues and obtained IC_{50} curves are shown. Primary screening hit compounds with respective macrodomain inhibition at 25 μM are indicated with a star. Grey circles highlight potential SAR information resulting in the observed differences on macrodomain inhibition based on IC_{50} values. All analogues with respective inhibitory activity are provided in Supplementary Table S4. (**D**) Molecular structure of the singleton compounds with their respective IC_{50} values and Hill slope parameters. (**E**) IC_{50} curves obtained for singletons in the HTRF assay.

To be unbiased for potential hit matter, we screened the "Manchester Institute Diversity Set" (MIDAS), comprising 10.1 k diverse, non-covalent, and tractable small molecules. The library was screened at a compound concentration of 25 μM in two batches, with an average assay performance of 0.87 for Z′ and an S/B ratio of 10.4 in the first run and of 0.83 for Z′ and an S/B ratio of 6.2 in the second run (Figure 2B). Setting a minimum NSP3 Mac1 inhibition of 40% and a maximum assay error of 10% to be defined as a hit compound, we obtained 10 primary hit compounds, resulting in an overall hit rate of 0.1%. More specifically, one compound showed complete NSP3 Mac1 inhibition at the screening concentration, four compounds were identified with 60–70% inhibition, and the remaining five compounds showed inhibitory activity in the range from 40% to 55%. Three out of these ten hit compounds were yet excluded from further hit confirmation based on their assay interference potential. Of note is that we judged—based on our experience in medicinal chemistry and FRET-based assays, as well as references in the literature (e.g., Baell and Walters, 2014 [41])—the screening compounds for the presence of chemical features known to cause assay interference and promiscuous binding behaviour. Compounds were excluded from the hit validation processes without further biophysical testing or computational predictions when certain motifs were identified. By considering the commercial availability of the remaining hit compounds, including of respective analogues along with the primary screening results, we determined to focus on four molecular scaffolds for follow-up (Figure 2C, Supplementary Table S3) and included three compounds, IAL-MD0017, IAL-MD0127 and IAL-MD0129, as singletons in the hit confirmation process. The hit compound defining scaffold type I with a furanyl-pyrrolo[2,3-*b*]pyridine structure (Compound **1**) could not be re-supplied; however, it was followed-up with seven close analogues to obtain initial structure–activity relationships (SAR) for its binding to NSP3 Mac1. The hit compounds defining scaffold type II with a pyridinyl-pyrrolo[2,3-*b*]pyridine (IAL-MD0128), scaffold type III with a thiophenyl-pyrazolo[3,4-*b*]pyridine attached to a piperazine substituent (IAL-MD0040), and scaffold type IV with a phenylquinoline-4-carboxylic acid (IAL-MD0031), were each followed-up with dose–response titrations along with 18 (type II and IV) or 55 (type III) analogues, respectively (Supplementary Table S4). The 98 analogues were selected based on having the core of their respective scaffold type group conserved with variations of the attached ring systems, functional groups, and additions of substituents, in order to explore the amenability and plasticity of the active site of NSP3 Mac1.

3.3. *NSP3 Macrodomain Inhibitors of Scaffold Type I*

All analogue compounds in the scaffold type I group were confirmed to inhibit NSP3 Mac1 with IC_{50} values between 4.9 μM and 25 μM (Supplementary Table S4). IAL-MD0148 that is structurally closest to the primary (non-purchasable) hit, i.e., Compound **1**, showed a macrodomain inhibition of 8.0 μM (Figure 2C top-left) and was slightly outcompeted in activity by compounds with smaller amid-containing substituents attached to the furan ring. Interestingly, IAL-MD0131 characterised by a variation of the methoxyethylamide to morpholine showed the best inhibitory activity (IC_{50} of 4.9 μM) (Supplementary Figure S3A) among this series. Furthermore, IAL-MD0134 stood out as the only compound of lower activity with an IC_{50} of 25 μM (Figure 2C top-left). Notably, its amino group being attached to the pyrrolo[2,3-*b*]pyridine core may lead to steric clashes within the active site of the macrodomain, resulting in the decrease in its inhibitory ability (Figure 2C, top-left). Of note is that all compounds in this series showed assay effects at higher compound concentrations, potentially indicating solubility-related issues, which is to be considered for compound optimisation.

3.4. NSP3 Macrodomain Inhibitors of Scaffold Type II

Scaffold type II is defined by the primary screening hit IAL-MD0128, whose NSP3 Mac1 inhibition was verified in the confirmatory dose–response titrations (Supplementary Figure S3B), albeit with assay interference effects at higher compound concentrations. Its IC_{50} activity was estimated with 3.1 μM and, as such, took the lead compared to the inhibitory activities of the analogue compounds in this series with determinable IC_{50} values ranging between 6.9 μM (IAL-MD0140) and 45 μM (IAL-MD0143) (Supplementary Table S4). The attachment of the piperidinyl ethenone substituent and variations in ortho instead of in para position on the pyridine (IAL-MD0138) most notably decreased NSP3 Mac1 inhibition and, particularly, meta position substituents were not tolerated (Figure 2C top-right, Supplementary Table S4), most likely by making the compound poorly fit into the active site of the macrodomain. Moreover, any of the tested variations of the pyridinyl substituents in para position including smaller (non-)aromatic ring systems or functional group extensions did not increase the inhibitory activity of the compounds compared to the primary screening hit (Supplementary Table S4). The para substituent seems yet to be involved in macrodomain interaction, since minor variations have notable effects. While a ring opening to *N,N*-dimethylacetamide (IAL-MD0140) is well tolerated, *N*-ethyl, *N*-methylacetamide (IAL-MD0144) or a simple reduction in ring size to pyrrolidine are less favoured (IAL-MD0142).

3.5. NSP3 Macrodomain Inhibitors of Scaffold Type III

Follow-up characterisation of the primary hit and 54 analogue compounds classifying to scaffold group type III provided further SAR information to target NSP3 Mac1. Only for six compounds IC_{50} values could be obtained ranging between 12.6 μM and 68 μM (Supplementary Table S4), while the primary hit (IAL-MD0040) was confirmed with an IC_{50} of 20 μM (Figure 2C bottom-left). IAL-MD0051, the best performing compound of this series, which has minor alterations of the core-attached ring systems, i.e.,thiophen (replaced with methyl-thiophen) and the piperazin-2-one (replaced with morpholine), showed slightly stronger macrodomain inhibition (IC_{50} of 12.6 μM) than the primary hit compound, yet accompanied by secondary assay effects at higher concentrations. Analogues IAL-MD0064 (IC_{50} of 25 μM) and IAL-MD0070 (IC_{50} of 16.3 μM), whose ethyl substituent is replaced with isopropyl or, additionally, the thiophen with a furan ring system, show similar activity on the macrodomain as the primary hit. In contrast, replacement of the thiophen with the slightly larger phenyl substituent (IAL-MD0123) as well as replacement of the ethyl substituent with a larger benzyl substituent (IAL-MD0124) is not tolerated, clearly showing the relevance of the size and nature of the substituents in these positions for the macrodomain binding. Moreover, linearisation of the piperazin-2-one (IAL-MD0074, IC_{50} of 25.3 μM) did not have any notable effects on macrodomain inhibitory activity. However, its replacement with an amide-linked piperidine (IAL-MD0108) was less favoured, decreasing the IC_{50} to 68 μM, while its replacement with larger substituents (which was sampled by the majority for the analogues) including dihydroquinoxalin-2-one (IAL-MD0049) was not tolerated, likely due to causing steric hindrance within the active site (Figure 2C bottom-left, Supplementary Table S4).

3.6. NSP3 Macrodomain Inhibitors of Scaffold Type IV and Singletons

Scaffold type IV compounds showed overall lower inhibitory activity on the macrodomain compared to the other groups. The primary hit IAL-MD0031 was characterised with an

IC_{50} of 19 μM, while derivative compounds showed either similar (IAL-MD0059 and IAL-MD0088) or notably decreased (IAL-MD0024 and IAL-MD0029) macrodomain activity (Figure 2C bottom-right, Supplementary Table S4). Interestingly, the exchange of the carboxylic acid group with carboxamide was well tolerated (IAL-MD0088, IC_{50} of 24 μM); however, its replacement with *N*-methoxy carboxamide (IAL-MD0024) was far less accepted, decreasing the IC_{50} to 68 μM, and replacement with any larger substituent (sampled by the majority of analogue compounds) resulted in a loss of inhibitory activity on the macrodomain. Furthermore, the addition of a methyl group to the quinoline core in position 7 also resulted in a decrease in macrodomain inhibitory activity (IAL-MD0029, IC_{50} of 76 μM), and even more so with a larger methoxy group (IAL-MD0030), similarly to its addition to the phenyl in position 3 (IAL-MD0094). In contrast, even bulky additions to position 2 of the quinoline core were tolerated (IAL-MD0059, IC_{50} of 22.8 μM) (Figure 2C bottom-right, Supplementary Table S4), indicating that substituents in position 2 may be directed outwards of the active site of the macrodomain.

Finally, the dose–response titrations of the singleton hit compounds, IAL-MD0017, IAL-MD0127, and MD0129 (Figure 2D), also confirmed their inhibitory activity on NSP3 Mac1, with IC_{50}s of 12.6 μM, 38.0 μM, and 14.2 μM, respectively, (Figure 2E, Supplementary Table S4) and may be considered as possible chemical matter accepted by the active site of the macrodomain for inhibitor development.

3.7. Screening of FDA-Approved Compounds Reveal Antibiotics as NSP3 Macrodomain Inhibitors

In light of the benefits of discovering FDA-approved drugs as inhibitors for the target of interest, we complemented our in vitro screening approach with screening a library of 1600 FDA-approved molecules. The screening was performed with a compound assay concentration of 50 μM in one run of single shot experiments, with an average assay performance of 0.73 for Z′ and an S/B ratio of 5.8 (Figure 3A). Applying the same criteria for hit compounds as for the MIDAS screen, i.e., NSP3 Mac1 inhibition over 40% and assay error less than 10%, we obtained 30 hit compounds, which corresponds to a hit rate of 1.9% (Supplementary Table S5). Thus, compared to the MIDAS library screen, notably more compounds were identified showing assay activity, yet including both compounds of direct target inhibition and potential assay interference. Interestingly, adenine was also included among the drug molecules, yet it did not show any inhibitory activity on NSP3 Mac1, indicating that hits to be identified in the screening with the chosen parameters are likely required to undergo more interactions with the domain than ADP-ribose targeting the adenine binding site. The selection of hit compounds for follow-up was guided by structural inspection, thereby excluding compounds with features known for likely assay interference. These included biotin (NSP3 Mac1 inhibition of 107%), due to its competition with the biotinylated peptide over binding to the streptavidin-conjugated XL665 fluorophore, and in particular, compounds with large multi-ring aromatic systems such as sennoside A (112% inhibition), chlorophyllide–copper complex (106% inhibition), methacycline (87% inhibition), protoporphyrin IX (75% inhibition) and candicidin (50%). The metal-complexed compounds pyrithione zinc, cisplatin, and zinc undecylenate (68–100% inhibition) were also not prioritised for follow-up; furthermore, the high inhibition of nadide (100%) was assumed to be based on the limited stability of NAD^+, resulting in its degradation to ADP-ribose and nicotinamide. Overall, the most active FDA-approved compounds identified from the primary screening (Supplementary Table S5) and selected for hit confirmation included the selenium and mercury-containing compounds ebselen and thiomersal (both showing 107% NSP3 Mac1 inhibition), thioctic acid (71% inhibition), avobenzone (62% inhibition), and oxantel pamoate (56% inhibition). Moreover and notably, several antibiotic compounds ranked upon the hits showing NSP3 Mac1 inhibitory activity between 108% (ceftazidime) to 43% (aztreonam). Although generally differing in structural makeup, the antibiotics were grouped into either anthracene scaffold-based compounds (methacycline and mitoxantrone) or beta-lactam-based antibiotics (ceftazidime, cephalosporin C, cefepime, ceftibuten and aztreonam).

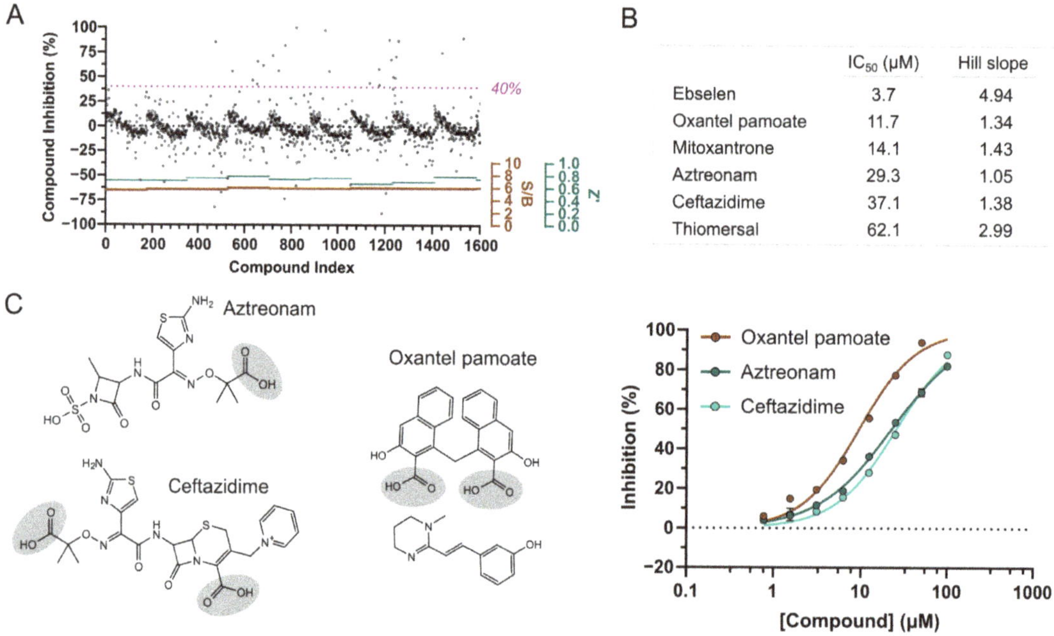

	IC_{50} (μM)	Hill slope
Ebselen	3.7	4.94
Oxantel pamoate	11.7	1.34
Mitoxantrone	14.1	1.43
Aztreonam	29.3	1.05
Ceftazidime	37.1	1.38
Thiomersal	62.1	2.99

Figure 3. NSP3 macrodomain hit discovery by screening a library of FDA-approved compounds. (**A**) Compound inhibition profile and assay quality monitoring of the FDA-approved compound library screen. The cut-off for hit compounds was set to a macrodomain inhibition of ≥40%. S/B: signal-to-background; Z′: assay quality parameter. (**B**) NSP3 Mac1 IC_{50} values and Hill slope parameters of best performing hit compounds in confirmatory HTRF assays. (**C**) Molecular structures and IC_{50} curves of the most promising hit compounds from the FDA-approved compound library. The carboxylic group commonly present in all three compounds is highlighted in grey circles.

The re-supply and testing of the 15 selected hit compounds in dose–response titration at a top assay concentration of 125 μM confirmed the activity of six drug molecules with IC_{50} values from 3.7 μM (ebselen) to 62 μM (thiomersal) (Figure 3B). The other characterised compounds did not reproduce the primary screening results, showing either inactivity over the tested concentration range or strong assay interference (carboplatin and methacycline). Ebselen showed the strongest inhibitory effect on NSP3 Mac1, yet its high Hill slope parameter and due to being generally known as a promiscuous binder indicated unspecific effects of ebselen on the target; therefore, it was not further pursued. Moreover, although an IC_{50} value of 14 μM could be estimated for mitoxantrone, strong assay interference likely based on aggregation effects could be observed. Moreover, considering the assay inactivity of the similar hit compound methacycline, mitoxantrone was also excluded from further follow-up characterisation. However, two of the three beta-lactam antibiotics, aztreonam and ceftazidime, had confirmed NSP3 Mac1 inhibition with IC_{50} values of 29 μM and 37 μM, respectively, along with oxantel pamoate with an IC_{50} value of 12 μM (Figure 3B,C). Of note is that oxantel pamoate is a two-component drug, using the embonate salt as a counterion for the oxantel base for controlling the dissolution rate of the formulation, and assuming that only one component is active on NSP3 Mac1, the IC_{50} of the latter may be around 6 μM. In all three compounds (pamoate, aztreonam, and ceftazidime), we also noted the presence of a carboxylic group (Figure 3C), potentially indicating a common motif that enables interaction with the active site of NSP3 Mac1.

3.8. Aztreonam Targets the NSP3 Macrodomain Active Site Similar to MIDAS Hit Compound

To confirm the binding of the hit compounds to SARS-CoV-2 NSP3 Mac1 and gain insights into the binding mode for structure-guided compound development, we performed co-crystallisation experiments of the macrodomain with MIDAS and FDA-approved hit compounds, whose inhibitory activity was confirmed in the dose–response titrations. For the MIDAS hit confirmation, we could determine the structure of NSP3 Mac1 in complex with IAL-MD0131 (Supplementary Table S1), which is one of the best performing MIDAS hits with an IC_{50} of 4.9 μM and belongs to the scaffold type I group. IAL-MD0131 was resolved in the crystallographic map, yet with low occupancy (Figure 4A, right). The ligand occupied the adenosine binding site of the ADP-ribose binding pocket, with its methyl-pyrrolo[2,3-*b*]pyridine moiety aligning with the adenine base of the ADP-ribose and the furan positioning in the ribose binding site (Figure 4A, left). The binding of IAL-MD0131 appears to be stabilised by hydrogen bonding of the pyrrolo[2,3-*b*]pyridine moiety to the backbone amine of Ile23 and its off-set π-π stacking to the Phe156 side chain, presenting interactions which are also established by the natural ligand ADP-ribose. Furthermore, the furan-carbonyl substructure of IAL-MD0131 enables targeting of the Phe156/Asp157 backbone amines, which is also defined as the "oxyanion" subsite of NSP3 Mac1 [36] (Figure 4A, middle). In the NSP3 Mac1:ADP-ribose-bound structure, the proximal ribose interacts with this oxyanion subsite via a bridging water molecule. Chemical matter exploring this subsite with direct interaction could, therefore, presenta valuable starting point for macrodomain inhibitor development. The morpholino ring system is directed outwards of the ADP-ribose pocket and does not engage with direct interactions to the macrodomain. Functionalising this moiety for NSP3 Mac1 binding, for instance, by targeting the side-chain of Asp157, could provide possibilities to improve inhibitor potency.

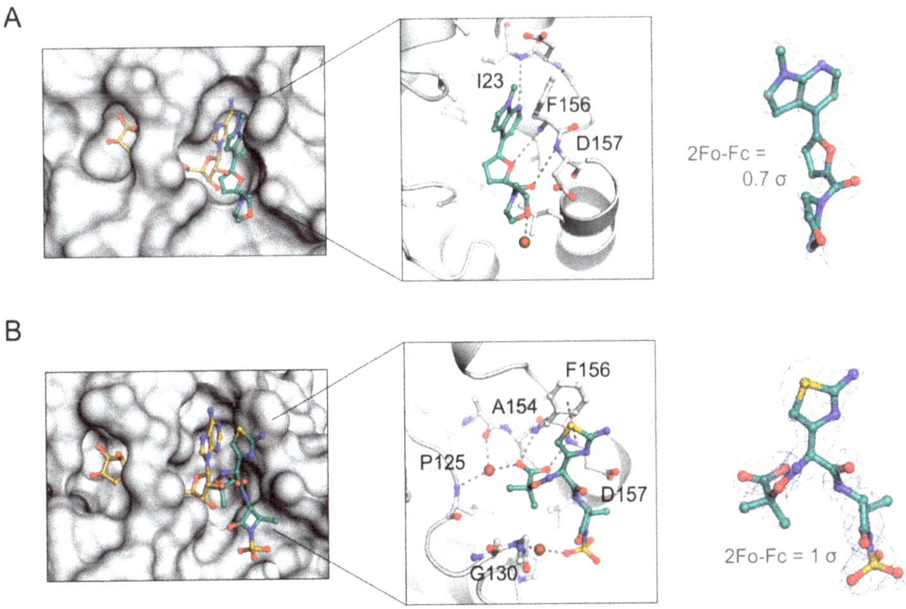

Figure 4. Crystallographic studies confirm the binding of the MIDAS library analogue hit IAL-MD0131 (**A**) and the FDA-approved compound aztreonam (**B**) in the active site of NSP3 Mac1. The hit compounds are shown as cyan stick model. Binding to NSP3 Mac1 in reference to ADP-ribose (brown stick model; generated by structure overlay with PDB ID 7KQP) is shown on the left, their molecular interactions with the macrodomain are shown in the middle, and the resolution of the compounds in the crystallographic map are presented on the right. Waters are shown as red spheres.

Moreover, we obtained a co-crystal structure of the macrodomain with the FDA-approved antibiotic drug aztreonam (Figure 4B, Supplementary Table S1). Aztreonam, confirming with an IC_{50} of 29 μM in dose–response titrations, was indeed identified to target the active site of the macrodomain by being well-resolved in the crystallographic map (Figure 4B, right). Compared to the ADP-ribose binding mode, aztreonam does not occupy the adenine-binding subsite and instead takes an arching confirmation into a groove adjacent below the adenosine binding pocket (Figure 4B, left). Similar to ADP-ribose and IAL-MD0131, aztreonam shows π-π stacking with Phe156, using its aminothiazol substituent. The amide beta-lactam bridges to the sulfonic acid, which is stabilised over a water-mediated contact to Gly130. Moreover, aztreonam hydrogen bonds to the oxyanion subsite of NSP3 Mac1 with its carboxylic acid that additionally coordinates over a water molecule to the backbone amines of Ala154 and Pro125 (Figure 4B, middle). The hydrogen bond interactions established by the carboxylic acid within the proximal ribose binding pocket may contribute to the binding affinity and, as a result, to the observed inhibitory activity of aztreonam. The carboxylic acid group was a common functional group among the discovered FDA-approved hit compounds which confirmed in dose–response titrations with confidence; therefore, it is tempting to assume that, also in these compounds, the carboxylic acid group contributes to NSP3 Mac1 inhibition via interaction with the oxyanion subsite.

4. Discussion

SARS-CoV-2 NSP3 Mac1 presents an alternative, promising target for the development of a new type of antiviral agent [26]. While characterised antiviral drugs acting against host-derived targets still need to prove their effectiveness in the clinic, several compounds targeting proteins involved in the viral life cycle and/or pathogenesis are approved for treatment, with the RNA-dependent RNA polymerase and viral protease inhibitors being most widely used at present [10,11]. However, recent studies demonstrated that resistances to remdesivir and nirmatrelvir (marketed as Paxlovid™) can arise via multiple pathways [49,50], rendering treatments ineffective and requiring the development of alternative strategies. We supported these efforts by performing a computational screen along with an in vitro compound library screen of experimental small molecules and FDA-approved drugs against NSP3 Mac1.

Virtual screening of a medium-sized library (125,000) of small molecules from BioAscent yielded two novel inhibitors with micromolar potency, IAL-MD0305 and IAL-MD0306. Efforts to improve the compound potency using structure-based drug design was hampered by the failure of either compound to yield ligand-bound crystal structures, and, as the hits from the in vitro screening of the MIDAS library were more attractive, effort shifted to those hits.

Complementation of this computational approach with the in vitro screening of ~12,000 compounds allowed us to gain more insights into the chemical matter amenable for the active site of NSP3 Mac1. The established HTRF assay [36] allowed a reliable identification of hit compounds, considering that we observed a high reproducibility of primary screening hits confirming in dose–response titrations of the re-supplied compounds. Taken together, the in vitro screening approach was performed with an overall hit rate of 0.3%, yielding diverse hits with inhibitory activity of both, experimentally screening molecules to explore the chemical space and developed FDA-approved drugs. Four MIDAS library-based compounds defined by different scaffold types were followed-up by the characterisation of close analogues. Despite limitations in compound availability that would allow a step-by-step analysis of the contributions of individual functional groups and ring-systems attached to the core scaffold motif to the inhibitory activity on NSP3 Mac1, the characterisation of the selected analogues provided valuable SAR information, and the tested substituents extending and/or modifying the core scaffolds allowed preliminary exploration of the active site for NSP3 Mac1 targeting. Consistent observations for related analogues regarding effects on the inhibitory activity of similar scaffold modifications also increased our confidence for the discovered hit compounds as being true NSP3 Mac1 inhibitors. Of all characterised analogues, compounds classified to the scaffold type I group generally showed the strongest NSP3 Mac1 inhibition, with IAL-MD0131 also being crystallographically confirmed for target binding. The IAL-MD0131 co-crystal structure may also allow us to infer the binding mode of the other characterised analogues within this group, considering their close structural similarity. As such, based on the orientation of the pyrrolo[2,3-*b*]pyridine in the crystal structure, attachment of the amino group in compound IAL-MD0134 leads as assumed to a sterical clash within the adenosine binding pocket, rationalising the observed drop of the IC_{50} compared to the primary hit. The scaffold type I core motif, i.e., the pyrrolo[2,3-*b*]pyridine, is moreover present in variations in scaffold types II and III, which eases molecular docking studies for defining the binding site and orientation of the molecules for the structure-guided design of follow-up compounds. Further structural and biophysical studies that generally confirm the binding along with providing NSP3 Mac1 targeting information of compounds belonging to the scaffold type II–IV groups would greatly foster and complement the in vitro-based analogue characterisation.

Re-purposing drugs for new targets has the potential to considerably accelerate the drug discovery and development process. Screening a library of FDA-approved molecules led to the discovery of several new hits which have not been described in the literature so far. This may partly be due to the different composition of the available library and partly due to the different screening assay format, which hampers comparability. Cefatrizine, dasatinib, and dihydralazine were previously described as NSP3 Mac1 inhibitors [30,51], yet this could not be confirmed since they were not included in our library. Interestingly, hydralazine was among the screened drug molecules that showed no NSP3 Mac1 inhibition at the set screening concentration, thus potentially indicating that the second hydrazine moiety notably contributes to the inhibitory activity. However, cefaclor, rabeprazole, and telmisartan were also included in our FDA-approved compound library, for which we could yet not measure any inhibitory activity on NSP3 Mac1 in our set-up compared to other studies [51]. Furthermore, oxaprozin may present an example that could not be identified in our screen for NSP3 Mac1 targeting, which is likely due to the assay format. The binding of oxaprozin to NSP3 Mac1 was discovered by protein-based nuclear magnetic resonance (NMR) screening experiments [37]. However, with the HTRF assay employed in this study particularly detecting compounds displacing the ADP-ribose-conjugated peptide from the active site of the macrodomain, the assay is less sensitive for identifying allosteric binders that do not induce notable conformational changes of the active site. However, we could confirm the inhibitory activity of suramin, which was identified as NSP3 Mac1 hit, also by using a FRET-based screening assay [52]; however, we did not follow suramin up due to its complex molecular structure and the potential overlap of these two studies. Furthermore, thiomersal inhibited NSP3 Mac1 with an IC_{50} of 62.1 μM, which was also identified as an inhibitor for MacroD1 with an IC_{50} of 5.2 μM in an AlphaScreen-based assay format [53]. Possibly, thiomersal is able to target both macrodomains due to the structural conservation of their folds.

Compared to the previous studies, our screen identified a series of antibiotic drug molecules as inhibitors for NSP3 Mac1. Although not all of them showed reproducibility with confidence in the confirmatory dose–response titrations, we verified ceftazidime and aztreonam with micromolar inhibitory activity on NSP3 Mac1. Moreover, aztreonam was confirmed for true NSP3 Mac1 targeting by crystallographic studies, providing insights into its binding mode, which would have been potentially challenging to predict by computational docking studies due to exploring regions of the macrodomain outside of the well-defined ADP-ribose binding pocket. Aztreonam uses for NSP3 Mac1 inhibition motifs, including Phe156 targeting with an aromatic system and the interaction with the oxyanion site using a carboxylate group, as has been previously observed by low-molecular fragments for targeting the domain [36]. Of note is that, albeit being a conserved fold, NSP3 Mac1 is subjected to evolutionary amino acid substitutions that have effects on ADP-ribose binding and enhance the ability of SARS-CoV-2 to counteract host immune response [54]. It is therefore conceivable that substitutions, e.g., of Phe156 or in the oxyanion site, can evolve, which may impair inhibitor binding as drug resistance strategy of the virus. Also of note is that Phe156 (also targeted by the MIDAS hit compound IAL-MD0131) is unique to SARS-CoV-2 amongst the β-coronaviruses; however, it is found present in human macrodomains including MacroD2 and PARP14 MD3. While being potentially advantageous for the development of SARS-CoV-2 NSP3 Mac1 inhibitors that are selective over other viral Mac1 variants, additional inhibitor interactions need to be considered and elaborated to achieve selectivity over human macrodomains. Compared to fragments, aztreonam also provides an underlying scaffold linking these motifs that additionally extends beyond the ADP-ribose binding site, which could potentially be exploited for selectivity purposes over other macrodomains [26]. As an FDA-approved drug, aztreonam furthermore possesses good pharmacokinetic properties, and merging it with the fragment chemical matter could be considered to develop potent NSP3 Mac1 inhibitors in future studies.

Supplementary Materials: The following supporting information can be downloaded at: https://www.mdpi.com/article/10.3390/pathogens12020324/s1, Figure S1: Redocking of ADP-ribose into SARS-CoV-2 NSP3 Mac1; Figure S2: ADP-ribose is used as positive control and reference compound in the HTRF-based assay for inhibitor screening and characterisation for SARS-CoV-2 NSP3 Mac1.; Figure S3: Dose-response titrations of MIDAS compound library hits leading regarding NSP3 Mac1 inhibitory activity.; Table S1: Data collection and refinement statistics for crystal structures described in this study.; Table S2: Top hit compounds from virtual screening approach.; Table S3: Hit compounds from MIDAS compound library screen.; Table S4: Confirmation of hit compounds and analogues for SAR information.; Table S5: Hit compounds from FDA-approved compound library screen. Excel file: Supplementary Tables S2–4 with molecular structures.

Author Contributions: I.A., P.E.B. and M.S. conceived the project and conceptualised the experiments; M.S. designed the screening and hit follow-up experiments, conducted biochemical and crystallographic studies including deposition into PDB, and performed data analysis and interpretation; T.Z.-T. designed the macrodomain hit follow-up studies; J.B. and O.F. performed HTRF assays for primary and confirmatory screening; S.D.C. performed virtual ligand screening; D.F. and F.v.D. supported crystallographic hit characterisation; M.S., P.E.B. and I.A. wrote the manuscript with the support of all other authors. All authors have read and agreed to the published version of the manuscript.

Funding: T.Z.T. and P.E.B. thank Alzheimer's Research UK for support (grant no. ARUK-2021DDI-OX). Work in the Ivan Ahel Laboratory is supported by the Biotechnology and Biological Sciences Research Council (BB/R007195/1 and BB/W016613/1), the Wellcome Trust (210634 and 223107), Oxford University Challenge Seed Fund (USCF 456), and by Ovarian Cancer Research Alliance (813369).

Institutional Review Board Statement: Not applicable.

Informed Consent Statement: Not applicable.

Data Availability Statement: Crystallography atomic coordinates and structure factors are deposited in the PDB (https://www.rcsb.org, accessed on 11 January 2023) under the following accession codes: 8C19 and 8C1A. All data supporting the findings of this study are available within the Article or the Supplementary Information. Any further information will be provided upon request to the corresponding authors.

Acknowledgments: We thank Daniel Ebner and his team from the High Throughput Screening Facility of the Nuffield Department of Medicine for providing access to the FDA-approved compound library. We are grateful to Karly Aitmakhanova at the CMD/TDI Oxford screening facility for her support performing HTRF assays. We thank Edward Lowe for his help with the X-ray diffraction data collection. We are grateful to Diamond Light Source for access to and assistance at beamline I03 throughout the project [proposal numbers mx23459].

Conflicts of Interest: The authors declare no conflict of interest.

References

1. Adam, D. True covid death toll could be more than double official count. *Nature* **2022**, *605*, 206. [CrossRef] [PubMed]
2. Gorbalenya, A.E.; Baker, S.C.; Baric, R.S.; de Groot, R.J.; Drosten, C.; Gulyaeva, A.A.; Haagmans, B.L.; Lauber, C.; Leontovich, A.M.; Neuman, B.W.; et al. The species Severe acute respiratory syndrome-related coronavirus: Classifying 2019-nCoV and naming it SARS-CoV-2. *Nat. Microbiol.* **2020**, *5*, 536–544.
3. Nicola, M.; Alsafi, Z.; Sohrabi, C.; Kerwan, A.; Al-jabir, A. The socio-economic implications of the coronavirus pandemic (COVID-19): A review. *Int. J. Surg. J.* **2020**, *78*, 185–193. [CrossRef] [PubMed]
4. Cui, J.; Li, F.; Shi, Z.L. Origin and evolution of pathogenic coronaviruses. *Nat. Rev. Microbiol.* **2019**, *17*, 181–192. [CrossRef]
5. Bai, C.; Zhong, Q.; Gao, G.F. Overview of SARS-CoV-2 genome-encoded proteins. *Sci. China Life Sci.* **2022**, *65*, 280–294. [CrossRef]
6. Chan, J.F.; Kok, K.H.; Zhu, Z.; Chu, H.; To, K.K.; Yuan, S.; Yuen, K.Y. Genomic characterization of the 2019 novel human-pathogenic coronavirus isolated from a patient with atypical pneumonia after visiting Wuhan. *Emerg. Microbes Infect.* **2020**, *9*, 540. [CrossRef]
7. Wu, F.; Zhao, S.; Yu, B.; Chen, Y.M.; Wang, W.; Song, Z.G.; Hu, Y.; Tao, Z.W.; Tian, J.H.; Pei, Y.Y.; et al. A new coronavirus associated with human respiratory disease in China. *Nature* **2020**, *579*, 265–269. [CrossRef]
8. Sheahan, T.P.; Sims, A.C.; Zhou, S.; Graham, R.L.; Pruijssers, A.J.; Agostini, M.L.; Leist, S.R.; Schafer, A.; Dinnon, K.H.; Stevens, L.J.; et al. An orally bioavailable broad-spectrum antiviral inhibits SARS-CoV-2 in human airway epithelial cell cultures and multiple coronaviruses in mice. *Sci. Transl. Med.* **2020**, *12*, eabb5883. [CrossRef]
9. Shyr, Z.A.; Gorshkov, K.; Chen, C.Z.; Zheng, W. Drug discovery strategies for sars-cov-2. *J. Pharmacol. Exp. Ther.* **2020**, *375*, 127–138. [CrossRef]
10. Aiello, T.F.; García-Vidal, C.; Soriano, A. Antiviral drugs against SARS-CoV-2. *Rev. Esp. Quimioter.* **2022**, *35*, 10–15. [CrossRef]
11. Cully, M. A tale of two antiviral targets—And the COVID-19 drugs that bind them. *Nat. Rev. Drug Discov.* **2022**, *21*, 3–5. [CrossRef]
12. Lei, J.; Kusov, Y.; Hilgenfeld, R. Nsp3 of coronaviruses: Structures and functions of a large multi-domain protein. *Antiviral Res.* **2020**, *149*, 58–74. [CrossRef] [PubMed]
13. Palazzo, L.; Mikolčević, P.; Mikoč, A.; Ahel, I. ADP-ribosylation signalling and human disease. *Open Biol.* **2019**, *9*, 190041. [CrossRef] [PubMed]
14. Rack, J.G.M.; Perina, D.; Ahel, I. Macrodomains: Structure, Function, Evolution, and Catalytic Activities. *Annu. Rev. Biochem.* **2016**, *85*, 431–454. [CrossRef] [PubMed]
15. Schuller, M.; Ahel, I. Beyond protein modification: The rise of non-canonical ADP-ribosylation. *Biochem. J.* **2022**, *479*, 463–477. [CrossRef]
16. Atasheva, S.; Frolova, E.I.; Frolov, I. Interferon-Stimulated Poly(ADP-Ribose) Polymerases Are Potent Inhibitors of Cellular Translation and Virus Replication. *J. Virol.* **2014**, *88*, 2116–2130. [CrossRef]
17. Fehr, A.R.; Singh, S.A.; Kerr, C.M.; Mukai, S.; Higashi, H.; Aikawa, M. The impact of PARPs and ADP-ribosylation on inflammation and host-pathogen interactions. *Genes Dev.* **2020**, *34*, 341–359. [CrossRef] [PubMed]
18. Li, L.; Zhao, H.; Liu, P.; Li, C.; Quanquin, N.; Ji, X.; Sun, N.; Du, P.; Qin, C.F.; Lu, N.; et al. PARP12 suppresses Zika virus infection through PARP-dependent degradation of NS1 and NS3 viral proteins. *Sci. Signal.* **2018**, *11*, eaas9332. [CrossRef]
19. Zhang, Y.; Mao, D.; Roswit, W.T.; Jin, X.; Patel, A.C.; Patel, D.A.; Agapov, E.; Wang, Z.; Tidwell, R.M.; Atkinson, J.J.; et al. PARP9-DTX3L ubiquitin ligase targets host histone H2BJ and viral 3C protease to enhance interferon signaling and control viral infection. *Nat. Immunol.* **2015**, *16*, 1215–1227. [CrossRef]
20. Goenka, S.; Boothby, M. Selective potentiation of Stat-dependent gene expression by collaborator of Stat6 (CoaSt6), a transcriptional cofactor. *PNAS* **2006**, *103*, 4210–4215. [CrossRef]
21. Iwata, H.; Goettsch, C.; Sharma, A.; Ricchiuto, P.; Goh, W.W.B.; Halu, A.; Yamada, I.; Yoshida, H.; Hara, T.; Wei, M.; et al. PARP9 and PARP14 cross-regulate macrophage activation via STAT1 ADP-ribosylation. *Nat. Commun.* **2016**, *7*, 12849. [CrossRef]
22. Caprara, G.; Prosperini, E.; Piccolo, V.; Sigismondo, G.; Melacarne, A.; Cuomo, A.; Boothby, M.; Rescigno, M.; Bonaldi, T.; Natoli, G. PARP14 Controls the Nuclear Accumulation of a Subset of Type I IFN–Inducible Proteins. *J. Immunol.* **2018**, *200*, 2439–2454. [CrossRef] [PubMed]
23. Grunewald, M.E.; Chen, Y.; Kuny, C.; Maejima, T.; Lease, R.; Ferraris, D.; Aikawa, M.; Sullivan, C.S.; Perlman, S.; Fehr, A.R. The coronavirus macrodomain is required to prevent PARP-mediated inhibition of virus replication and enhancement of IFN expression. *PLoS Pathog.* **2019**, *15*, e1007756. [CrossRef] [PubMed]
24. Parthasarathy, S.; Fehr, A.R. PARP14: A key ADP-ribosylating protein in host–virus interactions? *PLoS Pathog.* **2022**, *18*, 8–12. [CrossRef] [PubMed]
25. Hoch, N.C. Host ADP-ribosylation and the SARS-CoV-2 macrodomain. *Biochem. Soc. Trans.* **2021**, *49*, 1711–1721. [CrossRef]
26. Rack, J.M.G.; Zorzini, V.; Zhu, Z.; Schuller, M.; Ahel, D.; Ahel, I. Viral macrodomains: A structural and evolutionary assessment of the pharmacological potential. *Open Biol.* **2020**, *10*, 200237. [CrossRef]
27. Fehr, A.R.; Channappanavar, R.; Jankevicius, G.; Fett, C.; Zhao, J.; Athmer, J.; Meyerholz, D.K.; Ahel, I.; Perlman, S. The conserved coronavirus macrodomain promotes virulence and suppresses the innate immune response during severe acute respiratory syndrome coronavirus infection. *MBio* **2016**, *7*, 1–12. [CrossRef]
28. Alhammad, Y.M.O.; Fehr, A.R. The viral macrodomain counters host antiviral ADP-ribosylation. *Viruses* **2020**, *12*, 384. [CrossRef]
29. Leung, A.K.L.; Griffin, D.E.; Bosch, J.; Fehr, A.R. The Conserved Macrodomain Is a Potential Therapeutic Target for Coronaviruses and Alphaviruses. *Pathogens* **2022**, *11*, 94. [CrossRef]

30. Dasovich, M.; Zhuo, J.; Goodman, J.A.; Thomas, A.; McPherson, R.L.; Jayabalan, A.K.; Busa, V.F.; Cheng, S.J.; Murphy, B.A.; Redinger, K.R.; et al. High-Throughput Activity Assay for Screening Inhibitors of the SARS-CoV-2 Mac1 Macrodomain. *ACS Chem. Biol.* **2022**, *17*, 17–23. [CrossRef]
31. Gahbauer, S.; Correyb, G.J.; Schuller, M.; Ferlad, M.P.; Dorukf, Y.U.; Rachmana, M.; Wug, T.; Diolaitif, M.; Wangh, S.; Neitzi, R.J.; et al. Iterative computational design and crystallographic screening identifies potent inhibitors targeting the Nsp3 macrodomain of SARS-CoV-2. *Proc. Natl. Acad. Sci. USA* **2023**, *120*, e2212931120. [CrossRef]
32. Moiani, D.; Link, T.M.; Brosey, C.A.; Katsonis, P.; Lichtarge, O.; Kim, Y.; Joachimiak, A.; Ma, Z.; Kim, I.K.; Ahmed, Z.; et al. *An Efficient Chemical Screening Method for Structure-Based Inhibitors to Nucleic Acid Enzymes Targeting the DNA Repair-Replication Interface and SARS CoV-2*; Elsevier Inc.: Amsterdam, The Netherlands, 2021.
33. Roy, A.; Alhammad, Y.M.; Mcdonald, P.; Johnson, D.K.; Zhuo, J.; Leung, A.K.L.; Fehr, A.R. Discovery of compounds that inhibit SARS-CoV-2 Mac1-ADP-ribose binding by high-throughput screening. *Antiviral Res.* **2022**, *203*, 105344. [CrossRef] [PubMed]
34. Correy, G.J.; Kneller, D.W.; Phillips, G.; Pant, S.; Russi, S.; Cohen, A.E.; Meigs, G.; Holton, J.M.; Gahbauer, S.; Thompson, M.C.; et al. The mechanisms of catalysis and ligand binding for the SARS-CoV-2 NSP3 macrodomain from neutron and x-ray diffraction at room temperature. *Sci. Adv.* **2022**, *8*, eabo5083. [CrossRef] [PubMed]
35. Ni, X.; Der, M.S.; Olieric, V.; Sharpe, M.E.; Hernandez-olmos, V.; Proschak, E.; Merk, D.; Knapp, S.; Chaikuad, A. Structural Insights into Plasticity and Discovery of Remdesivir Metabolite GS-441524 Binding in SARS-CoV - 2 Macrodomain. *ACS Med. Chem. Lett.* **2021**, *12*, 603–609. [CrossRef] [PubMed]
36. Schuller, M.; Correy, G.J.; Gahbauer, S.; Fearon, D.; Wu, T.; Díaz, R.E.; Young, I.D.; Carvalho Martins, L.; Smith, D.H.; Schulze-Gahmen, U.; et al. Fragment binding to the Nsp3 macrodomain of SARS-CoV-2 identified through crystallographic screening and computational docking. *Sci. Adv.* **2021**, *7*, eabf8711. [CrossRef] [PubMed]
37. Li, J.; Zhong, F.; Li, M.; Liu, Y.; Wang, L.; Liu, M.; Li, F.; Zhang, J.; Wu, J.; Shi, Y.; et al. Two Binding Sites of SARS-CoV-2 Macrodomain 3 Probed by Oxaprozin and Meclomen. *J. Med. Chem.* **2022**, *65*, 15227–15237. [CrossRef]
38. Babar, Z.; Khan, M.; Zahra, M.; Anwar, M.; Noor, K.; Hashmi, H.F.; Suleman, M.; Waseem, M.; Shah, A.; Ali, S.; et al. Drug similarity and structure-based screening of medicinal compounds to target macrodomain-I from SARS-CoV-2 to rescue the host immune system: A molecular dynamics study. *J. Biomol. Struct. Dyn.* **2022**, *40*, 523–537. [CrossRef]
39. Patel, D.C.; Hausman, K.R.; Arba, M.; Tran, A.; Lakernick, P.M.; Wu, C. Novel inhibitors to ADP ribose phosphatase of SARS-CoV-2 identified by structure-based high throughput virtual screening and molecular dynamics simulations. *Comput. Biol. Med.* **2022**, *140*, 105084. [CrossRef]
40. Brosey, C.A.; Houl, J.H.; Katsonis, P.; Jones, D.E.; Ahmed, Z.; Tainer, J.A. Targeting SARS-CoV-2 Nsp3 macrodomain structure with insights from human poly(ADP-ribose) glycohydrolase (PARG) structures with inhibitors. *Prog. Biophys. Mol. Biol.* **2021**, *163*, 171–186. [CrossRef] [PubMed]
41. Baell, J.; Walters, M.A. Chemical con artists foil drug discovery. *Nature* **2014**, *513*, 481–483. [CrossRef] [PubMed]
42. Winter, G. *xia2*: An expert system for macromolecular crystallography data reduction. *J. Appl. Crystallogr.* **2010**, *43*, 186–190. [CrossRef]
43. Storoni, L.C.; McCoy, A.J.; Read, R.J. Likelihood-enhanced fast rotation functions. *Acta Crystallogr. Sect. D Biol. Crystallogr.* **2004**, *60*, 432–438. [CrossRef] [PubMed]
44. Emsley, P.; Cowtan, K. Coot: Model-building tools for molecular graphics. Acta Crystallogr. *Sect. D Biol. Crystallogr.* **2004**, *60*, 2126–2132. [CrossRef]
45. Murshudov, G.N.; Skubák, P.; Lebedev, A.A.; Pannu, N.S.; Steiner, R.A.; Nicholls, R.A.; Winn, M.D.; Long, F.; Vagin, A.A. REFMAC5 for the refinement of macromolecular crystal structures. Acta Crystallogr. *Sect. D Biol. Crystallogr.* **2011**, *67*, 355–367. [CrossRef] [PubMed]
46. Chen, V.B.; Arendall, W.B.; Headd, J.J.; Keedy, D.A.; Immormino, R.M.; Kapral, G.J.; Murray, L.W.; Richardson, J.S.; Richardson, D.C. MolProbity: All-atom structure validation for macromolecular crystallography. *Acta Crystallogr.* **2010**, *D66*, 12–21. [CrossRef]
47. Schuller, M.; Riedel, K.; Gibbs-Seymour, I.; Uth, K.; Sieg, C.; Gehring, A.P.; Ahel, I.; Bracher, F.; Kessler, B.M.; Elkins, J.M.; et al. Discovery of a Selective Allosteric Inhibitor Targeting Macrodomain 2 of Polyadenosine-Diphosphate-Ribose Polymerase 14. *ACS Chem. Biol.* **2017**, *12*, 2866–2874. [CrossRef] [PubMed]
48. Frick, D.N.; Virdi, R.S.; Vuksanovic, N.; Dahal, N.; Silvaggi, N.R. Molecular Basis for ADP-Ribose Binding to the Mac1 Domain of SARS-CoV-2 nsp3. *Biochemistry* **2020**, *59*, 2608–2615. [CrossRef] [PubMed]
49. Iketani, S.; Mohri, H.; Culbertson, B.; Hong, S.J.; Duan, Y.; Luck, M.I.; Annavajhala, M.K.; Guo, Y.; Sheng, Z.; Uhlemann, A.; et al. Multiple pathways for SARS-CoV-2 resistance to nirmatrelvir. *Nature* **2022**, *613*, 558–564. [CrossRef]
50. Stevens, L.J.; Pruijssers, A.J.; Lee, H.W.; Gordon, C.J.; Tchesnokov, E.P.; Gribble, J.; George, A.S.; Hughes, T.M.; Lu, X.; Li, J.; et al. Mutations in the SARS-CoV-2 RNA-dependent RNA polymerase confer resistance to remdesivir by distinct mechanisms. *Sci. Transl. Med.* **2022**, *14*, eabo0718. [CrossRef]
51. Virdi, R.S.; Bavisotto, R.V.; Hopper, N.C.; Vuksanovic, N.; Melkonian, T.R.; Silvaggi, N.R.; Frick, D.N. Discovery of Drug-Like Ligands for the Mac1 Domain of SARS-CoV-2 Nsp3. *SLAS Discov.* **2020**, *25*, 1162–1170. [CrossRef]
52. Sowa, S.T.; Galera-Prat, A.; Wazir, S.; Alanen, H.I.; Maksimainen, M.M.; Lehtiö, L. A molecular toolbox for ADP-ribosyl binding proteins. *Cell Rep. Methods* **2021**, *1*, 100121. [CrossRef] [PubMed]

53. Haikarainen, T.; Maksimainen, M.M.; Obaji, E.; Lehtiö, L. Development of an Inhibitor Screening Assay for Mono-ADP-Ribosyl Hydrolyzing Macrodomains Using AlphaScreen Technology. *SLAS Discov.* **2018**, *23*, 255–263. [CrossRef] [PubMed]
54. Hussain, I.; Pervaiz, N.; Khan, A.; Saleem, S.; Shireen, H.; Wei, D.Q.; Labrie, V.; Bao, Y.; Abbasi, A.A. Evolutionary and structural analysis of SARS-CoV-2 specific evasion of host immunity. *Genes Immun.* **2020**, *21*, 409–419. [CrossRef] [PubMed]

pathogens

Review

ADP-Ribosylation in Antiviral Innate Immune Response

Qian Du [1,2], Ying Miao [1,2], Wei He [1,2] and Hui Zheng [1,2,*]

1 Institutes of Biology and Medical Sciences, Soochow University, Suzhou 215123, China
2 Jiangsu Key Laboratory of Infection and Immunity, Soochow University, Suzhou 215123, China
* Correspondence: huizheng@suda.edu.cn

Abstract: Adenosine diphosphate (ADP)-ribosylation is a reversible post-translational modification catalyzed by ADP-ribosyltransferases (ARTs). ARTs transfer one or more ADP-ribose from nicotinamide adenine dinucleotide (NAD^+) to the target substrate and release the nicotinamide (Nam). Accordingly, it comes in two forms: mono-ADP-ribosylation (MARylation) and poly-ADP-ribosylation (PARylation). ADP-ribosylation plays important roles in many biological processes, such as DNA damage repair, gene regulation, and energy metabolism. Emerging evidence demonstrates that ADP-ribosylation is implicated in host antiviral immune activity. Here, we summarize and discuss ADP-ribosylation modifications that occur on both host and viral proteins and their roles in host antiviral response.

Keywords: ADP-ribosylation; PARylation; MARylation; post-translational modifications; viral infection; antiviral response; IFN-I; innate immunity

check for updates

Citation: Du, Q.; Miao, Y.; He, W.; Zheng, H. ADP-Ribosylation in Antiviral Innate Immune Response. *Pathogens* **2023**, *12*, 303. https://doi.org/10.3390/pathogens12020303

Academic Editors: Anthony K. L. Leung, Anthony Fehr and Rachy Abraham

Received: 17 January 2023
Revised: 9 February 2023
Accepted: 10 February 2023
Published: 12 February 2023

1. Introduction

Rapid and appropriate cellular responses are essential for organisms to respond to external stimuli. Post-translational modifications (PTMs) play an important role in this process by modulating intracellular signal transduction pathways [1]. Mechanically, PTMs regulate signaling pathways and gene expression mainly by affecting the catalytic activity of the target protein or the interaction of the target protein with other molecules [2]. Adenosine diphosphate (ADP)-ribosylation is an ancient PTM discovered around the 1960s [3]. ADP-ribosyltransferases (ARTs) are responsible for ADP-ribosylation [4]. ARTs are a superfamily consisting of 23 members, two of which are diphtheria toxin-like ADP-ribosyltransferase (ARTD) and cholera-like ADP-ribosyltransferase (ARTC) [5]. The ARTD family includes PARPs and TNKS (tankyrase), as detailed in Table 1. ARTs transfer one or more ADP-ribose (ADPr) units from nicotinamide adenine dinucleotide (NAD^+) to target proteins on a variety of amino acids, including lysine (Lys), arginine (Arg), glutamate (Glu), aspartate (Asp), and cysteine (Cys) [6], leading to mono-ADP-ribosylation (MARylation) or poly-ADP-ribosylation (PARylation) of substrates. The human genome encodes 17 PARPs [7], most of which share a common NAD^+ binding motif in their catalytic domain with a similar secondary structure [8,9]. In these NAD^+ binding motifs, a histidine (His) residue and a tyrosine (Tyr) residue are essential for positioning NAD^+ in a correct orientation, and a conserved Glu residue is crucial for ADP-ribose transfer. The histidine-tyrosine-glutamate motif (H–Y–E motif and variants thereof) is also known as the catalytic triad of PARPs [10,11]. However, not all PARPs reserve these critical residues. For example, PARP13 lacks catalytic activity due to the substitution of a His residue on the key NAD^+-binding motif. Therefore, PARP13 is the only PARP family member with a catalytically inactive domain [12,13]. In fact, only PARP1, 2, 3, TNKS1 and TNKS2 retain these residues and catalytic PARylation [14], while the rest of PARPs (except PARP13) possess the MARylation activity due to the substitution of the Glu residue [8,15,16]. In brief, the ability of PARPs to catalyze MARylation or PARylation of their protein substrates depends on conserved structural features such as the catalytic triad and the presence of certain cofactors.

As for degradation of the poly-ADP-ribose chain, poly-ADP-ribose glycohydrolase (PARG) and ADP-ribosyl-acceptor hydrolase (ARH) 3 are responsible for it. PARG can only hydrolyze ribose-ribosyl *O*-glycosidic bonds, and the final product of hydrolysis is a MARylated protein. While ARH3 can hydrolyze the bond between the amino acid side chain and the ribose, thereby completely reversing ADP-ribosylation. The chemical bond between the amino acid side chain and ribose is different from the ribose-ribose bond in the PAR chain, which explains why different enzymes are required to accomplish the two reactions. Moreover, ARH1 is also an ADP-ribosylhydrolase, which can specifically hydrolyze MARylated proteins at arginine residues [17,18]. In addition, three macrodomain-containing enzymes, MacroD1, MacroD2, and TARG1, can also act as mono-ADP-ribosylhydrolases [19–21]. Notably, ARH3 acts primarily at the terminal sites of the PAR chains and the serine residues of MARylated proteins, while MacroD1, D2, and TARG1 are involved in hydrolyzing ADP-ribosylation of acidic residues [2]. In a word, ADP-ribosylation is a fully reversible PTM. The dynamic balance of ADP-ribosylation synthesis and degradation provides plasticity for rapid cellular response to external signals.

ADP-ribosylation occurs in both prokaryotes and eukaryotes and is particularly prevalent in stress responses requiring rapid adaptation [22,23]. MARylation and PARylation are involved in a variety of physiological and pathophysiological processes, including DNA damage recognition, chromatin regulation, and oxidative stress response [24–30]. During this last decade, several studies uncovered the roles of ADP-ribosylation in antiviral immunity [31,32]. Of note is that there are many PARPs that perform antiviral functions without their catalytic activity, and will not be discussed in detail here [15]. In this review, we mainly summarize and discuss new findings on the role of ADP-ribosylation occurred on the host and viral proteins in antiviral response. As to other substrates, such as nucleotides or other small molecules, we will only give a brief introduction here (Table 1).

Table 1. The enzymes that catalyze ADP-ribosylation and their main functions in antiviral innate immune response.

Name	ADP-Ribosylation Activity	Other Names	Roles in Antiviral Innate Immune Response
PARP1	PARylation	ARTD1	PARylating EBNA1 and LANA [33,34] Degradation of IFNAR1 [35]
PARP2	PARylation	ARTD2	Not known
PARP3	PARylation	ARTD3	Regulating PARP1 [31]
PARP4	MARylation	ARTD4	Not known
PARP5a	PARylation	ARTD5 Tankyrase 1	PARylating MAVS and promoting its degradation [36] PARylating EBNA1 [37]
PARP5b	PARylation	ARTD6 Tankyrase 2	PARylating MAVS and promoting its degradation [36] PARylating EBNA1 [37]
PARP6	MARylation	ARTD17	Not known
PARP7	MARylation	ARTD14, BAL3	MARylating and inhibiting TBK1 [38] MARylating PARP13 [23]
PARP8	MARylation	ARTD16	Not known
PARP9	MARylation	ARTD9, BAL1	Forming a complex with DTX3L and MARylating ubiquitin at Gly76 [39] Suppressing PARP14-mediated MARylation of STAT1 [40]
PARP10	MARylation	ARTD10	MARylating nsP2 of CHIKV and promoting its degradation [41]
PARP11	MARylation	ARTD11	Targeting β-TrCP and promoting IFNAR1 degradation [42] Enhancing PARP12-mediated NS1 and NS3 ADP-ribosylation [43]

Table 1. *Cont.*

Name	ADP-Ribosylation Activity	Other Names	Roles in Antiviral Innate Immune Response
PARP12	MARylation	ARTD12	Targeting NS1 and NS3 of ZIKV and promoting its degradation [44]
PARP13	Inactive	ARTD13	Restricts viral replication [45–48]
PARP14	MARylation	ARTD8	MARylating STAT1 and preventing its phosphorylation [40] MARylating PARP13 [49]
PARP15	MARylation	ARTD7	Not known
PARP16	MARylation	ARTD15	Not known

2. ADP-Ribosylation of Host Proteins

The innate immune system is the main defense mechanism of higher organisms against pathogens such as viruses. It senses and responds to pathogen-associated molecular patterns (PAMPs) through pattern recognition receptors (PRRs). During this process, interferons (IFNs) are produced [50–52], enabling a rapid host response to viral infection.

Interferons are commonly classified into three types according to their receptor complex, designated types I to III [53,54]. Type I IFN (IFN-I) comprises IFNα and IFNβ, and most virus-infected cells are able to produce IFN-I to resist viruses. Type II IFN (IFN-II) consists only of IFN-γ, which is synthesized by certain immune cells. However, it can offer resistance to a wide range of pathogens [55]. Type III IFN (IFN-III) includes IFN-λ, which can be produced by most cell types. IFN-III plays an important role in the innate immune response of the intestinal and respiratory mucosal barriers [56–58]. Among the three types of IFNs, IFN-I is well characterized and plays an important role in the host response against viral infection [59–61]. After IFN-I is produced, it first binds to its membrane receptors (IFNAR1 and IFNAR2) to induce their dimerization and then initiates the autophosphorylation of Janus kinase 1 (JAK1) and tyrosine kinase 2 (TYK2). Phosphorylated JAK1 and TYK2 recruit and activate the signal transducers and activators of transcription 1 (STAT1) and STAT2. Phosphorylated STAT1 and STAT2, together with IFN-regulatory factor 9 (IRF9), form a well-characterized complex, IFN-stimulated gene factor 3 (ISGF3). Then, ISGF3 translocates to the nucleus and binds to IFN-stimulated response elements (ISRE), the promoter of IFN-stimulated genes (ISGs), leading to the expression of ISGs [62–66]. Eventually, these ISGs perform antiviral functions, and therefore host establishes the antiviral status [59,67,68]. In addition, IFN affects several other processes, including cell growth, differentiation, and apoptosis, as well as immune regulation [69–71].

IFN and IFN-induced ISGs are so important for the maintenance of host antiviral status that their production undergoes complex and delicate regulation. ADP-ribosylation plays an important role in this process. Here, we first focus on ADP-ribosylation modification that occurs on proteins associated with IFN-I production and IFN-I-induced JAK-STAT signaling pathway (Figure 1). These include several proteins such as Mitochondrial antiviral signaling protein (MAVS), TANK-binding kinase 1 (TBK1) and STAT1. In addition, lots of ADP-ribosylation that occurs on other host antiviral factors can also affect host antiviral response. For example, PARP11 MARylates β-transducin repeat-containing protein (β-TrCP) and promote IFNAR1 ubiquitination and degradation, resulting in the downregulation of IFN-I signaling and antiviral activity. Next, we will describe these contents in detail.

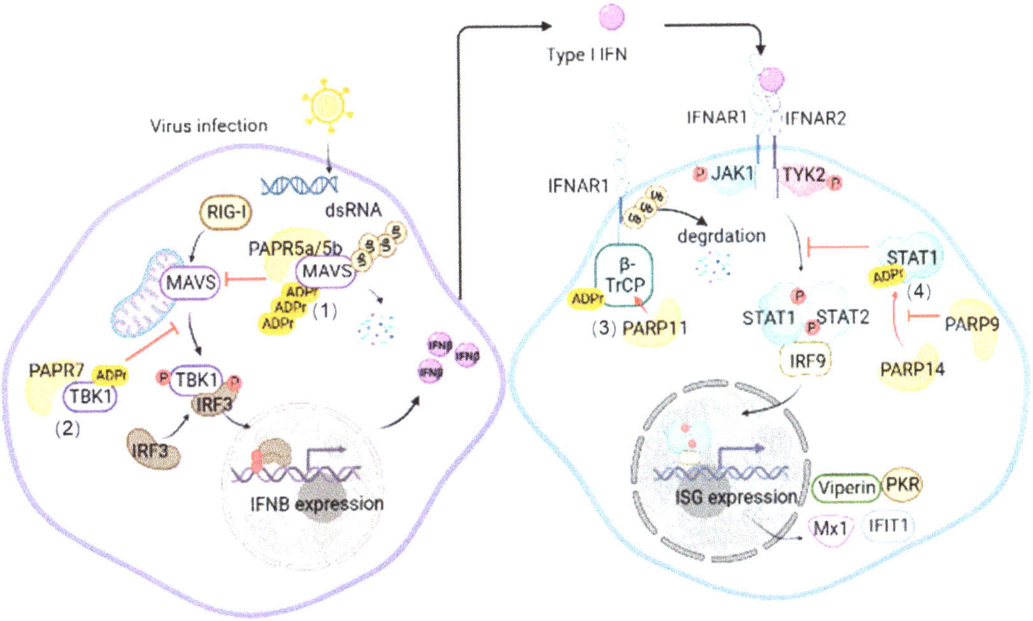

Figure 1. ADP-ribosylation in IFN-I production and IFN-I-induced JAK-STAT signaling pathway. During viral infection, (1) TNKS1 and TNKS2 interact with MAVS and catalyze its PARylation, leading to the degradation of MAVS and impairment of IFN production. (2) PARP7 targets TBK1 and catalyzes it mono-ADP-ribosylation, which suppresses the activation of TBK1 and downstream events. (3) PARP11 mono-ADP-ribosylates β-TrCP and mediates IFNAR1 ubiquitination and degradation, thus acting as a negative regulator of IFN-I response. (4) PARP14 catalyzes mono-ADP-ribosylation of STAT1 and prevents its phosphorylation, thus repressing the IFN-I response. PARP9 suppresses PARP14-mediated MARylation of STAT1, which sustains STAT1 phosphorylation. (Created with BioRender.com (accessed on 16 January 2023).

2.1. *ADP-Ribosylation of Proteins Associated with the Signaling Pathway of IFN-I Production*

MAVS (also known as virus-induced signal adaptor [VISA]) is a key signaling molecule that mediates antiviral innate immune response initiated by RNA viruses [72–75]. In this process, retinoic acid-induced gene I (RIG-I) acts as a pattern recognition receptor to recognize and bind viral RNA in response to RNA virus infection [76]. Subsequent conformational change of RIG-I exposes its N-terminal caspase recruitment domain (2CARD), which binds to the N-terminal CARD domain of MAVS and induces MAVS polymerization at the mitochondrial outer membrane [77]. Active MAVS polymers recruit the tumor necrosis factor receptor-associated factor (TRAF) family to synthesize polyubiquitin chains that activate TBK1 and IκB-kinase (IKK). Activated TBK1 and IKK first phosphorylate MAVS so that phosphorylated MAVS can recruit IFN-regulatory factor 3 (IRF3) by binding to its conserved positively charged surface. When TBK1 and IRF3 are in close proximity, IRF3 is phosphorylated by TBK1. Finally, phosphorylated IRF3 dissociates from MAVS and translocates to the nucleus after forming a dimer, where it binds to the promoter of the IFN-I gene and drives type I IFN production [50]. Therefore, the strict and subtle dynamic regulation of MAVS is very important to antiviral immune response. PARP5a/Tankyrase 1 (TNKS1) and its homolog PARP5b/TNKS2 are known to catalyze the PARylation of their substrates. The five ankyrin repeat (ANK) units at the N-terminus of TNKS1 and TNKS2 are the structural basis for their substrate recognition. The C-terminal catalytic domain is responsible for the ADP-ribosylation of their substrates [78]. A recent study showed that TNKS1 and TNKS2 can poly-ADP-ribosylate MAVS at Glu137 residue. After viral infection,

TNKS1 and TNKS2 are upregulated and translocate from the cytosol to mitochondria, interacting with MAVS and catalyzing its PARylation [36]. The PARylation of MAVS serves as a signal for the ubiquitin E3 ligase Ring figure protein 146 (RNF146)-mediated K48-linked polyubiquitination and subsequent degradation of MAVS [79], thereby negatively regulating the innate immune response to RNA viruses.

As mentioned above, TBK1 is a key kinase that induces IFN-I production. PARP7 (TIPARP) is an ADP-ribosylase whose expression is upregulated by aryl hydrocarbon receptor (AHR) [38,80]. AHR is a ligand-activated transcription factor that can be activated by a variety of environmental xenobiotics. Therefore, there is a close connection between AHR and innate immune signaling [80–83]. During viral infection, AHR-induced PARP7 can interact with TBK1 and catalyze its mono-ADP-ribosylation, which suppresses the activation of TBK1 and subsequent phosphorylation of IRF3. Consistent with this, AHR-deficient ($Ahr^{-/-}$) MEFs and MEFs with PARP7-knockdown by siPARP7 show stronger antiviral effects [38]. Thus, the AHR-PARP7 axis is a negative regulator of interferon signaling. However, the ADP ribosylation site of TBK1 remains to be determined. It is reported that PARP7 mainly modifies cysteines and acidic amino acids (glutamates and aspartates), which also provides a basis for the determination of the ADP-ribosylation site of TBK1 [84,85].

2.2. *ADP-Ribosylation of Proteins Associated with IFN-I-Induced JAK-STAT Signaling Pathway*

STAT1 stands out as a key functional component of the interferon signaling pathway, and its post-translational modification profoundly affects signal transduction [86]. For example, linear ubiquitination of STAT1 inhibits STAT1 activation and thus lowers the strength of IFN-I antiviral signaling [66]. The ADP-ribosylation of STAT1 was first reported by Iwata et al. [40]. It is PARP14 that catalyzes mono-ADP-ribosylation of STAT1 and prevents its phosphorylation. Consistent with this, PARP14 silencing and STAT1 ribosylation site mutation promote STAT1 phosphorylation and STAT1-driven ISG expression, thus enhancing interferon signaling. Interestingly, PARP9 and PARP14 physically and functionally interact with each other. Protein ribosylation assay and mass spectrometry showed that PARP9 suppresses PARP14-mediated MARylation of STAT1, which sustains STAT1 phosphorylation. In other words, PARP14 and PARP9 play opposite roles in this process [40]. However, STAT1 SUMOylation also affects its phosphorylation, which in turn affects interferon downstream events [87]. The ADP-ribosylation of STAT1 has therefore been questioned and requires further evidence. It is worth noting that in mouse bone marrow-derived macrophages (BMDMs), the mRNA level of PARP14 is significantly upregulated when stimulated by IFNβ or toll-like receptor (TLR) agonists such as polyinosinic-polycytidylic acid (Poly (I: C)). Depletion of PARP14 results in a decrease in IRF3-mediated IFNβ production and ISG expression, thereby impairing the interferon response [88]. Taken together, PARP14 plays a complex role in the process of interferon resistance to pathogens.

2.3. *ADP-Ribosylation of Other Host Antiviral Factors*

β-TrCP is the ubiquitin E3 ligase of interferon-alpha/beta receptor 1 (IFNAR1) and mediates IFNAR1 ubiquitination and degradation [89]. β-TrCP belongs to the F-box/WD40 repeats family, which contains an F-box motif and seven WD40 motifs [90]. Among them, WD40 repeats are responsible for the binding of β-TrCP to its protein substrate. PARP11 can mono-ADP-ribosylate β-TrCP in the WD40 repeats. The MARylatopn of β-TrCP inhibits its ubiquitin-proteasome degradation and enhances the ability of β-TrCP to interact with IFNAR1, which in turn promotes IFNAR1 ubiquitination and degradation. The above process ultimately results in the downregulation of IFN-I signaling and antiviral activity. Moreover, viral infection could lead to the upregulation of PARP11, thus restricting IFN-I-induced expression of ISGs and promoting ADP-ribosylation-mediated immune evasion of the virus. Similarly, PARP11 knockdown significantly downregulates the protein level of β-TrCP and upregulates the protein level of IFNAR1, therefore enhancing the IFN-I-activated signaling pathway and antiviral activity. Taken together, PARP11 and PARP11-

mediate-MARylation are highly effective targets for improving the antiviral efficacy of type-I interferon [42]. Consistent with this, a recent study shows that during Influenza A viruses (IAV) infection, PARP1 activity could facilitate IAV-induced IFNAR1 degradation, thus promoting virus propagation [35]. However, the substrate for this ADP-ribosylation process remains to be determined.

Initially, PARP9 was also thought to be enzymatically inactive. PARP9 was reported to form a heterodimer complex with Deltex E3 ubiquitin ligase 3L (DTX3L) [91]. DTX3L and PARP9 are relatively highly expressed in prostate cancer and breast cancer and share a common promoter. Both genes are responsive to IFNγ and activated in cells that express a dominant active form of STAT1 [92]. In addition, the complex can interact with STAT1 to promote STAT1-mediated ISG expression and thus enhance antiviral response [93]. Interestingly, in 2017, a study found that PARP9 can display mono-ADP-ribosylation activity in the case of DTX3L as its chaperone. DTX3L is an E3 ligase containing a RING domain, and PARP9/DTX3L heterodimer still has an E3 ligase activity. In the presence of high levels of NAD^+, E1 and E2 enzymes, PARP9/DTX3L heterodimer exhibits ADP-ribosyltransferase activity and catalyzes mono-ADP-ribosylation of ubiquitin (Ub) at Gly76, which is involved in Ub conjugation to substrates. Therefore, the ADP-ribosylation of the Ub restricts the E3 function of DTX3L. In this process, both the RING domain of DTX3L and the catalytic domain of PARP9 are indispensable [39]. Although it has been reported that PARP9/DTX3L can target host histone H2BJ to enhance ISG expression and target encephalomyocarditis virus (EMCV) 3C protease to disrupt viral assembly as an E3 ubiquitin ligase [93], current evidence shows that PARP9 can only ADP-ribosylate Ub. It highlights the selectivity of ADP-ribosylation. However, the effect of PARP9-mediated MARylation on the antiviral activity of PARP9/DTX3L remains to be further investigated.

In addition, PARP7 can MARylate other proteins that function in antiviral immunity, such as PARP13. Recent studies have shown that PARP7 MARylates PARP13 at cysteines (Cys), rather than glutamates (Glu) or aspartates (Asp) [23]. Of course, there are many examples of PARP7 modifying these two amino acids, such as PARP7 MARylates α-tubulin at glutamic acid and aspartic acid residues [84]. PARP13 is also a member of the PARP family, which is considered to be catalytically inactive [13]. Even so, PARP13 is still known for its antiviral activity [48]. PARP13 has shown antiviral activity against a variety of DNA and RNA viruses, including influenza A virus, alphaviruses, filoviruses, and human immunodeficiency virus 1 (HIV-1) [45–47,94]. Mechanistically, PARP13 can promote interferon signaling by interacting with RIG-I and other ISGs. In addition, PARP13 can also target viral RNA to induce its degradation [95–97]. Since the Cys residues in the zinc-finger domains of PARP13 are responsible for the coordination of Zn^{2+} [23,85], the MARylation of these Cys residues may prevent RNA binding. Thus, the Cys MARylation of PARP13 appears to limit the antiviral response. Of note is that the MARylation of Cys in cells is more stable than Glu/Asp, suggesting that the MARylation site of protein targets regulates the duration of the signal. Moreover, PARP14 can also MARylate PARP13 on several Glu/Asp residues, providing a link for crosstalk among PARP family members [49].

3. ADP-Ribosylation of Virus-Encoded Proteins

Recent studies have found that viral proteins can also undergo ADP-ribosylation modification. These include the Epstein-Barr nuclear antigen 1 (EBNA1) of Epstein-Barr virus (EBV), the latency-associated nuclear antigen (LANA) of Kaposi's sarcoma-associated herpesvirus (KSHV), the nonstructural proteins NS1 and NS3 of Zika virus (ZIKV), the nonstructural polyprotein nsP2 of Chikungunya virus (CHIKV), the nucleocapsid (N) protein of coronavirus (CoVs) and the core proteins of adenovirus.

EBV is a human herpesvirus, and EBNA1 is the only viral protein required for the stable maintenance of the viral genome [98,99]. It is reported that PARP1 and TNKS can interact directly with the EBNA1 protein and catalyze its poly-ADP-ribosylation, resulting in the inhibition of OriP replication [33,37]. In the same way, PARP1 can also PARylate LANA of KSHV. KSHV, another human herpesvirus, is usually in the latent phase of disease as reported for EVB [34,100,101]. In latent replication, the KSHV genome is considered to replicate once per cell cycle, and several potentially pathogenic genes such as LANA, K-cyclin (ORF72), K15, and vFLIP (ORF71) are expressed [102]. Thus, the poly-ADP-ribosylation of LANA appears to affect the maintenance of the viral genome. Taking the above discussion together, PARP1- and TNKS-mediated ADP-ribosylation of viral proteins inhibit viral replication and infection using a similar mechanism.

ZiKV is a mosquito-borne flavivirus with a single-stranded positive RNA genome [103,104], and its nonstructural viral proteins NS1 and NS3 have been reported to be targeted by PARP12 [44]. In this process, PARP12 is responsible for the initial MARylation reaction of NS1 and NS3. Subsequent PARylation is performed by other PARPs. This modification of NS1 and NS3 triggers their K48-linked ubiquitination and proteasome-mediated degradation. Since NS1 and NS3 are involved in viral replication and immune evasion [105,106], the PARP-dependent degradation of NS1 and NS3 suppresses Zika virus infection and immune evasion [44]. In addition, a recent study finds that PARP11 can enhance PARP12-mediated NS1 and NS3 ADP-ribosylation and degradation, thus inhibiting Zika virus infection [43]. However, the exact mechanism remains to be elucidated. Again, the ADP-ribosylation of ZiKV proteins shows the potent antiviral activity of this modification.

CHIKV is a mosquito-borne virus with a genome of approximately 11,800 nucleotides [107]. Early in infection, CHIKV encodes a nonstructural polyprotein (nsP). Subsequently, nsP is cleaved into four separate nsP1–4 that together form a replication complex. In this process, nsP2 functions as a key protease. Therefore, it is an important target for antiviral drugs [108–111]. PARP10 can MARylate nsP2 and its protease domain and inhibit its proteolytic activity, thus restricting the processing of nsP and suppressing CHIKV replication. Interestingly, nsP3 is responsible for MAR hydrolase activity and can remove the MARylation modification from nsP2, thereby reactivating its proteolytic activity and promoting CHIKV infection [41,112,113]. In summary, ADP-ribosylation plays an indispensable role in host antiviral response.

In addition, it has been reported that some viral proteins can undergo ADP-ribosylation, but which PARP catalyzes this process, and the effect of ADP-ribosylation on viral infection are still unclear. For example, the coronavirus (CoVs) nucleocapsid (N) protein is ADP-ribosylated in cells during coronavirus infection, and the nsp3 macrodomain does not affect ADP-ribosylation of the N protein. N protein ADP-ribosylation can only be detected in the context of viral infection and cannot be detected in mock-infected cells. Interestingly, nucleocapsid protein ADP-ribosylation is conserved in both α- and β-coronaviruses [114]. This suggests that it is important for viral replication in the host or host resistance to the virus. Future experiments are needed to explore the function of N protein ADP-ribosylation. Another example is that adenovirus core proteins also undergo ADP-ribosylation during viral infection and may play a role in virus decapsidation [115]. However, more details remain to be explored.

4. ADP-Ribosylation of Nucleic Acid Molecules

It is of interest to mention that in addition to proteins, many nucleic acid molecules can also undergo MARylation or PARylation modification [116–120], but we still know relatively little about them. The first nucleic acid discovered to undergo ADP-ribosylation was bacterial DNA, which results in an inhibition of bacterial DNA replication. The antitoxin DarG, a macrodomain protein, catalyzes this process [121]. This suggests that ADP-ribosylation of nucleic acids may also be present in mammals. In fact, it has been shown that PARP1 and PARP2 can PARylate the phosphorylated ends of double-stranded or single-stranded DNA in vitro [122,123]. Similarly, PARP3 can MARylate DNA substrates [124]. In addition, RNA can also serve as a substrate for ADP-ribosylation. In vitro experiments showed that PARP10 can modify the 5′-phosphorylated termini of RNA [119]. Post-translational modifications of these nucleic acid molecules provide new insights into the molecular mechanisms of ADP-ribosylation modifications mediated by PARPs. However, these in vitro experiments still need more validation, especially in vivo validation.

We do not yet know whether PARPs are also sequence-specific when recognizing nucleic acid molecules, which still requires more exploration. The next step is to explore whether ADP-ribosylation occurs on viral genomes and whether it is related to antiviral immunity response. This might provide a seed for subsequent studies.

5. Conclusions and Perspectives

As discussed above, ADP-ribosylation plays a complex function in antiviral response. ADP-ribosylation of proteins involved in IFN-I signaling pathways generally weakens antiviral responses, whereas inhibition of this process enhances antiviral signaling. Interestingly, several of the rapidly evolving PARP genes, including PARP9, PARP10, PARP12, PARP13, and PARP14, are upregulated by IFN, and overexpression of them upregulates several antiviral effectors. Therefore, these PARPs are considered as ISGs [125–128]. ADP-ribosylation of viral proteins often promotes their degradation and thus enhances the host antiviral response, while viruses have also evolved strategies to reverse ADP-ribosylation modifications. One common strategy is that viral macrodomains can degrade ADP-ribosylation modification. As previously described, the nsP3 macrodomain of CHIKV is able to remove the MARylation modification of nsP2. In addition, severe acute respiratory syndrome coronavirus (SARS-CoV) macrodomain mutations significantly increase the virus's sensitivity to interferon. Similarly, mutations in the ADP-ribose binding region of the Sindbis virus (SINV) macrodomain impair viral replication [129,130]. All these suggest that viral macrodomains may play an important role in antiviral immune escape. It appears that viral macrodomains are phylogenetically and structurally closely related to cellular macrodomains. It is, therefore, not surprising that viral macrodomains can hydrolyze ADP-ribosylation [41,131–133]. Together, these observations offer mechanisms of how ADP-ribosylation functions in host-virus conflicts. It seems that ADP-ribosylation can both enhance and restrict the antiviral response, depending on the specificity of the substrate. Therefore, a thorough understanding of the role of ADP-ribosylation in the antiviral response is crucial for the prevention and treatment of virus-associated diseases.

ADP-ribosylation plays such an important role in many biological processes, such as innate immunity, that PARP inhibitors (PARPi) have also become a research hotspot in recent years. For example, rucaparib inhibits PARP11-induced β-TrCP ADP-ribosylation, leading to a decrease in β-TrCP level and an increase in IFNAR1 level [42]. Therefore, rucaparib can effectively promote type-I IFN signaling pathway transduction and enhance host antiviral activity. Other PARP inhibitors, such as olaparib, veliparib, and niraparib, are still under experimental or clinical investigation with rapid progress [134,135]. However, there is still a long way to go in the development of PARP inhibitors due to the difficulty of developing specific inhibitors for individual PARPs and the drug resistance of PARP inhibitors.

Despite these notions, we would like to point out that only a few direct substrates of ADP-ribosylation have been identified so far. Thus far, we do not have a deep understanding of how PARPs select and interact with target substrates and how they are activated [116]. It might be interesting to define more broadly relevant substrates (both host and viral factors), which requires further exploration. Certainly, the identification of specific sites of ADP-ribosylation is equally important. With the development of mass spectrometry techniques and detection tools, it is possible to identify specific substrates for ADP-ribosylation and map modification sites. This will contribute to a further understanding of the role of ADP-ribosylation in antiviral response.

Author Contributions: Conceptualization, Q.D. and H.Z.; writing—review and editing, Q.D., H.Z., Y.M. and W.H. All authors have read and agreed to the published version of the manuscript.

Funding: This work is supported by the National Natural Science Foundation of China (32241009, 31970846) and the Priority Academic Program Development of Jiangsu Higher Education Institutions (PAPD).

Conflicts of Interest: The authors declare no conflict of interest.

References

1. Liu, J.; Qian, C.; Cao, X. Post-Translational Modification Control of Innate Immunity. *Immunity* **2016**, *45*, 15–30. [CrossRef] [PubMed]
2. Luscher, B.; Butepage, M.; Eckei, L.; Krieg, S.; Verheugd, P.; Shilton, B.H. ADP-Ribosylation, a Multifaceted Posttranslational Modification Involved in the Control of Cell Physiology in Health and Disease. *Chem. Rev.* **2018**, *118*, 1092–1136. [CrossRef]
3. Chambon, P.; Weill, J.D.; Mandel, P. Nicotinamide mononucleotide activation of new DNA-dependent polyadenylic acid synthesizing nuclear enzyme. *BioChem. Biophys Res. Commun.* **1963**, *11*, 39–43. [CrossRef]
4. Schreiber, V.; Dantzer, F.; Ame, J.C.; de Murcia, G. Poly(ADP-ribose): Novel functions for an old molecule. *Nat. Rev. Mol. Cell Biol.* **2006**, *7*, 517–528. [CrossRef] [PubMed]
5. Lüscher, B.; Ahel, I.; Altmeyer, M.; Ashworth, A.; Bai, P.; Chang, P.; Cohen, M.; Corda, D.; Dantzer, F.; Daugherty, M.D.; et al. ADP-ribosyltransferases, an update on function and nomenclature. *FEBS J.* **2022**, *289*, 7399–7410. [CrossRef] [PubMed]
6. Gibson, B.A.; Kraus, W.L. New insights into the molecular and cellular functions of poly(ADP-ribose) and PARPs. *Nat. Rev. Mol. Cell Biol.* **2012**, *13*, 411–424. [CrossRef] [PubMed]
7. Hassa, P.O.; Hottiger, M.O. The diverse biological roles of mammalian PARPS, a small but powerful family of poly-ADP-ribose polymerases. *Front. Biosci.* **2008**, *13*, 3046–3082. [CrossRef] [PubMed]
8. Hottiger, M.O.; Hassa, P.O.; Luscher, B.; Schuler, H.; Koch-Nolte, F. Toward a unified nomenclature for mammalian ADP-ribosyltransferases. *Trends BioChem. Sci.* **2010**, *35*, 208–219. [CrossRef]
9. Gagne, J.P.; Ethier, C.; Defoy, D.; Bourassa, S.; Langelier, M.F.; Riccio, A.A.; Pascal, J.M.; Moon, K.M.; Foster, L.J.; Ning, Z.; et al. Quantitative site-specific ADP-ribosylation profiling of DNA-dependent PARPs. *DNA Repair. (Amst.)* **2015**, *30*, 68–79. [CrossRef]
10. Otto, H.; Reche, P.A.; Bazan, F.; Dittmar, K.; Haag, F.; Koch-Nolte, F. In silico characterization of the family of PARP-like poly(ADP-ribosyl)transferases (pARTs). *BMC Genom.* **2005**, *6*, 139. [CrossRef]
11. Bell, C.E.; Eisenberg, D. Crystal structure of diphtheria toxin bound to nicotinamide adenine dinucleotide. *Biochemistry* **1996**, *35*, 1137–1149. [CrossRef]
12. Todorova, T.; Bock, F.J.; Chang, P. Poly(ADP-ribose) polymerase-13 and RNA regulation in immunity and cancer. *Trends Mol. Med.* **2015**, *21*, 373–384. [CrossRef]
13. Karlberg, T.; Klepsch, M.; Thorsell, A.G.; Andersson, C.D.; Linusson, A.; Schuler, H. Structural basis for lack of ADP-ribosyltransferase activity in poly(ADP-ribose) polymerase-13/zinc finger antiviral protein. *J. Biol. Chem.* **2015**, *290*, 7336–7344. [CrossRef]
14. Rippmann, J.F.; Damm, K.; Schnapp, A. Functional characterization of the poly(ADP-ribose) polymerase activity of tankyrase 1, a potential regulator of telomere length. *J. Mol. Biol.* **2002**, *323*, 217–224. [CrossRef]
15. Lüscher, B.; Verheirstraeten, M.; Krieg, S.; Korn, P. Intracellular mono-ADP-ribosyltransferases at the host-virus interphase. *Cell Mol. Life Sci.* **2022**, *79*, 288. [CrossRef]
16. Kleine, H.; Poreba, E.; Lesniewicz, K.; Hassa, P.O.; Hottiger, M.O.; Litchfield, D.W.; Shilton, B.H.; Luscher, B. Substrate-assisted catalysis by PARP10 limits its activity to mono-ADP-ribosylation. *Mol. Cell* **2008**, *32*, 57–69. [CrossRef]
17. Moss, J.; Stanley, S.J.; Nightingale, M.S.; Murtagh, J.J., Jr.; Monaco, L.; Mishima, K.; Chen, H.C.; Williamson, K.C.; Tsai, S.C. Molecular and immunological characterization of ADP-ribosylarginine hydrolases. *J. Biol. Chem.* **1992**, *267*, 10481–10488. [CrossRef]
18. Kato, J.; Zhu, J.; Liu, C.; Moss, J. Enhanced sensitivity to cholera toxin in ADP-ribosylarginine hydrolase-deficient mice. *Mol. Cell Biol.* **2007**, *27*, 5534–5543. [CrossRef]
19. Bonicalzi, M.E.; Haince, J.F.; Droit, A.; Poirier, G.G. Regulation of poly(ADP-ribose) metabolism by poly(ADP-ribose) glycohydrolase: Where and when? *Cell Mol. Life Sci.* **2005**, *62*, 739–750. [CrossRef]
20. Oka, S.; Kato, J.; Moss, J. Identification and characterization of a mammalian 39-kDa poly(ADP-ribose) glycohydrolase. *J. Biol. Chem.* **2006**, *281*, 705–713. [CrossRef]
21. Mashimo, M.; Kato, J.; Moss, J. Structure and function of the ARH family of ADP-ribosyl-acceptor hydrolases. *DNA Repair. (Amst.)* **2014**, *23*, 88–94. [CrossRef]
22. Cho, C.C.; Chien, C.Y.; Chiu, Y.C.; Lin, M.H.; Hsu, C.H. Structural and biochemical evidence supporting poly ADP-ribosylation in the bacterium Deinococcus radiodurans. *Nat. Commun.* **2019**, *10*, 1491. [CrossRef]
23. Rodriguez, K.M.; Buch-Larsen, S.C.; Kirby, I.T.; Siordia, I.R.; Hutin, D.; Rasmussen, M.; Grant, D.M.; David, L.L.; Matthews, J.; Nielsen, M.L.; et al. Chemical genetics and proteome-wide site mapping reveal cysteine MARylation by PARP-7 on immune-relevant protein targets. *Elife* **2021**, *10*, e60480. [CrossRef]
24. Audebert, M.; Salles, B.; Calsou, P. Involvement of poly(ADP-ribose) polymerase-1 and XRCC1/DNA ligase III in an alternative route for DNA double-strand breaks rejoining. *J. Biol. Chem.* **2004**, *279*, 55117–55126. [CrossRef]
25. Langelier, M.F.; Planck, J.L.; Roy, S.; Pascal, J.M. Structural basis for DNA damage-dependent poly(ADP-ribosyl)ation by human PARP-1. *Science* **2012**, *336*, 728–732. [CrossRef]
26. Kraus, W.L.; Lis, J.T. PARP goes transcription. *Cell* **2003**, *113*, 677–683. [CrossRef]
27. Luo, X.; Kraus, W.L. On PAR with PARP: Cellular stress signaling through poly(ADP-ribose) and PARP-1. *Genes Dev.* **2012**, *26*, 417–432. [CrossRef]

28. Hassa, P.O.; Hottiger, M.O. The functional role of poly(ADP-ribose)polymerase 1 as novel coactivator of NF-kappaB in inflammatory disorders. *Cell Mol. Life Sci.* **2002**, *59*, 1534–1553. [CrossRef]
29. Bai, P.; Canto, C.; Oudart, H.; Brunyanszki, A.; Cen, Y.; Thomas, C.; Yamamoto, H.; Huber, A.; Kiss, B.; Houtkooper, R.H.; et al. PARP-1 inhibition increases mitochondrial metabolism through SIRT1 activation. *Cell Metab.* **2011**, *13*, 461–468. [CrossRef]
30. Pellegrino, S.; Altmeyer, M. Interplay between Ubiquitin, SUMO, and Poly(ADP-Ribose) in the Cellular Response to Genotoxic Stress. *Front. Genet.* **2016**, *7*, 63. [CrossRef]
31. Rosado, M.M.; Pioli, C. ADP-ribosylation in evasion, promotion and exacerbation of immune responses. *Immunology* **2021**, *164*, 15–30. [CrossRef] [PubMed]
32. Zhu, H.; Zheng, C. When PARPs Meet Antiviral Innate Immunity. *Trends MicroBiol.* **2021**, *29*, 776–778. [CrossRef]
33. Tempera, I.; Deng, Z.; Atanasiu, C.; Chen, C.J.; D'Erme, M.; Lieberman, P.M. Regulation of Epstein-Barr virus OriP replication by poly(ADP-ribose) polymerase 1. *J. Virol.* **2010**, *84*, 4988–4997. [CrossRef] [PubMed]
34. Ohsaki, E.; Ueda, K.; Sakakibara, S.; Do, E.; Yada, K.; Yamanishi, K. Poly(ADP-ribose) polymerase 1 binds to Kaposi's sarcoma-associated herpesvirus (KSHV) terminal repeat sequence and modulates KSHV replication in latency. *J. Virol.* **2004**, *78*, 9936–9946. [CrossRef]
35. Xia, C.; Wolf, J.J.; Sun, C.; Xu, M.; Studstill, C.J.; Chen, J.; Ngo, H.; Zhu, H.; Hahm, B. PARP1 Enhances Influenza A Virus Propagation by Facilitating Degradation of Host Type I Interferon Receptor. *J. Virol.* **2020**, *94*, e01572-19. [CrossRef]
36. Xu, Y.R.; Shi, M.L.; Zhang, Y.; Kong, N.; Wang, C.; Xiao, Y.F.; Du, S.S.; Zhu, Q.Y.; Lei, C.Q. Tankyrases inhibit innate antiviral response by PARylating VISA/MAVS and priming it for RNF146-mediated ubiquitination and degradation. *Proc. Natl. Acad. Sci. USA* **2022**, *119*, e2122805119. [CrossRef]
37. Deng, Z.; Atanasiu, C.; Zhao, K.; Marmorstein, R.; Sbodio, J.I.; Chi, N.W.; Lieberman, P.M. Inhibition of Epstein-Barr virus OriP function by tankyrase, a telomere-associated poly-ADP ribose polymerase that binds and modifies EBNA1. *J. Virol.* **2005**, *79*, 4640–4650. [CrossRef]
38. Yamada, T.; Horimoto, H.; Kameyama, T.; Hayakawa, S.; Yamato, H.; Dazai, M.; Takada, A.; Kida, H.; Bott, D.; Zhou, A.C.; et al. Constitutive aryl hydrocarbon receptor signaling constrains type I interferon-mediated antiviral innate defense. *Nat. Immunol.* **2016**, *17*, 687–694. [CrossRef]
39. Yang, C.S.; Jividen, K.; Spencer, A.; Dworak, N.; Ni, L.; Oostdyk, L.T.; Chatterjee, M.; Kusmider, B.; Reon, B.; Parlak, M.; et al. Ubiquitin Modification by the E3 Ligase/ADP-Ribosyltransferase Dtx3L/Parp9. *Mol. Cell* **2017**, *66*, 503–516.e5. [CrossRef]
40. Iwata, H.; Goettsch, C.; Sharma, A.; Ricchiuto, P.; Goh, W.W.; Halu, A.; Yamada, I.; Yoshida, H.; Hara, T.; Wei, M.; et al. PARP9 and PARP14 cross-regulate macrophage activation via STAT1 ADP-ribosylation. *Nat. Commun.* **2016**, *7*, 12849. [CrossRef]
41. Krieg, S.; Pott, F.; Eckei, L.; Verheirstraeten, M.; Bütepage, M.; Lippok, B.; Goffinet, C.; Lüscher, B. Mono-ADP-ribosylation by ARTD10 restricts Chikungunya virus replication by interfering with the proteolytic activity of nsP2. *bioRxiv* **2020**. [CrossRef]
42. Guo, T.; Zuo, Y.; Qian, L.; Liu, J.; Yuan, Y.; Xu, K.; Miao, Y.; Feng, Q.; Chen, X.; Jin, L.; et al. ADP-ribosyltransferase PARP11 modulates the interferon antiviral response by mono-ADP-ribosylating the ubiquitin E3 ligase β-TrCP. *Nat. MicroBiol.* **2019**, *4*, 1872–1884. [CrossRef] [PubMed]
43. Li, L.; Shi, Y.; Li, S.; Liu, J.; Zu, S.; Xu, X.; Gao, M.; Sun, N.; Pan, C.; Peng, L.; et al. ADP-ribosyltransferase PARP11 suppresses Zika virus in synergy with PARP12. *Cell Biosci.* **2021**, *11*, 116. [CrossRef] [PubMed]
44. Li, L.; Zhao, H.; Liu, P.; Li, C.; Quanquin, N.; Ji, X.; Sun, N.; Du, P.; Qin, C.F.; Lu, N.; et al. PARP12 suppresses Zika virus infection through PARP-dependent degradation of NS1 and NS3 viral proteins. *Sci. Signal.* **2018**, *11*, eaas9332. [CrossRef] [PubMed]
45. Bick, M.J.; Carroll, J.W.; Gao, G.; Goff, S.P.; Rice, C.M.; MacDonald, M.R. Expression of the zinc-finger antiviral protein inhibits alphavirus replication. *J. Virol.* **2003**, *77*, 11555–11562. [CrossRef] [PubMed]
46. Müller, S.; Möller, P.; Bick, M.J.; Wurr, S.; Becker, S.; Günther, S.; Kümmerer, B.M. Inhibition of filovirus replication by the zinc finger antiviral protein. *J. Virol.* **2007**, *81*, 2391–2400. [CrossRef] [PubMed]
47. Tang, Q.; Wang, X.; Gao, G. The Short Form of the Zinc Finger Antiviral Protein Inhibits Influenza A Virus Protein Expression and Is Antagonized by the Virus-Encoded NS1. *J. Virol.* **2017**, *91*, e01909-16. [CrossRef]
48. Xue, G.; Braczyk, K.; Gonçalves-Carneiro, D.; Dawidziak, D.M.; Sanchez, K.; Ong, H.; Wan, Y.; Zadrozny, K.K.; Ganser-Pornillos, B.K.; Bieniasz, P.D.; et al. Poly(ADP-ribose) potentiates ZAP antiviral activity. *PLoS Pathog.* **2022**, *18*, e1009202. [CrossRef]
49. Carter-O'Connell, I.; Vermehren-Schmaedick, A.; Jin, H.; Morgan, R.K.; David, L.L.; Cohen, M.S. Combining Chemical Genetics with Proximity-Dependent Labeling Reveals Cellular Targets of Poly(ADP-ribose) Polymerase 14 (PARP14). *ACS Chem. Biol.* **2018**, *13*, 2841–2848. [CrossRef]
50. Liu, S.; Cai, X.; Wu, J.; Cong, Q.; Chen, X.; Li, T.; Du, F.; Ren, J.; Wu, Y.T.; Grishin, N.V.; et al. Phosphorylation of innate immune adaptor proteins MAVS, STING, and TRIF induces IRF3 activation. *Science* **2015**, *347*, aaa2630. [CrossRef]
51. Park, A.; Iwasaki, A. Type I and Type III Interferons—Induction, Signaling, Evasion, and Application to Combat COVID-19. *Cell Host Microbe* **2020**, *27*, 870–878. [CrossRef] [PubMed]
52. Zhang, H.G.; Wang, B.; Yang, Y.; Liu, X.; Wang, J.; Xin, N.; Li, S.; Miao, Y.; Wu, Q.; Guo, T.; et al. Depression compromises antiviral innate immunity via the AVP-AHI1-Tyk2 axis. *Cell Res.* **2022**, *32*, 897–913. [CrossRef] [PubMed]
53. Pestka, S.; Krause, C.D.; Walter, M.R. Interferons, interferon-like cytokines, and their receptors. *Immunol. Rev.* **2004**, *202*, 8–32. [CrossRef] [PubMed]

54. Zhang, S.Y.; Boisson-Dupuis, S.; Chapgier, A.; Yang, K.; Bustamante, J.; Puel, A.; Picard, C.; Abel, L.; Jouanguy, E.; Casanova, J.L. Inborn errors of interferon (IFN)-mediated immunity in humans: Insights into the respective roles of IFN-alpha/beta, IFN-gamma, and IFN-lambda in host defense. *Immunol. Rev.* **2008**, *226*, 29–40. [CrossRef] [PubMed]
55. Muller, U.; Steinhoff, U.; Reis, L.F.; Hemmi, S.; Pavlovic, J.; Zinkernagel, R.M.; Aguet, M. Functional role of type I and type II interferons in antiviral defense. *Science* **1994**, *264*, 1918–1921. [CrossRef] [PubMed]
56. Ye, L.; Schnepf, D.; Staeheli, P. Interferon-λ orchestrates innate and adaptive mucosal immune responses. *Nat. Rev. Immunol.* **2019**, *19*, 614–625. [CrossRef]
57. Kotenko, S.V.; Rivera, A.; Parker, D.; Durbin, J.E. Type III IFNs: Beyond antiviral protection. *Semin. Immunol.* **2019**, *43*, 101303. [CrossRef]
58. Ank, N.; West, H.; Bartholdy, C.; Eriksson, K.; Thomsen, A.R.; Paludan, S.R. Lambda interferon (IFN-lambda), a type III IFN, is induced by viruses and IFNs and displays potent antiviral activity against select virus infections in vivo. *J. Virol.* **2006**, *80*, 4501–4509. [CrossRef]
59. Schneider, W.M.; Chevillotte, M.D.; Rice, C.M. Interferon-Stimulated Genes: A Complex Web of Host Defenses. *Annu. Rev. Immunol.* **2014**, *32*, 513–545. [CrossRef]
60. Sadler, A.J.; Williams, B.R. Interferon-inducible antiviral effectors. *Nat. Rev. Immunol.* **2008**, *8*, 559–568. [CrossRef]
61. Zuo, Y.; He, J.; Liu, S.; Xu, Y.; Liu, J.; Qiao, C.; Zang, L.; Sun, W.; Yuan, Y.; Zhang, H.; et al. LATS1 is a central Signal. transmitter for achieving full type-I interferon activity. *Sci. Adv.* **2022**, *8*, eabj3887. [CrossRef] [PubMed]
62. Mohr, A.; Chatain, N.; Domoszlai, T.; Rinis, N.; Sommerauer, M.; Vogt, M.; Muller-Newen, G. Dynamics and non-canonical aspects of JAK/STAT signalling. *Eur. J. Cell Biol.* **2012**, *91*, 524–532. [CrossRef] [PubMed]
63. Levy, D.E.; Kessler, D.S.; Pine, R.; Reich, N.; Darnell, J.E., Jr. Interferon-induced nuclear factors that bind a shared promoter element correlate with positive and negative transcriptional control. *Genes Dev.* **1988**, *2*, 383–393. [CrossRef] [PubMed]
64. Yuan, Y.; Miao, Y.; Qian, L.; Zhang, Y.; Liu, C.; Liu, J.; Zuo, Y.; Feng, Q.; Guo, T.; Zhang, L.; et al. Targeting UBE4A Revives Viperin Protein in Epithelium to Enhance Host Antiviral Defense. *Mol. Cell* **2020**, *77*, 734–747.e737. [CrossRef]
65. Yuan, Y.; Miao, Y.; Ren, T.; Huang, F.; Qian, L.; Chen, X.; Zuo, Y.; Zhang, H.G.; He, J.; Qiao, C.; et al. High salt activates p97 to reduce host antiviral immunity by restricting Viperin induction. *EMBO Rep.* **2022**, *23*, e53466. [CrossRef]
66. Zuo, Y.; Feng, Q.; Jin, L.; Huang, F.; Miao, Y.; Liu, J.; Xu, Y.; Chen, X.; Zhang, H.; Guo, T.; et al. Regulation of the linear ubiquitination of STAT1 controls antiviral interferon signaling. *Nat. Commun.* **2020**, *11*, 1146. [CrossRef]
67. Liu, S.Y.; Aliyari, R.; Chikere, K.; Li, G.; Marsden, M.D.; Smith, J.K.; Pernet, O.; Guo, H.; Nusbaum, R.; Zack, J.A.; et al. Interferon-inducible cholesterol-25-hydroxylase broadly inhibits viral entry by production of 25-hydroxycholesterol. *Immunity* **2013**, *38*, 92–105. [CrossRef]
68. Raftery, N.; Stevenson, N.J. Advances in anti-viral immune defence: Revealing the importance of the IFN JAK/STAT pathway. *Cell Mol. Life Sci.* **2017**, *74*, 2525–2535. [CrossRef]
69. Minegishi, Y.; Saito, M.; Morio, T.; Watanabe, K.; Agematsu, K.; Tsuchiya, S.; Takada, H.; Hara, T.; Kawamura, N.; Ariga, T.; et al. Human tyrosine kinase 2 deficiency reveals its requisite roles in multiple cytokine signals involved in innate and acquired immunity. *Immunity* **2006**, *25*, 745–755. [CrossRef]
70. Essers, M.A.; Offner, S.; Blanco-Bose, W.E.; Waibler, Z.; Kalinke, U.; Duchosal, M.A.; Trumpp, A. IFNalpha activates dormant haematopoietic stem cells in vivo. *Nature* **2009**, *458*, 904–908. [CrossRef]
71. González-Navajas, J.M.; Lee, J.; David, M.; Raz, E. Immunomodulatory functions of type I interferons. *Nat. Rev. Immunol.* **2012**, *12*, 125–135. [CrossRef] [PubMed]
72. Seth, R.B.; Sun, L.; Ea, C.K.; Chen, Z.J. Identification and characterization of MAVS, a mitochondrial antiviral signaling protein that activates NF-kappaB and IRF 3. *Cell* **2005**, *122*, 669–682. [CrossRef] [PubMed]
73. Xu, L.G.; Wang, Y.Y.; Han, K.J.; Li, L.Y.; Zhai, Z.; Shu, H.B. VISA is an adapter protein required for virus-triggered IFN-beta signaling. *Mol. Cell* **2005**, *19*, 727–740. [CrossRef] [PubMed]
74. Liu, S.; Chen, J.; Cai, X.; Wu, J.; Chen, X.; Wu, Y.T.; Sun, L.; Chen, Z.J. MAVS recruits multiple ubiquitin E3 ligases to activate antiviral signaling cascades. *Elife* **2013**, *2*, e00785. [CrossRef]
75. Hou, F.; Sun, L.; Zheng, H.; Skaug, B.; Jiang, Q.X.; Chen, Z.J. MAVS forms functional prion-like aggregates to activate and propagate antiviral innate immune response. *Cell* **2011**, *146*, 448–461. [CrossRef]
76. Kawasaki, T.; Kawai, T.; Akira, S. Recognition of Nucleic. acids by pattern-recognition receptors and its relevance in autoimmunity. *Immunol. Rev.* **2011**, *243*, 61–73. [CrossRef] [PubMed]
77. Ren, Z.; Ding, T.; Zuo, Z.; Xu, Z.; Deng, J.; Wei, Z. Regulation of MAVS Expression and Signaling Function in the Antiviral Innate Immune Response. *Front. Immunol.* **2020**, *11*, 1030. [CrossRef]
78. Guettler, S.; LaRose, J.; Petsalaki, E.; Gish, G.; Scotter, A.; Pawson, T.; Rottapel, R.; Sicheri, F. Structural basis and sequence rules for substrate recognition by Tankyrase explain the basis for cherubism disease. *Cell* **2011**, *147*, 1340–1354. [CrossRef]
79. DaRosa, P.A.; Wang, Z.; Jiang, X.; Pruneda, J.N.; Cong, F.; Klevit, R.E.; Xu, W. Allosteric activation of the RNF146 ubiquitin ligase by a poly(ADP-ribosyl)ation signal. *Nature* **2015**, *517*, 223–226. [CrossRef] [PubMed]
80. Diani-Moore, S.; Ram, P.; Li, X.; Mondal, P.; Youn, D.Y.; Sauve, A.A.; Rifkind, A.B. Identification of the aryl hydrocarbon receptor target gene TiPARP as a mediator of suppression of hepatic gluconeogenesis by 2,3,7,8-tetrachlorodibenzo-p-dioxin and of nicotinamide as a corrective agent for this effect. *J. Biol. Chem.* **2010**, *285*, 38801–38810. [CrossRef]

81. Linden, J.; Lensu, S.; Tuomisto, J.; Pohjanvirta, R. Dioxins, the aryl hydrocarbon receptor and the central regulation of energy balance. *Front. Neuroendocr.* **2010**, *31*, 452–478. [CrossRef]
82. Denison, M.S.; Nagy, S.R. Activation of the aryl hydrocarbon receptor by structurally diverse exogenous and endogenous chemicals. *Annu. Rev. Pharm. Toxicol.* **2003**, *43*, 309–334. [CrossRef]
83. Stockinger, B.; Di Meglio, P.; Gialitakis, M.; Duarte, J.H. The aryl hydrocarbon receptor: Multitasking in the immune system. *Annu. Rev. Immunol.* **2014**, *32*, 403–432. [CrossRef]
84. Palavalli Parsons, L.H.; Challa, S.; Gibson, B.A.; Nandu, T.; Stokes, M.S.; Huang, D.; Lea, J.S.; Kraus, W.L. Identification of PARP-7 substrates reveals a role for MARylation in microtubule control in ovarian cancer cells. *Elife* **2021**, *10*, e60481. [CrossRef]
85. Gomez, A.; Bindesbøll, C.; Satheesh, S.V.; Grimaldi, G.; Hutin, D.; MacPherson, L.; Ahmed, S.; Tamblyn, L.; Cho, T.; Nebb, H.I.; et al. Characterization of TCDD-inducible poly-ADP-ribose polymerase (TIPARP/ARTD14) catalytic activity. *BioChem. J.* **2018**, *475*, 3827–3846. [CrossRef]
86. Stark, G.R.; Darnell, J.E., Jr. The JAK-STAT pathway at twenty. *Immunity* **2012**, *36*, 503–514. [CrossRef]
87. Begitt, A.; Cavey, J.; Droescher, M.; Vinkemeier, U. On the role of STAT1 and STAT6 ADP-ribosylation in the regulation of macrophage activation. *Nat. Commun.* **2018**, *9*, 2144. [CrossRef]
88. Caprara, G.; Prosperini, E.; Piccolo, V.; Sigismondo, G.; Melacarne, A.; Cuomo, A.; Boothby, M.; Rescigno, M.; Bonaldi, T.; Natoli, G. PARP14 Controls the Nuclear Accumulation of a Subset of Type I IFN-Inducible Proteins. *J. Immunol.* **2018**, *200*, 2439–2454. [CrossRef]
89. Kumar, K.G.; Krolewski, J.J.; Fuchs, S.Y. Phosphorylation and specific ubiquitin acceptor sites are required for ubiquitination and degradation of the IFNAR1 subunit of type I interferon receptor. *J. Biol. Chem.* **2004**, *279*, 46614–46620. [CrossRef]
90. Fuchs, S.Y.; Spiegelman, V.S.; Kumar, K.G. The many faces of beta-TrCP E3 ubiquitin ligases: Reflections in the magic mirror of cancer. *Oncogene* **2004**, *23*, 2028–2036. [CrossRef]
91. Takeyama, K.; Aguiar, R.C.; Gu, L.; He, C.; Freeman, G.J.; Kutok, J.L.; Aster, J.C.; Shipp, M.A. The BAL-binding protein BBAP and related Deltex family members exhibit ubiquitin-protein isopeptide ligase activity. *J. Biol. Chem.* **2003**, *278*, 21930–21937. [CrossRef]
92. Juszczynski, P.; Kutok, J.L.; Li, C.; Mitra, J.; Aguiar, R.C.; Shipp, M.A. BAL1 and BBAP are regulated by a gamma interferon-responsive bidirectional promoter and are overexpressed in diffuse large B-cell lymphomas with a prominent inflammatory infiltrate. *Mol. Cell Biol.* **2006**, *26*, 5348–5359. [CrossRef]
93. Zhang, Y.; Mao, D.; Roswit, W.T.; Jin, X.; Patel, A.C.; Patel, D.A.; Agapov, E.; Wang, Z.; Tidwell, R.M.; Atkinson, J.J.; et al. PARP9-DTX3L ubiquitin ligase targets host histone H2BJ. and viral 3C protease to enhance interferon signaling and control viral infection. *Nat. Immunol.* **2015**, *16*, 1215–1227. [CrossRef]
94. Zhu, Y.; Chen, G.; Lv, F.; Wang, X.; Ji, X.; Xu, Y.; Sun, J.; Wu, L.; Zheng, Y.T.; Gao, G. Zinc-finger antiviral protein inhibits HIV-1 infection by selectively targeting multiply spliced viral mRNAs for degradation. *Proc. Natl. Acad. Sci. USA* **2011**, *108*, 15834–15839. [CrossRef]
95. Luo, X.; Wang, X.; Gao, Y.; Zhu, J.; Liu, S.; Gao, G.; Gao, P. Molecular Mechanism of RNA Recognition by Zinc-Finger Antiviral Protein. *Cell Rep.* **2020**, *30*, 46–52.e4. [CrossRef]
96. Hayakawa, S.; Shiratori, S.; Yamato, H.; Kameyama, T.; Kitatsuji, C.; Kashigi, F.; Goto, S.; Kameoka, S.; Fujikura, D.; Yamada, T.; et al. ZAPS is a potent stimulator of signaling mediated by the RNA helicase RIG-I during antiviral responses. *Nat. Immunol.* **2011**, *12*, 37–44. [CrossRef]
97. Guo, X.; Ma, J.; Sun, J.; Gao, G. The zinc-finger antiviral protein recruits the RNA processing exosome to degrade the target mRNA. *Proc. Natl. Acad. Sci. USA* **2007**, *104*, 151–156. [CrossRef]
98. Kennedy, G.; Sugden, B. EBNA-1, a bifunctional transcriptional activator. *Mol. Cell Biol.* **2003**, *23*, 6901–6908. [CrossRef]
99. Hayman, I.R.; Temple, R.M.; Burgess, C.K.; Ferguson, M.; Liao, J.; Meyers, C.; Sample, C.E. New insight into Epstein-Barr Virus infection using models of stratified epithelium. *PLoS Pathog.* **2023**, *19*, e1011040. [CrossRef]
100. Imai, S.; Koizumi, S.; Sugiura, M.; Tokunaga, M.; Uemura, Y.; Yamamoto, N.; Tanaka, S.; Sato, E.; Osato, T. Gastric carcinoma: Monoclonal epithelial malignant cells expressing Epstein-Barr virus latent infection protein. *Proc. Natl. Acad. Sci. USA* **1994**, *91*, 9131–9135. [CrossRef]
101. Hu, J.; Garber, A.C.; Renne, R. The latency-associated nuclear antigen of Kaposi's sarcoma-associated herpesvirus supports latent DNA replication in dividing cells. *J. Virol.* **2002**, *76*, 11677–11687. [CrossRef]
102. Chang, Y.; Moore, P.S. Kaposi's Sarcoma (KS)-associated herpesvirus and its role in KS. *Infect. Agents Dis.* **1996**, *5*, 215–222.
103. Campos, G.C.; Sardi, S.I.; Sarno, M.; Brites, C. Zika virus infection, a new public health challenge. *Braz. J. Infect. Dis.* **2016**, *20*, 227–228. [CrossRef]
104. Ong, C.W. Zika virus: An emerging infectious threat. *Intern. Med. J.* **2016**, *46*, 525–530. [CrossRef]
105. Akey, D.L.; Brown, W.C.; Dutta, S.; Konwerski, J.; Jose, J.; Jurkiw, T.J.; DelProposto, J.; Ogata, C.M.; Skiniotis, G.; Kuhn, R.J.; et al. Flavivirus NS1 structuRes. reveal surfaces for associations with membranes and the immune system. *Science* **2014**, *343*, 881–885. [CrossRef]
106. Sironi, M.; Forni, D.; Clerici, M.; Cagliani, R. Nonstructural Proteins Are Preferential Positive Selection Targets in Zika Virus and Related Flaviviruses. *PLoS Negl. Trop. Dis.* **2016**, *10*, e0004978. [CrossRef]
107. Silva, L.A.; Dermody, T.S. Chikungunya virus: Epidemiology, replication, disease mechanisms, and prospective intervention strategies. *J. Clin. Investig.* **2017**, *127*, 737–749. [CrossRef]

108. Kril, V.; Aïqui-Reboul-Paviet, O.; Briant, L.; Amara, A. New Insights into Chikungunya Virus Infection and Pathogenesis. *Annu. Rev. Virol.* **2021**, *8*, 327–347. [CrossRef]
109. Göertz, G.P.; McNally, K.L.; Robertson, S.J.; Best, S.M.; Pijlman, G.P.; Fros, J.J. The Methyltransferase-Like Domain of Chikungunya Virus nsP2 Inhibits the Interferon Response by Promoting the Nuclear Export of STAT1. *J. Virol.* **2018**, *92*. [CrossRef]
110. Utt, A.; Das, P.K.; Varjak, M.; Lulla, V.; Lulla, A.; Merits, A. Mutations conferring a noncytotoxic phenotype on chikungunya virus replicons compromise enzymatic properties of nonstructural protein 2. *J. Virol.* **2015**, *89*, 3145–3162. [CrossRef]
111. Webb, L.G.; Veloz, J.; Pintado-Silva, J.; Zhu, T.; Rangel, M.V.; Mutetwa, T.; Zhang, L.; Bernal-Rubio, D.; Figueroa, D.; Carrau, L.; et al. Chikungunya virus antagonizes cGAS-STING mediated type-I interferon responses by degrading cGAS. *PLoS Pathog.* **2020**, *16*, e1008999. [CrossRef] [PubMed]
112. Abraham, R.; Hauer, D.; McPherson, R.L.; Utt, A.; Kirby, I.T.; Cohen, M.S.; Merits, A.; Leung, A.K.L.; Griffin, D.E. ADP-ribosyl-binding and hydrolase activities of the alphavirus nsP3 macrodomain are critical for initiation of virus replication. *Proc. Natl. Acad. Sci. USA* **2018**, *115*, E10457–E10466. [CrossRef] [PubMed]
113. McPherson, R.L.; Abraham, R.; Sreekumar, E.; Ong, S.E.; Cheng, S.J.; Baxter, V.K.; Kistemaker, H.A.; Filippov, D.V.; Griffin, D.E.; Leung, A.K. ADP-ribosylhydrolase activity of Chikungunya virus macrodomain is critical for virus replication and virulence. *Proc. Natl. Acad. Sci. USA* **2017**, *114*, 1666–1671. [CrossRef] [PubMed]
114. Grunewald, M.E.; Fehr, A.R.; Athmer, J.; Perlman, S. The coronavirus nucleocapsid protein is ADP-ribosylated. *Virology* **2018**, *517*, 62–68. [CrossRef]
115. Déry, C.V.; de Murcia, G.; Lamarre, D.; Morin, N.; Poirier, G.G.; Weber, J. Possible role of ADP-ribosylation of adenovirus core proteins in virus infection. *Virus Res.* **1986**, *4*, 313–329. [CrossRef]
116. Suskiewicz, M.J.; Palazzo, L.; Hughes, R.; Ahel, I. Progress and outlook in studying the substrate specificities of PARPs and related enzymes. *FEBS J.* **2021**, *288*, 2131–2142. [CrossRef]
117. Weixler, L.; Schäringer, K.; Momoh, J.; Lüscher, B.; Feijs, K.L.H.; Žaja, R. ADP-ribosylation of RNA and DNA: From in vitro characterization to in vivo function. *Nucleic. Acids Res.* **2021**, *49*, 3634–3650. [CrossRef]
118. Schuller, M.; Butler, R.E.; Ariza, A.; Tromans-Coia, C.; Jankevicius, G.; Claridge, T.D.W.; Kendall, S.L.; Goh, S.; Stewart, G.R.; Ahel, I. Molecular basis for DarT ADP-ribosylation of a DNA base. *Nature* **2021**, *596*, 597–602. [CrossRef]
119. Munnur, D.; Bartlett, E.; Mikolčević, P.; Kirby, I.T.; Rack, J.G.M.; Mikoč, A.; Cohen, M.S.; Ahel, I. Reversible ADP-ribosylation of RNA. *Nucleic. Acids Res.* **2019**, *47*, 5658–5669. [CrossRef]
120. Kim, D.S.; Challa, S.; Jones, A.; Kraus, W.L. PARPs and ADP-ribosylation in RNA biology: From RNA expression and processing to protein translation and proteostasis. *Genes Dev.* **2020**, *34*, 302–320. [CrossRef]
121. Jankevicius, G.; Ariza, A.; Ahel, M.; Ahel, I. The Toxin-Antitoxin System DarTG Catalyzes Reversible ADP-Ribosylation of DNA. *Mol. Cell* **2016**, *64*, 1109–1116. [CrossRef] [PubMed]
122. Matta, E.; Kiribayeva, A.; Khassenov, B.; Matkarimov, B.T.; Ishchenko, A.A. Insight into DNA substrate specificity of PARP1-catalysed DNA poly(ADP-ribosyl)ation. *Sci. Rep.* **2020**, *10*, 3699. [CrossRef]
123. Talhaoui, I.; Lebedeva, N.A.; Zarkovic, G.; Saint-Pierre, C.; Kutuzov, M.M.; Sukhanova, M.V.; Matkarimov, B.T.; Gasparutto, D.; Saparbaev, M.K.; Lavrik, O.I.; et al. Poly(ADP-ribose) polymerases covalently modify strand break termini in DNA fragments in vitro. *Nucleic. Acids Res.* **2016**, *44*, 9279–9295. [CrossRef] [PubMed]
124. Zarkovic, G.; Belousova, E.A.; Talhaoui, I.; Saint-Pierre, C.; Kutuzov, M.M.; Matkarimov, B.T.; Biard, D.; Gasparutto, D.; Lavrik, O.I.; Ishchenko, A.A. Characterization of DNA ADP-ribosyltransferase activities of PARP2 and PARP3: New insights into DNA ADP-ribosylation. *Nucleic. Acids Res.* **2018**, *46*, 2417–2431. [CrossRef]
125. Shaw, A.E.; Hughes, J.; Gu, Q.; Behdenna, A.; Singer, J.B.; Dennis, T.; Orton, R.J.; Varela, M.; Gifford, R.J.; Wilson, S.J.; et al. Fundamental properties of the mammalian innate immune system revealed by multispecies comparison of type I interferon responses. *PLoS Biol.* **2017**, *15*, e2004086. [CrossRef]
126. Eckei, L.; Krieg, S.; Bütepage, M.; Lehmann, A.; Gross, A.; Lippok, B.; Grimm, A.R.; Kümmerer, B.M.; Rossetti, G.; Lüscher, B.; et al. The conserved macrodomains of the non-structural proteins of Chikungunya virus and other pathogenic positive strand RNA viruses function as mono-ADP-ribosylhydrolases. *Sci. Rep.* **2017**, *7*, 41746. [CrossRef] [PubMed]
127. Schoggins, J.W. Interferon-stimulated genes: Roles in viral pathogenesis. *Curr. Opin. Virol.* **2014**, *6*, 40–46. [CrossRef]
128. Karki, S.; Li, M.M.; Schoggins, J.W.; Tian, S.; Rice, C.M.; MacDonald, M.R. Multiple interferon stimulated genes synergize with the zinc finger antiviral protein to mediate anti-alphavirus activity. *PLoS ONE* **2012**, *7*, e37398. [CrossRef]
129. Kuri, T.; Eriksson, K.K.; Putics, A.; Züst, R.; Snijder, E.J.; Davidson, A.D.; Siddell, S.G.; Thiel, V.; Ziebuhr, J.; Weber, F. The ADP-ribose-1″-monophosphatase domains of severe acute respiratory syndrome coronavirus and human coronavirus 229E mediate resistance to antiviral interferon responses. *J. Gen. Virol.* **2011**, *92*, 1899–1905. [CrossRef]
130. Park, E.; Griffin, D.E. The nsP3 macro domain is important for Sindbis virus replication in neurons and neurovirulence in mice. *Virology* **2009**, *388*, 305–314. [CrossRef]
131. Fu, W.; Yao, H.; Butepage, M.; Zhao, Q.; Luscher, B.; Li, J. The search for inhibitors of macrodomains for targeting the readers and erasers of mono-ADP-ribosylation. *Drug. Discov. Today* **2021**, *26*, 2547–2558. [CrossRef] [PubMed]
132. Alhammad, Y.M.O.; Fehr, A.R. The Viral Macrodomain Counters Host Antiviral ADP-Ribosylation. *Viruses* **2020**, *12*, 384. [CrossRef] [PubMed]

133. Grunewald, M.E.; Chen, Y.; Kuny, C.; Maejima, T.; Lease, R.; Ferraris, D.; Aikawa, M.; Sullivan, C.S.; Perlman, S.; Fehr, A.R. The coronavirus macrodomain is required to prevent PARP-mediated inhibition of virus replication and enhancement of IFN expression. *PLoS Pathog.* **2019**, *15*, e1007756. [CrossRef] [PubMed]
134. Li, H.; Liu, Z.Y.; Wu, N.; Chen, Y.C.; Cheng, Q.; Wang, J. PARP inhibitor resistance: The underlying mechanisms and clinical implications. *Mol. Cancer* **2020**, *19*, 107. [CrossRef]
135. Curtin, N.J.; Szabo, C. Poly(ADP-ribose) polymerase inhibition: Past, present and future. *Nat. Rev. Drug. Discov.* **2020**, *19*, 711–736. [CrossRef]

pathogens MDPI

Perspective

The DarT/DarG Toxin–Antitoxin ADP-Ribosylation System as a Novel Target for a Rational Design of Innovative Antimicrobial Strategies

Giuliana Catara [1], Rocco Caggiano [2] and Luca Palazzo [2,3,*]

1 Institute of Biochemistry and Cell Biology, National Research Council of Italy, CNR, 80131 Naples, Italy
2 Institute for the Experimental Endocrinology and Oncology, National Research Council of Italy, CNR, 80131 Naples, Italy
3 Department of Molecular Medicine and Medical Biotechnology, University of Naples "Federico II", 80131 Naples, Italy
* Correspondence: luca.palazzo@ieos.cnr.it or luca.palazzo@unina.it

Abstract: The chemical modification of cellular macromolecules by the transfer of ADP-ribose unit(s), known as ADP-ribosylation, is an ancient homeostatic and stress response control system. Highly conserved across the evolution, ADP-ribosyltransferases and ADP-ribosylhydrolases control ADP-ribosylation signalling and cellular responses. In addition to proteins, both prokaryotic and eukaryotic transferases can covalently link ADP-ribosylation to different conformations of nucleic acids, thus highlighting the evolutionary conservation of archaic stress response mechanisms. Here, we report several structural and functional aspects of DNA ADP-ribosylation modification controlled by the prototype DarT and DarG pair, which show ADP-ribosyltransferase and hydrolase activity, respectively. DarT/DarG is a toxin–antitoxin system conserved in many bacterial pathogens, for example in *Mycobacterium tuberculosis*, which regulates two clinically important processes for human health, namely, growth control and the anti-phage response. The chemical modulation of the DarT/DarG system by selective inhibitors may thus represent an exciting strategy to tackle resistance to current antimicrobial therapies.

Keywords: ADP-ribosylation; toxin–antitoxin; DarT/DarG; DNA modification; cell growth; phage defence; antimicrobial resistance

check for updates

Citation: Catara, G.; Caggiano, R.; Palazzo, L. The DarT/DarG Toxin–Antitoxin ADP-Ribosylation System as a Novel Target for a Rational Design of Innovative Antimicrobial Strategies. *Pathogens* **2023**, *12*, 240. https://doi.org/10.3390/pathogens12020240

Academic Editors: Anthony K L Leung, Anthony Fehr and Rachy Abraham

Received: 30 December 2022
Revised: 27 January 2023
Accepted: 31 January 2023
Published: 2 February 2023

1. Introduction

ADP-ribosylation is a reversible post-translational modification (PTM) found in all three domains of life, as well as in several viruses [1–4]. It was identified in the 1970s as a key enzymatic activity required for cholera and diphtheria toxin pathogenesis [5,6]. Since then, the understanding of ADP-ribosylation has increased, from bacteria to mammals. Today, it is mainly known in the scientific community for its key role in DNA damage repair [7–10] and for being the target of tailored cancer therapies [11–14]. However, the functions of ADP-ribosylation are also vital for controlling many additional physiological processes, such as transcription and translation [15–21], cell proliferation [22,23], and cell death [24–26] along with stress and immune responses [27–33] and many others [34–36]. The control of cell homeostasis in both prokaryotes and eukaryotes by ADP-ribosylation [19,34,35,37–41] makes this a PTM of great interest in many fields of human health.

ADP-ribosylation is characterised by the addition of ADP-ribose unit(s) from nicotinamide adenine dinucleotide (NAD^+) onto cellular substrates with the release of nicotinamide [42,43]. Consistent with the number of ADP-ribose moieties attached, single or multiple, the reaction is further differentiated into mono-ADP-ribosylation (MARylation) and poly-ADP-ribosylation (PARylation), respectively [44]. It was originally discovered as a PTM mainly targeting proteins [45], but it can also be covalently attached to additional macromolecules, including DNA [38,46–49] and RNA [48,50], as well as to small molecules such as antibiotics [51], ATP, and ADP [36].

There is a balanced interplay between specialised enzymes, namely, ADP-ribosyltransferases and ADP-ribosylhydrolases, which are responsible for the attachment and removal of the modification from cellular targets, respectively. This interplay shapes ADP-ribosylation signalling [43,44,52–54]. The dysregulation of these enzymatic activities in humans thus has implications in the pathogenicity of several diseases, above all, neurological disorders [55,56], cancer [57,58], and bacterial- and viral-mediated infections, as discussed here [59–62].

ADP-ribosylation is currently known to be involved in a large number of infectious diseases worldwide [63–66], including COVID-19 [33,60,61,67–72], Legionnaires' disease [73–77], and the infectious diseases caused by the virulent *M. tuberculosis*. From a pathogenic point of view, the mechanisms of ADP-ribosylation in *M. tuberculosis* infection involve the modification of their own/endogenous targets rather than the host proteins, ultimately enabling the cell to adapt within the host and to improve the biological fitness. Among these mechanisms, the activity of the DarT/DarG toxin–antitoxin (TA) ADP-ribosylation system in *M. tuberculosis* targets bacterial genomic DNA. As a result of DNA modification, ADP-ribosylation slows growth and potentially induces bacterial persistence, a phenotypic state that correlates with antibiotic resistance [38,40,78].

The excessive use of antibiotics to counteract pathogen infections has led to the spread of antibiotic resistant "superbugs", which currently represent a major public health threat [79,80]. Antibiotic resistance occurs in a wide range of bacterial infections and is determined by the activation of pathogen resistance/defence mechanisms, which also enable the cells to become persistent and tolerant to antibiotics [81,82]. Bacterial TA systems are widespread in Gram-negative and Gram-positive bacteria, and are involved in cell regulatory mechanisms in response to stress stimuli [83–85], including antibiotic resistance. Targeting TA modules such as the ADP-ribosylation DarT/DarG system can thus act as a blueprint for designing alternative drugs to current therapeutic treatments of antibiotic-resistant pathogens [86,87].

In this perspective, we discuss the structural and mechanistic aspects of DarT/DarG toxin–antitoxin-mediated control of DNA ADP-ribosylation. In addition, we then address the pathogenic role of the DarT/DarG TA pair as well as the therapeutic perspectives that the modulation of this specific ADP-ribosylation signalling may have.

2. ADP-Ribosylation in Bacteria

2.1. NAD^+-Dependent Endotoxins and Exotoxins Involved in ADP-Ribosylation Signalling

ADP-ribosylation sustains prokaryotic cells in both cell metabolic processes and pathogenic mechanisms through the activity of NAD^+-dependent enzymes, namely, endotoxins and exotoxins, respectively (Figure 1A–C).

The majority of endotoxins and exotoxins belong to the superfamily of ADP-ribosyltra nsferase (ART) enzymes, which, despite limited sequence similarity, share a conserved secondary structure and protein fold [2,88]. ARTs fall into three phylogenetically distinct clades according to the catalytic triad composition: (*i*) the diphtheria toxin-like ARTs (ARTDs), characterised by the catalytic H-Y-[EDQ] triad; (*ii*) the cholera toxin-like ARTs (ARTCs), harbouring the R-S-E motif in the catalytic domain; and (*iii*) the tRNA 2′-phosphotransferase (TpT1/KptA) containing the H-H-h motif, with h containing any hydrophobic residues [2,43,52]. In addition to these subgroups, in *Staphylococcus aureus* and *Streptococcus pyogenes*, the microbial SirTMs, which belong to the Sirtuin superfamily of enzymes, catalyse the lipoyl-dependent ADP-ribosylation of proteins following a non-canonical deacylation reaction [89].

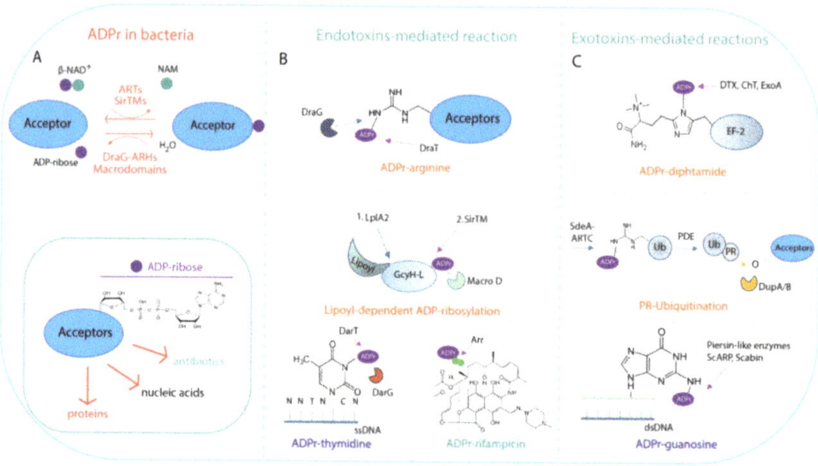

Figure 1. Schematic representation of ADP-ribosylation reaction in bacterial meta-cell. (**A**) ADP-ribosylation (ADPr) reaction is catalysed by NAD$^+$-dependent diphtheria toxin-like ARTs ARTDs, the cholera toxin-like ARTs ARTCs, or SirTMs, which transfer a single ADP-ribose unit on acceptors. Macrodomain-containing hydrolases (Macrodomains) or DraG-related ADP-ribosylhydrolases reverse the reaction by generating free ADP-ribose and unmodified acceptor. Inset: ADP-ribose moiety linked to acceptor substrates, which can be proteins, nucleic acids, or antibiotics. (**B**) Endotoxin-mediated reaction. Endotoxins can modify proteins, nucleic acids, or antibiotics. ARTDs, ARTCs, and SirTM modify endogenous bacterial substrates on different residues as indicated. ADP-ribosylated (ADPr)-arginine: MARylation of arginine residue is performed by the DraT enzyme, and is reversed by the cognate DraG ADP-ribosylhydrolase. Further examples are provided in the text. Lipoyl-dependent MARylation is carried out by SirTM and is dependent on prior lipoylation of the lipoyl-carrier protein GcyH-L, by the lipoate-protein ligase A (LplA2). The modification is reversed by the MacroD hydrolase, which is encoded within the same SirTM operon; ADP-ribosylated (ADPr)-thymidine: the reaction is performed by the endotoxin DarT that modifies thymidine base on ssDNA; the cognate DarG antitoxin reverses the modification; ADP-ribosylated (ADPr)-rifampicin: MARylation of the rifampin antibiotic is catalysed by Arr toxin. (**C**) Exotoxin-mediated reactions. ARTDs and ARTCs modify host targets on different residues as indicated. ADP-ribosylated (ADPr)-diphthamide: the reaction is catalysed by the toxins DTX, ChT, and ExoA, which irreversibly transfer ADP-ribose on the residue diphthamide on the elongation factor 2; PR-Ubiquitination. SdeA toxin catalyses the ADP-ribosylation (ADPr)-dependent ubiquitination of host proteins in a two-step reaction as detailed in the text. The reaction is reversed by the phosphodiesterases DubA/B; ADP-ribosylated (ADPr)-guanosine. The irreversible ADP-ribosylation on guanosine in dsDNA is performed by the pierisin-like enzymes ScARP and Scabin.

The substrate selectivity of ARTs is provided by two conserved functional motifs called the ARTT loop (ADP-ribosylating turn-turn), which is also known as the acceptor-loop (A-loop), and the donor-loop (D-loop). The D-loop is exclusive to ARTDs. Both loop structures are evolutionarily highly conserved, although their amino acid sequence and length vary greatly among the ARTDs [43]. In comparison with eukaryotic ARTs, bacterial enzymes show narrow amino acid residues specificity in host targets. Bacterial ARTs are in fact able to MARylate target proteins on several amino acid residues (i.e., arginine, cysteine, threonine, asparagine, and glutamine for ARTCs, diphthamide for ARTDs) (Figure 1B,C). Unlike some mammalian ART homologues (namely, PARP1 and PARP2) [90–92], bacterial ARTs do not need specificity factors [43]. In addition, bacterial ADP-ribosyltransferases can also modify nucleic acids (Figure 1B,C) [38,93]. Some bacterial ARTs, such as the ARTD homologue Arr-ms from *Mycobacterium smegmatis,* can also ADP-ribosylate the hydroxyl group at C23 of rifamycin and derivatives, thus inactivating antibiotics [51,94] (Figure 1B).

ADP-ribosylation is a reversible PTM. Two structurally distinct families of ADP-ribosylhydrolases, namely, DraG-related ADP-ribosylhydrolases (ARHs) and macrodomain-containing ADP-ribosylhydrolases, revert ADP-ribosylation signalling in bacterial cells (Figure 1A) [4].

DraG-related ARHs, from the founder DraG protein found in the nitrogen-fixing bacterial species *R. rubrum* and *A. brasilense* [37,95], are compact and globular modules with a typical domain length of 290–360 residues, with a central core motif comprising 13 orthogonal α-helices and a variable number of supplementary helices depending on the organism and type. The divalent metal ions enable the correct positioning of the substrate in the active site [53]. Structural studies on *R. rubrum* DraG hydrolase show that the de-MARylation of substrates occurs via the opening of the ribose ring and the formation of a protonated Schiff base. This substrate opening leads to a shift in metal coordination, allowing a nucleophilic attack by a water molecule activated by Mn^{2+} and resulting in a tetrahedral intermediate. The proton transfer via D97 promotes intermediate collapse and the release of arginine [96].

Macrodomain-containing hydrolases, harbouring the ADP-ribose-binding domain known as the macrodomain, share a conserved α/β/α fold consisting of a six-stranded mixed β-sheet surrounded by five α-helices [53]. Substrate binding takes place in a deep cleft on the top of the domain and several conserved interactions contribute to stabilise the ligand–macrodomain complex [53,54]. Based on ADP-ribosylhydrolase activity, macrodomains are further classified into mono-ADP-ribosylhydrolases (including MacroD-type and ALC1-like enzymes) and poly-ADP-ribosylhydrolases (PARG). Several bacterial macrodomains have been characterised as belonging to MacroD-, ALC1-, and PARG-like phylogenetically distinct groups that regulate a variety of cellular processes by deacetylating O-acetyl-ADP-ribose, and by hydrolysing MARylated targets, which include proteins and RNA [50,97,98]. In addition, several enzymes such as the TARG1-type macrodomain enzyme from *Fusobacterium mortiferum* ATCC 9817 [99] and the bacterial PARG from *Thermomonospora curvata* [97] have been reported to hydrolyse chains of ADP-ribose in vitro. The finding of an endogenous bacterial PARG-processing enzyme in *Deinococcus radiodurans* would seem to indicate an active prokaryotic PARylation machinery that may be involved in the stress response, given that PARG disruption leads to PAR enrichment in treated cells and loss of recovery after UV irradiation [100].

In a similar way to what happened for cancer treatment with the discovery of specific inhibitors of PARP1/PARP2 and PARG, the in-depth understanding of the enzymatic reactions and structural features of both bacterial ARTs and hydrolases promises important advances in antimicrobial therapies, which may eventually help to tackle antibiotic resistance.

2.2. Functional Aspects of ADP-Ribosylation in Bacteria

From a functional point of view, endotoxins modify endogenous targets, thus regulating the oxidative stress response [89], morphological differentiation and antibiotic production [101,102], and the maintenance of cell homeostasis in response to environmental stimuli, as exemplified by the *Rhodospirillum rubrum* and *Azospirillum brasilense* DraT/DraG system that regulates nitrogen fixation [37,95,96,103]. On the other hand, exotoxins promote pathogenic mechanisms through the transfer of ADP-ribose onto host targets, which alters signal transduction (e.g., CTX from *Vibrio. cholerae*; ETEC from *Escherichia coli*), cellular cytoskeleton organisation along with membrane trafficking (e.g., C2 toxin from *Clostridium botulinum*; ExoS toxin from *Pseudomonas. aeruginosa*), and protein synthesis (e.g., DTX from *Corynebacterium diphtheriae;* ChT from *Vibrio cholerae*) [4,45,62]. Bacterial exotoxins appear to be involved in the aetiology of important diseases [4,62,104]. Of these, SidE family effectors regulate the pathogenicity of *Legionella pneumophila* by non-canonical phosphoribosyl ubiquitination (Figure 1C), which interferes with the host physiological ubiquitination machinery [73,76,105,106], ultimately leading to the impairment of mitophagy and the secretory pathway [107]. The SdeA toxin, which is one of SidE family effectors released by the pathogenic *L. pneumophila*, catalyses the ADP-ribosylation-dependent ubiquitination of host proteins in a two-step reaction. Firstly, SdeA transfers the ADP-ribose on arginine 42 of a ubiquitin (Ub) molecule to generate an ADP-ribosylated-Ub intermediate due to the presence of an ARTC domain; the ADP-ribosylated-Ub intermediate is then converted to phosphoribosyl-Ub by the SdeA phosphodiesterase activity and is then conjugated through an ester linkage to a serine residue to target protein. Cognate phosphodiesterases DubA/B revert the reaction (Figure 1C).

Of particular interest from this perspective is that recent discoveries have established nucleic acids, such as genomic DNA and RNAs, as novel ADP-ribosylation targets [48–50], which, although involved in crucial physiological processes, are not yet fully understood in either mammals or prokaryotes [47,48,50,108]. To date, the ADP-ribosylation of DNA has only been characterised in a few bacterial systems including pierisin-like members and the DarT/DarG system. Pierisin and the pierisin-like ARTs represent a small group of ARTC toxins that prevalently ADP-ribosylate DNA. Pierisin, which is the founder of the family, has been identified in the cabbage butterfly species, *Pieris rapae*, where it counteracts the non-habitual parasitoids [109]. Extracts from *P. rapae* are highly cytotoxic toward insect and mammalian cells. In fact, pierisin induces irreversible host apoptosis by ADP-ribosylating N2 amino groups of 2′-deoxyguanosine into double-stranded DNA (dsDNA) in a non-conserved sequence manner, and as such, pierisin likely plays a role in the defence mechanism [110]. Similarly, the bacterial pierisin-like Scabin from *Streptomyces scabies* [111] and ScARP from *Streptomyces coelicolor* [112] are able to modify DNA on the exocyclic amino group on guanine bases and most of its derivatives in either single-stranded (ssDNA) or dsDNA. The disruption of ScARP affects *S. coelicolor* morphological differentiation, sporulation, and increased antibiotic production [101].

Unlike pierisin-like ARTCs, the DarT toxin from *Thermus aquaticus* and *M. tuberculosis* can specifically modify genomic ssDNA on thymidine in the conserved nucleotide sequence NNTNTCN, which can strongly hinder bacterial cell growth and, in turn, can have implications in antimicrobial susceptibility. Importantly, the cellular effects on the bacterial growth induced by DarT can be neutralised by DarG hydrolase, which, by reverting DNA-ADP-ribosylation, acts as an antitoxin [38]. Section 3 details the DarT/DarG system.

3. The DarT/DarG ADP-Ribosylation-Dependent System

3.1. DarT Is a New PARP-Like Toxin and a Potential Molecular Target for Antimicrobial Therapy

ADP-ribosylation catalysed by DarT specifically targets the thymidine bases present in conserved ssDNA sequence NNTNTCN in *T. aquaticus* and TTTT/A in *M. tuberculosis*, respectively, thereby showing no activity on dsDNA and RNA or protein targets. Substrate specificity toward a thymidine base also takes place for DarT toxin homologues, as highlighted in enteropathogenic *E. coli* DarT, which shows a preference for TCT/TTT sequences by modifying the third thymidine base of these motifs [40]. From a structural point of view, DarT can be considered a divergent ART enzyme given that it lacks the canonical catalytic triad residues found in ARTD and ARTC bacterial endotoxins (Figure 2A). Compared to bacterial ARTDs and ARTCs, DarT is very variable in terms of primary structure and motifs in comparison with bacterial ARTDs and ARTCs (Figure 2A). Nevertheless, the overall ART fold is maintained, as revealed by 3D resolution structural insights into *Thermus* sp. 2.9 DarT [78]. In fact, DarT is a PARP-like enzyme [78], as also predicted by phylogenetically analyses, as it is closer to human ARTDs than bacterial ART counterparts (Figure 2B).

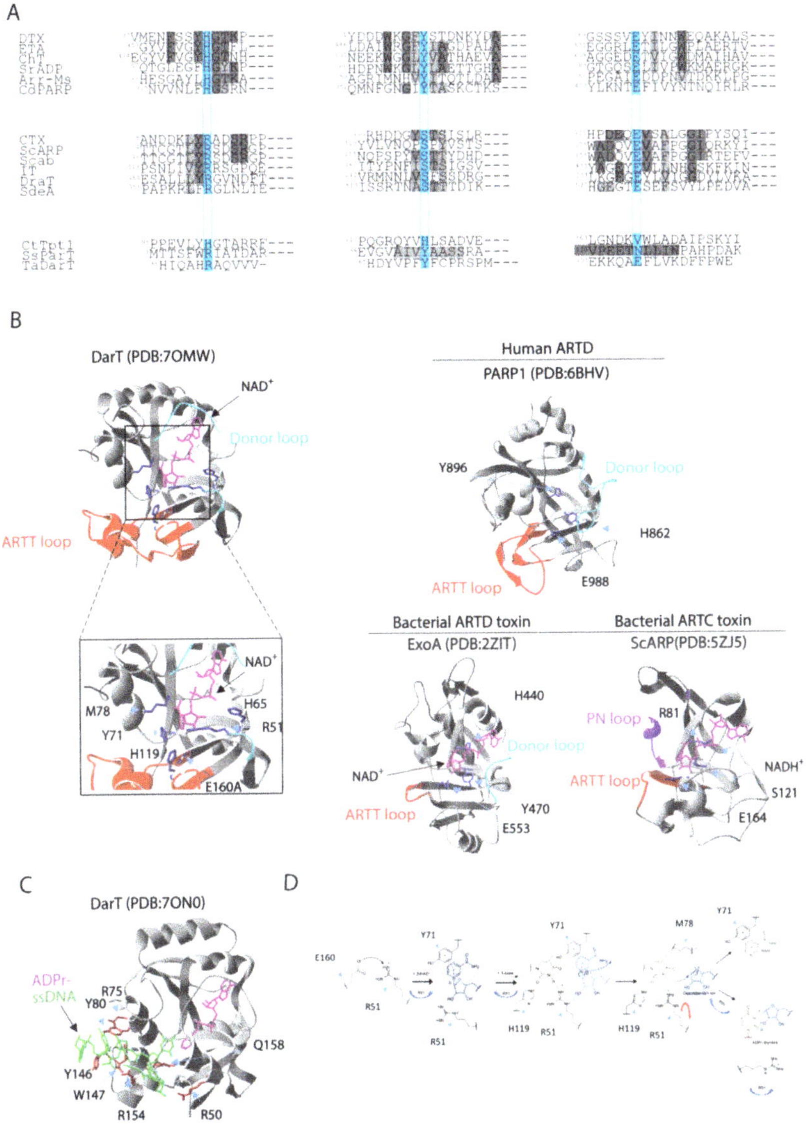

Figure 2. Comparison of amino acid sequences and 3D structures of representative ARTs. (**A**) Alignment of the partial sequences of the bacterial ARTs. ARTD members, which harbour the H-Y-E catalytic residues, include: DTX, diphtheria toxin from *C. diphtheriae;* ETA, exotoxin A from *P. aeruginosa;* Ch toxin, cholix toxin from *V. cholerae;* SrADP, toxin from *Streptomyces roseifaciens;* Arr-Ms, rifamycin ADP-ribosylation toxin from *Mycobacterium smegmatis;* Cd-PARP, toxin from *Clostridium perfringens* CD 160. ARTC members, which enclose R-S-E catalytic residues, include: CTX, cholera toxin from *V. cholerae;* ScARP, toxin from *S. coelicolor;* Scabin from *S. scabies;* IT, iota toxin from *C. perfringens;* DraT, dinitrogenase reductase ADP-ribosyltransferase from *R. rubrum;* SdeA, ADP-ribosylation-dependent ubiquitination toxin from *L. pneumophila*. Divergent enzymes include: CtTpt1, Tpt1 RNA-phosphotransferase enzyme from *Clostridium thermocellum;* ParT, ADP-ribosylating toxin of ParT/ParS TA system from *Sphingobium* sp. YBL2; TaDarT, DNA ADP-ribosylating toxin of DarT/DarG TA system from *T. aquaticus*. The residues involved in catalysis are boxed on a light blue

background. Identities or accepted amino acid substitutions are indicated in dark and light grey, respectively. (**B**) Cartoon–stick model of *Thermus* sp. 2.9 DarT(E160A) (PDB:7OMW, [78]) showing the NAD$^+$ binding site in complex with NAD$^+$, the amino acid residues involved in the catalytic activity (blue), the regulatory ARTT-loop (red) and the donor-loop (light blue). Inset: the catalytic site residues R51, H65, Y71, M78, H119, and E160A localise in proximity of nicotinamide in the active site. Cartoon–stick models of the 3D structure of the human ARTD PARP1 (PDB:6BHV, [113]), the bacterial ARTD-toxin ExoA (PDB:2ZIT, [114]), and the bacterial ARTC-toxin ScARP (PDB:5ZJ5, [115]) are shown as exemplars. (**C**) Cartoon–stick model of *Thermus* sp. 2.9 DarT(E160A) (PDB:7ON0, [78]) in complex with ADP-ribosylated ssDNA showing the residues (dark red) involved in the interaction with ssDNA (green). (**D**) Catalytic mechanism proposed for DarT-mediated ADP-ribosylation reaction of DNA.

Secondary structure elements are found conserved such as the fold of the central six-stranded b-sheet core and the helices between strands β1 and β2 (β1–2) and β2 and β3 (β2–3). Target DNA binds to a solvent-accessible channel placed orthogonally to the NAD$^+$ molecule (Figure 2B,C) and is stabilised by the ARTT loop, which is known to affect substrate specificity in ARTDs as mentioned before. The length of the ARTT loop in DarT exceeds the ARTT loop of bacterial ARTDs and is instead comparable to loops found in human ARTDs, including PARP1, PARP2, and PARP3, thus forming a stable scaffold for the DNA target.

Given that the ARTT loop is found conserved in human ARTDs and that several ARTDs also ADP-ribosylate DNA [47,116–118], it is tempting to speculate that the ARTT loop plays a role in the interaction with the DNA target. The DNA binding site is located in a groove enriched with basic amino acid residues that enable the formation of a series of interactions that serve the sequence-specific ADP-ribosylation by DarT, with thymidine targeted for ADP-ribosylation pointing orthogonally to the DNA backbone deep inside the active site of DarT. Additional interactions between the DNA fragment and DarT side chains and main chains, in addition to structural waters, stabilise the phosphate–ribose backbone. DarT does not exhibit any NADase activity or auto-ADP-ribosylation activity and shows a distinct catalytic mechanism in comparison with other ARTDs. DarT binds the NAD$^+$ substrate within a large pocket with key interactions resulting in a binding mode of constrained conformation. The adenine moiety is stabilised by hydrogen bonding to the K28 and L30 backbone amides, the adenine–proximal ribose bonds with its 2′ and 3′ hydroxyl groups to T15/H13 and N19, respectively. On the other hand, the NAM moiety is permanently maintained in position by π–π interaction with Y71 and hydrogen bonding of its primary amide to I14 and intra-molecularly to the beta-phosphate [78]. DarT shows a diverse arrangement of the catalytic site, wherein a key role is played by R51 residue, which expands the canonical ART catalytic motifs' repertoire. ADP-ribosylation reaction occurs in several steps including: (1) locking of the thymidine base in plane for ADP-ribosylation by H119; (2) polarisation of the NAD$^+$ molecule for cleavage sustained by Y71 and R51; (3) stabilisation of the oxocarbenium ion resulting from NAD$^+$ cleavage by M78; and (4) proton abstraction from N3 of the thymidine base by R51. The latter step represents a new mechanism of ADP-ribosylation that has not been reported for other ARTs (Figure 3C). In fact, DarT-mediated ADP-ribosylation requires the presence of both R51 and E160 residues to perform the reaction, as NAD$^+$ polarisation for cleavage is promoted by R51, which, when mutated, results in a loss of DarT cytotoxicity and enzymatic activity. This mechanism is different from the canonical NAD$^+$ polarisation generally found in ARTs, where it is mediated by the interaction of the 2′′ hydroxyl group of the NAM ribose with the conserved catalytic glutamate residue. These data show that a new motif also supports the ADP-ribosylation reaction, which prompts the question as to whether DarT is an early version of ARTDs or a more evolved form that specialised in ADP-ribosylation of DNA. The advance in the understanding of such peculiar DarT enzymatic catalytic mechanisms will surely help in designing specific small molecules able to modulate DarT activity. This would represent an interesting molecular target for designing future antimicrobial strategies (please see Section 4).

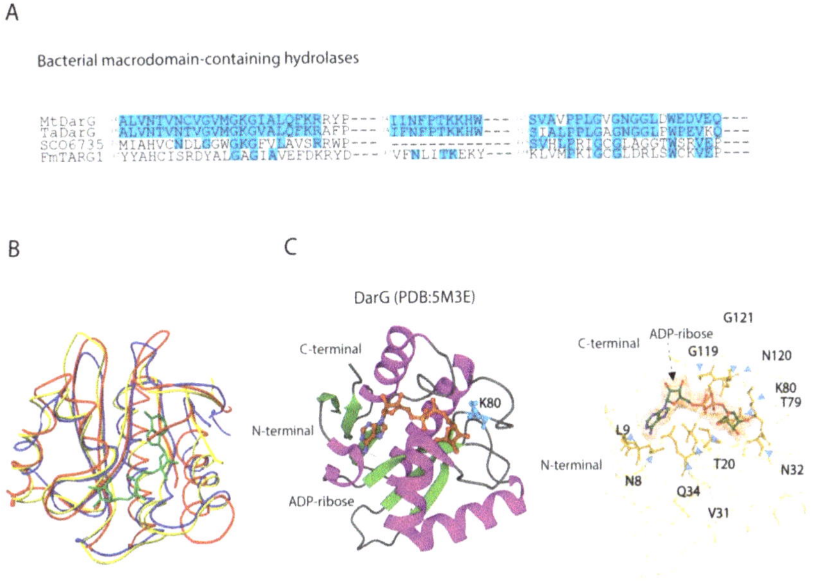

Figure 3. Comparison of amino acid sequences and 3D structures of macrodomain-containing hydrolases belonging to ALC1-like group. (**A**) Alignment of partial sequences of ALC1-like hydrolases from bacteria. MtDarG, DarG from *M. tuberculosis;* TaDarG, DarG from *T. aquaticus*; SCO6735, macrodomain-containing hydrolase from *S. coelicolor;* FmTARG1, TARG1 homologue from *F. mortiferum*. Identities are indicated in light blue. (**B**) Structural comparison between DarG from *T. aquaticus* in complex with ADP-ribose (yellow line, PDB: 5M3E, [38]), human TARG1 in complex with ADP-ribose (blue line, PDB:4J5S, [55]) and SCO6735 (red line, PDB:5E3B, [119]). (**C**) DarG from *T. aquaticus* (cartoon) in complex with ADP-ribose (ball and stick). The catalytic K80 is shown in light blue (left panel). Close up of the *T. aquaticus* DarG active site showing the residues involved in ADP-ribose binding (right panel).

3.2. DarG Macrodomain-Containing Hydrolase Counteracts DarT Toxicity

The macrodomain-containing hydrolase DarG from *T. aquaticus* reverts DNA-ADP-ribosylation adduct on the thymidine base as observed in overexpressing *E. coli* cells with the consequent rescue of cell growth [38]. DarG antitoxins, which were characterised in *T. aquaticus* and *M. tuberculosis,* show a 56.4% sequence identity and a low sequence similarity to other bacterial macrodomain-containing hydrolases (Figure 3A).

DarG antitoxins share a resemblance to human terminal ADP-ribose glycohydrolase 1 (TARG1), and thus belong to the phylogenetically distinct ALC1-like sub-class of macrodomains. The ALC1-like macrodomain-containing enzymes bear similarity to the macrodomain fold found in the human chromatin-remodelling enzyme, ALC1 (Amplified Liver in Cancer1), which does not possess enzymatic activity, but interacts with PAR and catalyses PARP1-dependent nucleosome remodelling upon DNA damage [120]. From a functional point of view, members of the ALC1-like class display mono-ADP-ribosylhydrolase activity by hydrolysing the acyl-ADP-ribose ester bond by lysine residue, also exemplified by TARG1. The K84 nucleophilically attacks the C1′ atom of the distal ribose, leading to the formation of a lysyl-ADP-ribose intermediate with the release of the de-ADP-ribosylated E/D residue. The lysyl-intermediate is then resolved by residue D125, which frees the ADP-ribose, and restores the K84 residue [55].

The DarG macrodomain adopts a typical macrodomain fold composed of a six-stranded mixed β sheet arranged between four α helices and one 3_{10}-helical element. The ligand-binding pocket of the DarG macrodomain is made up of four surface loops where the bound ADP-ribose is located, and it interacts with neighbouring amino acid residues by forming hydrogen bonds (Figure 3B). The finding of W83 at the entrance of the cleft to stack with thymidine base for a correct position and K80 mostly involved in catalysis reflects the corresponding residues found in TARG1. The ligand-binding pocket is stabilised by the formation of hydrogen bonds. Key residues are found conserved, including K80, which is free to access the thymidine–ribose bond, and which is located in an equivalent position of K84 in TARG1. In fact, the mutation of K80A results in inactive DarG with a loss of hydrolase activity, similarly to the corresponding mutation on catalytic lysine in TARG1 [55]. However, the DarG catalytic mechanism remains unclear and needs further investigations.

The great similarity in the structural fold encountered between bacterial DarG and human TARG1 (Figure 3B) suggests that TARG1 plays a role in reversing ADP-ribosylation from DNA. Experimental evidence shows that the overexpression of DarT in human TARG1 knockout cell lines induces a strong DNA damage response due to replication fork progression arrest and cell death, and that the reintroduction of TARG1 activity is required for the reversal of DarT genotoxic effects. This indicates that TARG1 is the main macrodomain enzyme in human cells that acts as a DNA repair factor analogously to DarG [121].

Similar reversal activity has been described for the macrodomain hydrolase SCO6735, known for its regulatory role in antibiotic production in *S. coelicolor* [102]. SCO6735 has been identified as a functional homologue of DarG as it neutralises DarT activity by hydrolysing the thymidine-linked DNA-ADP-ribosylation [119]. Structural studies have shown that SCO6735 has a notable structural similarity to *T. aquaticus* DarG and human TARG1, even though TARG1 and SCO6735 also de-MARylate protein targets while DarG does not [119]. The overall macrodomain fold in SCO6735 is conserved (Figure 3B); the superimposition of the SCO6735 crystal structure with TARG1 and DarG in complex with ADP-ribose revealed a putative active site confined by three loops. The diphosphate and distal ribose are located between two loops, namely, the phosphate-binding (PB) and substrate-binding (SB) loop. The central loop in SCO6735 is five amino acids longer than DarG and TARG1 and provides *Streptomyces* hydrolase the ability to reverse ADP-ribose from thymidine-linked ADP-ribosylation and from aspartate/glutamate-linked proteins. The catalytic mechanism relies on the correct arrangement of the V25 and Q85 residues and a catalytic water molecule within the active site that sits between these residues and the diphosphate of the ADP-ribose. The mutation of Q85, located in an equivalent position to the catalytic lysine in DarG and TARG1 (K80 and K84, respectively), leads to a complete loss of activity. These observations suggest a diversification of catalytic reaction in this sub-class of macrodomain hydrolases, providing the rationale for the design of selective inhibitors or even agonists [119].

3.3. DarT/DarG TA System Molecular Mechanisms and Biological Functions

Bacterial genes encoding toxin and cognate antitoxin are frequently organised into operons, whose gene expression is regulated at a transcriptional and translational level. Under certain physiological conditions, the antitoxin protects the cell from the harmful effects of the toxin through a blockade or neutralising toxin activity [83,122,123]. Under stress conditions, the toxin is released and free to specifically impair one or more of several different cell events including DNA replication, translation, cytoskeleton formation, and membrane integrity [85,124,125]. More than 10,000 putative TA systems have been predicted by bioinformatic analyses [126,127], which can be classified into different types based on the nature of the antitoxin (non-coding RNA or protein) and on the interaction mode between the toxin and antitoxin components (Table 1) [85,125].

Table 1. Classification of TA systems with the related targets and affected cellular functions.

TA Types	Toxin	Antitoxin	Interaction Mode	Main Targets	Affected Cellular Processes	References
Type –I	Protein	Noncoding RNA	Interference with toxin mRNA	Bacterial membrane	Biosynthesis of cell membrane	[128]
Type –II	Protein	Protein	Protein–protein interaction	DNA gyrase, EF-Tu elongation factor, genomic DNA, phosphoribosyl pyrophosphate synthetase	DNA replication, translation, nucleotide metabolism	[38,39,129,130]
Type III	Protein	Noncoding RNA	Sequestering of the toxin	mRNA	Translation	[131]
Type IV	Protein	Protein	Competition for cellular targets	Cytoskeletal proteins	Cell morphology	[132]
Type V	RNA	Protein	Hydrolysis of toxin mRNA	Cell membrane	Biosynthesis of cell membrane	[133]
Type VI	Protein	Protein	Complex formation resistant to proteolysis	β-sliding clamp	DNA replication	[134]
Type VII	Protein	Protein	Chemical modification of the toxin	Biofilm	Biofilm	[135]
Type VIII	Noncoding RNA	mRNAs	Targeting of mRNAs	YhcB inner membrane protein	Cell morphology	[136]

In type II TA systems, the toxin effects are mainly counteracted by the direct binding of antitoxin to cognate toxin through protein–protein interaction, forming a stable toxin–antitoxin complex [83,87,137]. As summarised in Table 2, type II toxins include endoribonucleases that target mRNA, rRNA, and tRNA; ribosome-poisoning protein acetyltransferases that target tRNAs; topoisomerase inhibitors; cell wall inhibitors; and enzymes generating PTMs that target a diverse array of cellular targets, with the majority of them involved in the downregulation of cell metabolism [138,139].

Table 2. Bacterial toxins displaying post-translational modification activity in type II TA systems.

Toxin	Bacterium	PTM Targets	Affected Biological Functions	References
HipA	*E. coli* K12	Phosphorylation of Gltx	Persistence induction	[140,141]
HipT	*E. coli* O127: H6	Phosphorylation of TrpS	Cell growth inhibition	[142]
Doc	*E. coli*	Phosphorylation of EF-Tu	Persistence induction	[130]
FicT	*P. aeruginosa*, *E. coli* and *Yersinia enterocolitica*	Adenylylation of DNA-gyrase and TopoIV	Cell growth inhibition	[143]
Fic-1	*P. fluorescens* 2P24	Adenylylation of DNA gyrase GyrB	Persistence induction	[144]
VbhT	*Bartonella schoenbuchensis*	T4SS effector	Secretion of virulence factors	[145]
DarT	*M. tuberculosis*	ADP-ribosylation of DNA	Cell growth inhibition Phage defence	[38,40,78,146]
ParT	*Sphingobium* sp. YBL2	ADP-ribosylation of Prs	Cell growth inhibition	[39]
Tre1	*Serratia proteamaculans*	ADP-ribosylation of FtsZ	Cell death	[147]
MbcT	*M. tuberculosis*	NAD^+ degradation	Cell death	[148]
TacT	*Salmonella typhimurium*	Acetylation of tRNAs	Translation inhibition	[149]
AtaT/AtaT2	*E. coli* O157:H7	Acetylation of fMet-tRNAs	Translation inhibition	[150]
KacT	*Klebsiella pneumoniae*	Acetylation of tRNA	Translation inhibition	[151]
ItaT	*E. coli* HS	Acetylation of tRNA	Translational inhibition	[152]

ADP-ribosylation is a new player in TA biology. DarT/DarG TA was initially ascribed to type II, but it is now recognised as a type II/IV hybrid system, as DarT toxicity is mainly abrogated by DarG enzymatic activity by removing the DNA-ADP-ribose adduct rather than by TA complex assembly, as detailed below.

DarT catalyses the MARylation of NNTNTCN and TTTT/A motifs in ssDNA genomic sequences in *T. aquaticus* and *M. tuberculosis*, respectively. This enzymatic activity results in the formation of a thymidine ADP-ribose adduct that is recognised by the DNA damage repair system as a DNA lesion [38,78] (Figure 4A). It seems that *M. tuberculosis* uses this system to introduce adducts at the origin of DNA replication (OriC), which controls replication and cell growth. DarT highly ADP-ribosylates genomic DNA in DarG-depleted *M. tuberculosis* cells, leading to the activation of DNA damage response (Figure 4A). As a final outcome, the recruitment of DNAB, the replicative helicase interacting with ssDNA at the OriC and driving DNA branch migration during replication, may be impaired at cell division [38,78].

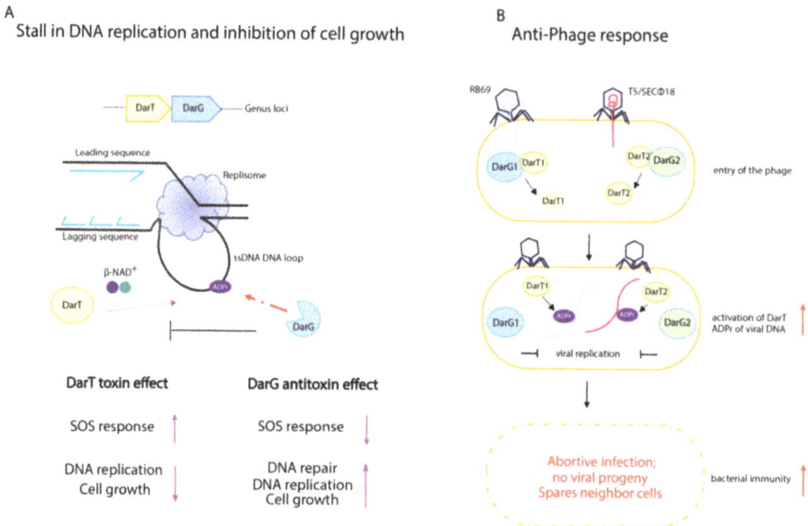

Figure 4. Schematic representation of DarT/DarG TA system biological functions. (**A**) DarT/DarG system in the regulation of bacterial cell growth. DarT-mediated ADP-ribosylation of ssDNA on thymidine found in consensus sequences causes a stall of DNA replication and concomitant arrest of cell growth. The activity of the DarG antitoxin counteracts DarT activity through the removal of ADP-ribose from the marked thymidine on ssDNA: DarG-mediated removal of ADP-ribose enables the replication to proceed and cell growth to re-establish. (**B**) The DarT/DarG system and the anti-phage response. Upon entry of the phages, the DarT1 and DarT2 endotoxins ADP-ribosylate viral DNA, which is unable to replicate. The overall downregulation of cell metabolic processes triggers the abortive infection programme, which leads to the host cell death and prevents viral progeny spreading in order to protect the bacterial cell population.

The ADP-ribosylation of genomic DNA can be counteracted by the DarG antitoxin, which reverts DNA-ADP-ribose adducts [38] (Figure 4A), thus acting as a non-canonical DNA repair factor specific for ADP-ribosyl-thymidine adducts and re-establishing bacterial cell replication. The exquisite specificity of DarG reversal has been confirmed by in vitro experiments, where human macrodomain-containing hydrolases such as MacroD1 or PARG and DraG-related ADP-ribosylhydrolase ARH3 were unable to remove ADP-ribose from genomic DNA [78]. Further investigations support the protective role of DarG against DarT-mediated toxic effects, as the activation of the DNA damage response leads to cell death in *M. tuberculosis* DarG-depleted cells [153]. The same molecular mechanisms are shared by DarT/DarG TA systems from other species, such as in the enteropathogenic *E. coli*, where the ADP-ribosylation of genomic TCT or TTT DNA sequences can affect bacterial growth and viability [40]. Notably, complementation studies show that *T. aquaticus* DarG bears mutations in the hydrolase domain, namely, in the catalytic K80 residue, yet negatively affects DarT activity, thus suggesting that the antitoxin effect of DarG can also pass through additional mechanisms [38]. Consequently, the DarT/DarG TA pair can be considered a novel hybrid TA system.

In addition to the DarG antitoxin, the DarT-mediated DNA adducts can be also repaired by the sequential action of RecF-mediated homologous recombination, which leads to the conversion of a single-strand lesion into a double-strand lesion, which is then repaired by the nucleotide excision repair (NER) pathway [40].

The finding that the DarT/DarG system is also present in other bacterial species including the pathogenic *P. aeruginosa*, *Acinetobacter baumannii*, and *K. pneumoniae* [40] suggests a conserved role in pathogenic bacteria. However, the triggers that specifically induce DarT toxin activity remain unknown.

3.4. Functional Outcomes of DarT/DarG System in Prokaryotic Immunity

TA systems regulate several physiological processes including plasmid stabilisation and cell viability [154], persister cell formation [82,155], stress response [156], and biofilms [157] as well as multidrug resistance [86], pathogenicity [158], and defence from bacteriophages [139,159]. Overall, TA systems behave like versatile modulators of bacterial physiology that exploit the same biochemical mechanism to regulate a wide range of different activities.

With regard to phage defence, several bacterial strains harness diverse anti-phage defence systems relying on: (*i*) the degradation of phage nucleic acids by acting through restriction endonucleases and the CRISPR-Cas system; (*ii*) abortive infection-activating mechanisms that kill the bacterial population before phage replication; and (*iii*) the inhibition of DNA and RNA synthesis through the production of small molecules with inhibitory activity, which in turn sustains bacterial immunity [160,161].

The DarT/DarG TA system modulates the anti-phage response through the ADP-ribosylation of viral DNA and the consequent induction of the abortive infection mechanism [146] (Figure 4B). As already mentioned above, abortive infection is a well-conserved mechanism within bacterial immunity, and is widespread in Gram-positive and Gram-negative bacteria, where cell death takes place prior to the maturation of the phage progeny, thus preventing the spread of phages to neighbouring cells and protecting the bacterial population.

TA systems are known to have a pivotal role in the immunity of bacteria against phages, as their activation upon phage infection leads to cell death or growth arrest [131,162–165]. The role of ADP-ribosylation in bacterial immunity is exemplified by the DarT1/DarG1 and DarT2/DarG2 TA systems, which are found in the defence islands of the *E. coli* MG1655 bacterial genome, which are homologues of *T. aquaticus* DarT/DarG [146]. DarT toxin is conserved in both systems, whereas DarG1 and DarG2 belong to two different subfamilies; DarG1 encloses a putative YbiA-like fold, while DarG2 harbours a highly conserved macrodomain. Both DarT1/DarG1 and DarT2/DarG2 protect *E. coli* cells from natural bacteriophage infections, given that DarT1 and DarT2 are involved in ADP-ribosylation of viral DNA, with the consequent hindering of the phage's genome replication, RNA synthesis, and assembly of mature/infective virions (Figure 4B).

DarT1/DarG1 and DarT2/DarG2 appear to target different phages (i.e., RB69 and T5/SEC 18, respectively) and are active under different growing conditions (DarTG1 during fast growth, DarTG2 during slow growth). These findings may suggest that a different regulatory mechanism activates the two DarT/DarG systems [146]. The molecular mechanism that unleashes the DarT toxin remains elusive; yet, though a still-unknown phage-derived trigger that frees the DarT toxin to exert an anti-phage response may perhaps explain it. The finding of phage mutants that exhibit spontaneous resistance to this immunity system by interfering with DarT/DarG activity adds another layer of complexity to the bacteria–phage conflicts [146].

A more comprehensive understanding of DarT/DarG biology may also result in the rational design of selective phage-based therapies as an alternative to antibiotics for treating resistant pathogens by manipulating endogenous anti-phage responses [166,167]. As such, small molecules inhibiting DarT may be exploited to counteract bacterial defence mechanisms against phages, which, to date, represent the most real alternative to antibiotics.

4. Exploitation of DarT/DarG Biology for a Rational Design of Antimicrobial Agents

Antibiotic resistance and the recurrence of bacterial infections are two of the most urgent threats to future global public health, with implications for all areas of medicine [168]. Antibiotic treatment misuse in humans and animals has accelerated the generation of antibiotic-resistant bacterial strains. In addition, the lack of novel effective antibacterial compounds, also due to poor investment in antimicrobial research, has increased this concern [168,169]. In fact, in the past 15 years, only one new class of antibiotics against Gram-positive bacteria has been introduced into clinical practice. The majority of antibiotics on the market are based on existing drugs selected to overcome the resistance acquired by bacteria against their related compounds [170]. Therefore, in order to tackle antibiotic resistance and recurring infections, it is imperative to search for antibacterial agents that rely on innovative modes of action.

Current antibiotics mostly target bacterial enzymes, ribosomal RNA, cell wall construction, and cell membrane function. Despite being widely used for the treatment of diverse infectious diseases, antibiotic treatments are not effective enough to eradicate highly resistant pathogens such as those referred as to ESKAPE. These resistant pathogens include *E. coli*, *S. aureus*, *K. pneumoniae*, *A. baumanii*, *P. aeruginosa*, and *Enterobacter* species, which are considered a priority by the World Health Organization for the urgent need of alternative therapeutics to antibiotic treatments. Therefore, alternative approaches to eradicate bacterial infection have been explored to deliver new therapies with clinical utility [171].

Mono-ADP-ribosyltransferase toxins are produced by pathogenic bacteria to infect the host cell with the impairment of vital molecular pathways [62]. These exotoxins exploit the host intracellular NAD^+ to accomplish bacterial infection, which, in turn, causes a decrease in the level of NAD^+ in the host, resulting in energy store depletion, immune evasion, or cell death [172]. In addition, some pathogens can also modulate NAD^+ metabolism to support their fitness through the activity of pathogenic-specific enzymes such as NADases, or by the modulation of the activity of host NAD^+-dependent enzymes (i.e., Sirtuins, PARPs, and CD38) [172,173]. Very recently, the pharmacological inhibition of PARPs in patients affected by diabetes mellitus has been reported to decrease intracellular *M. tuberculosis* (Mtb) in human macrophages, identifying PARP targeting as a potential novel strategy for host-directed therapy against *M. tuberculosis* and possibly against other infectious diseases [174].

With regard to NAD^+-targeting toxins, the therapeutic inhibition of NAD^+-dependent reactions in bacteria is still in its infancy and mainly relies on the chemical modulation of the NAD^+ interaction pocket within the ART domain in order to block enzymatic activity. In the last two decades, antimicrobial strategies against ADP-ribosylating toxins have been proposed given that they are expected to provide new drug targets to disarm antibiotic-resistant bacteria. Different strategies, starting from the combination of PARP inhibitors, have been tested on *P. aeruginosa* Exotoxin A, *V. cholerae* Cholix toxin, *V. splendidus* Vis toxin, *S. scabies* Scabin toxin, *Bacillus cereus* Certhrax toxin, *Paenibacillus larvae* C3larvin, and Plx2A ([175] and the references therein). Such strategies have been searched for using ARTD non-specific inhibitors such as PJ34 [176], largely known for targeting human ARTDs (i.e., PARPs), polyphenolic extracts [177], and small molecules from the screening of synthetic libraries [178–180]. These attempts have led to the identification of lead compounds that can be further modified and explored in drug design. A promising approach relies on the use of natural compounds from plant origin that can hinder bacterial infections [181,182].

Protein–protein interactions (PPIs) have emerged as promising drug targets [183–186] and intensive efforts have led to the clinical use of PPI modulators as next-generation therapeutics in cancer treatments [187]. With regard to infectious diseases, the treatment of HIV/AIDS with the antiretroviral drugs enfuvirtide and maraviroc, which target host–pathogen interactions, provides a successful example of PPI-based drugs [188]. Since PPIs naturally occur in bacteria and regulate a multitude of cellular processes, bacterial protein–protein interactions are considered to be good candidates as a target for antibiotic drug discovery; however, to date, they are still underexplored [189,190].

Toxin–antitoxin systems represent a substantial pool of interactions within bacteria [83,191] that can be exploited for the development of advanced antibiotics [189–191]. Diverse PPI-based approaches have led to the discovery of peptides and small-molecule compounds that interfere with PPIs in TA systems, with the impairment of translation, cell wall synthesis, and lipase activity. However, no inhibitors are currently used in clinics. Given that DarG counteracts DarT activity even through the formation of a DarT–DarG complex, a PPI-based approach may be also considered in order to interfere with DarT function.

From this perspective, we have discussed the recent advances in the regulatory role exerted by the DarT/DarG hybrid TA system in the control of cell growth and abortive infection, strictly relying on ADP-ribosylation signalling. Within this framework, targeting DarT activity may represent a valuable alternative strategy for the therapeutic treatment of highly resistant pathogenic bacteria, such as *M. tuberculosis,* by preventing persistence activation. In addition, the availability of high-resolution-structure DarT provides the rationale for designing selective drugs to use as antimicrobial agents with less daunting side effects for the host.

PARP inhibitors for the therapeutic manipulation of ADP-ribosylation have been proposed for a wide range of disorders both in human and animal models, including cardiovascular, inflammatory, autoimmune, and neurological disorders [57]. In contrast, targeting ADP-ribosylation as a therapeutic intervention to counteract infectious diseases and related antibiotic resistant bacterial strains has been much less explored, with the exception of viral-mediated disease, where ADP-ribosylation is emerging as a crucial process in host–pathogen conflicts [69,71]. The growing body of evidence for the critical role of NAD^+ as a co-factor of enzymes involved in bacterial physiology and the pathogenic mechanism as well as in host–pathogen interactions, also including viral-mediated diseases, highlights the importance of investigating these molecular pathways in order to find novel therapeutic strategies.

5. Conclusions

Recent discoveries have established DNA and RNA as the novel ADP-ribosylated substrates [48–50]. In mammals, the reversible ADP-ribosylation of DNA is mediated by PARP1, PARP2, and PARP3, which can ADP-ribosylate phosphorylated DNA termini on ds-DNA in vitro following a phosphorylation-dependent ADP-ribosylation mechanism; however, the functional outcomes remain unknown [116–118,192].

Recently, the reversible PARP1-mediated PARylation of ssDNA that targets adenosine residues has also been identified both in vitro and in vivo, where it does not seem to be activated by DNA strand breaks [193]. Other human PARPs, such as TRPT1, PARP10, PARP11, PARP12, and PARP15, appear to target the 5′-phosphorylated end of single-stranded RNA in vitro [47–50], giving rise to a novel RNA capping mechanism.

Several PARPs also have a role in the antiviral response through the inhibition of the virus life cycle at different stages, from transcription to translation, as exemplified by PARP7 and PARP13, which are involved in the exosome degradation of viral RNAs, and PARP12, which is responsible for the impairment of viral translation through the direct binding of viral RNA, and by the downregulation of cellular processes such as translation ([194] and the references therein). The importance of ADP-ribosylation-dependent systems in the antiviral response is highlighted by the fact that several viruses, such as alphaviruses and coronaviruses, have evolved macrodomains to counter hosts' defensive processes controlled by ADP-ribosylation [33,53,60,195–198]. Viral macrodomains represent potential targets of antiviral drugs [69–71,199–201]. The role of ADP-ribosylation in antiviral and stress response, for instance, involving the ADP-ribosylation of viral genetic material, are reminiscent of the DNA modifications observed in lower organisms, where the transfer of ADP-ribose on nucleic acids results in the defence mechanism's response. DarT/DarG represents one of the most ancient ADP-ribosylation-dependent systems with a role in bacterial immunity (e.g., against viral infections) as well as in the stress response. The modulation of the DarT/DarG system may also help the design of new effective antimicrobial agents in this context.

Anti-phage defence mechanisms have been extensively studied. However, many aspects still need clarification [202]. Several molecular processes underlie the anti-phage defence, which is mostly based on the degradation of the viral genome (e.g., restriction/modification enzymes, CRISPR-Cas systems, Argonaute proteins), the inhibition of DNA and RNA synthesis (e.g., chemical defence, secondary metabolite, nucleotide depletion), and abortive infection [160]. Abortive infection takes place through several molecular mechanisms, which include CRISPR-Cas and TA systems, among others [160]. A new group of retrons, belonging to prokaryotic reverse transcriptases, have been characterised to confer resistance to a wide range of phages [203,204]. Intriguingly, they share a tripartite module organisation reminiscent of TA systems, and are composed of reverse transcriptase, a multicopy single-stranded DNA (msDNA) and RcaT, an additional element protein [205]. Retrons can also be potentially used in genome editing, as they catalyse the polymerisation of DNA from an RNA template [204].

The systems mentioned above are just a few examples of the great diversity of defence systems found or predicted in bacterial cells to resist phage attack. In fact, a large number of genes encoding for different defence systems are found on the bacterial genome within chromosome regions known as "defence islands", some of which are estimated to contain more than 100 defence genes [202]. Overall, such co-localisation of different defence genes suggests a functional link between the defence systems, including a possible coregulation mechanism.

More than 10,000 TA systems have been found on bacterial genomes, with many bacterial species encoding dozens of TA modules. For instance, *E. coli* K12 and *M. tuberculosis* encode more than 30, and at least 80, different TA systems [146], thus suggesting that different molecular activities support TA systems in their functional outcomes. DarT/DarG represents the first TA system that induces the stress response by growth control and abortive infection by ADP-ribosylating host genomic DNA and viral DNA with the concomitant inhibition of host DNA replication and cell growth. More recently, a ParT/ParS TA system from *Sphingobium* sp. YBL2 was found to hinder nucleotide metabolism with the induction of a persistence state by ADP-ribosylating target protein [39], highlighting NAD^+ as a key component for triggering the prokaryotic immune response [173].

The wide distribution of ADP-ribosylation systems across all domains of life highlights the importance of this modification throughout evolution [1–4,62,206,207]. Overall, we believe that our review highlights the emergence of a new and exciting research area in the ADP-ribosylation field with implications in the regulation of cellular functions still to be discovered.

Author Contributions: G.C. and L.P. conceived and co-wrote the manuscript. R.C. performed supporting studies and contributed to the preparation of the figures. All authors have read and agreed to the published version of the manuscript.

Funding: Work in the L.P. group is funded by the Ovarian Cancer Research Alliance (OCRA, Collaborative Research Development Grant, n.813369), the POR Campania FESR 2014/2020 (projects SATIN and RECOVER-COVID19), and by NUTRIAGE-CNR. G.C.'s research activity is supported by "One

Health Basic and Translational Research Actions addressing Unmet Needs on Emerging Infectious Diseases—INFACT"—CUP B53C20040570005.

Institutional Review Board Statement: Not applicable.

Informed Consent Statement: Not applicable.

Data Availability Statement: Not applicable.

Acknowledgments: We thank Ivan Ahel, William Dunn School of Pathology, University of Oxford; Giuseppe Manco, Institute of Biochemistry and Cell Biology, National Research Council of Italy (CNR); and Rosa Marina Melillo, Department of Molecular Medicine and Medical Biotechnology, University of Naples "Federico II" for the critical reading of the manuscript.

Conflicts of Interest: The authors declare no conflict of interest.

References

1. Perina, D.; Mikoč, A.; Ahel, J.; Ćetković, H.; Žaja, R.; Ahel, I. Distribution of Protein Poly(ADP-Ribosyl)Ation Systems across All Domains of Life. *DNA Repair* **2014**, *23*, 4–16. [CrossRef] [PubMed]
2. Aravind, L.; Zhang, D.; de Souza, R.F.; Anand, S.; Iyer, L.M. The Natural History of ADP-Ribosyltransferases and the ADP-Ribosylation System. *Curr. Top. Microbiol. Immunol.* **2015**, *384*, 3–32. [CrossRef] [PubMed]
3. Palazzo, L.; Mikoč, A.; Ahel, I. ADP-Ribosylation: New Facets of an Ancient Modification. *FEBS J.* **2017**, *284*, 2932–2946. [CrossRef] [PubMed]
4. Mikolčević, P.; Hloušek-Kasun, A.; Ahel, I.; Mikoč, A. ADP-Ribosylation Systems in Bacteria and Viruses. *Comput. Struct. Biotechnol. J.* **2021**, *19*, 2366–2383. [CrossRef]
5. Jørgensen, R.; Purdy, A.E.; Fieldhouse, R.J.; Kimber, M.S.; Bartlett, D.H.; Merrill, A.R. Cholix Toxin, a Novel ADP-Ribosylating Factor from Vibrio Cholerae. *J. Biol. Chem.* **2008**, *283*, 10671–10678. [CrossRef]
6. Collier, R.J.; Pappenheimer, A.M. Studies on the mode of action of diphtheria toxin. I. Phosphorylated intermediates in normal and intoxicated hela cells. *J. Exp. Med.* **1964**, *120*, 1007–1018. [CrossRef]
7. D'Amours, D.; Desnoyers, S.; D'Silva, I.; Poirier, G.G. Poly(ADP-Ribosyl)Ation Reactions in the Regulation of Nuclear Functions. *Biochem. J.* **1999**, *342*, 249–268. [CrossRef]
8. De Vos, M.; Schreiber, V.; Dantzer, F. The Diverse Roles and Clinical Relevance of PARPs in DNA Damage Repair: Current State of the Art. *Biochem. Pharmacol.* **2012**, *84*, 137–146. [CrossRef]
9. Gupte, R.; Liu, Z.; Kraus, W.L. PARPs and ADP-Ribosylation: Recent Advances Linking Molecular Functions to Biological Outcomes. *Genes Dev.* **2017**, *31*, 101–126. [CrossRef]
10. Schützenhofer, K.; Rack, J.G.M.; Ahel, I. The Making and Breaking of Serine-ADP-Ribosylation in the DNA Damage Response. *Front. Cell Dev. Biol.* **2021**, *9*, 745922. [CrossRef]
11. Lord, C.J.; Ashworth, A. PARP Inhibitors: Synthetic Lethality in the Clinic. *Science* **2017**, *355*, 1152–1158. [CrossRef]
12. Palazzo, L.; Ahel, I. PARPs in Genome Stability and Signal Transduction: Implications for Cancer Therapy. *Biochem. Soc. Trans.* **2018**, *46*, 1681–1695. [CrossRef]
13. Slade, D. PARP and PARG Inhibitors in Cancer Treatment. *Genes Dev.* **2020**, *34*, 360–394. [CrossRef]
14. Curtin, N.J.; Szabo, C. Poly(ADP-Ribose) Polymerase Inhibition: Past, Present and Future. *Nat. Rev. Drug Discov.* **2020**, *19*, 711–736. [CrossRef]
15. Krishnakumar, R.; Kraus, W.L. PARP-1 Regulates Chromatin Structure and Transcription through a KDM5B-Dependent Pathway. *Mol. Cell* **2010**, *39*, 736–749. [CrossRef]
16. Gibson, B.A.; Zhang, Y.; Jiang, H.; Hussey, K.M.; Shrimp, J.H.; Lin, H.; Schwede, F.; Yu, Y.; Kraus, W.L. Chemical Genetic Discovery of PARP Targets Reveals a Role for PARP-1 in Transcription Elongation. *Science* **2016**, *353*, 45–50. [CrossRef]
17. Kim, D.-S.; Camacho, C.V.; Nagari, A.; Malladi, V.S.; Challa, S.; Kraus, W.L. Activation of PARP-1 by SnoRNAs Controls Ribosome Biogenesis and Cell Growth via the RNA Helicase DDX21. *Mol. Cell* **2019**, *75*, 1270–1285.e14. [CrossRef]
18. Gupte, R.; Nandu, T.; Kraus, W.L. Nuclear ADP-Ribosylation Drives IFNγ-Dependent STAT1α Enhancer Formation in Macrophages. *Nat. Commun.* **2021**, *12*, 3931. [CrossRef]
19. Challa, S.; Khulpateea, B.R.; Nandu, T.; Camacho, C.V.; Ryu, K.W.; Chen, H.; Peng, Y.; Lea, J.S.; Kraus, W.L. Ribosome ADP-Ribosylation Inhibits Translation and Maintains Proteostasis in Cancers. *Cell* **2021**, *184*, 4531–4546.e26. [CrossRef]
20. Jones, A.; Kraus, W.L. Multiomics Analysis of the NAD+-PARP1 Axis Reveals a Role for Site-Specific ADP-Ribosylation in Splicing in Embryonic Stem Cells. *Genes Dev.* **2022**, *36*, 601–617. [CrossRef]
21. Manco, G.; Lacerra, G.; Porzio, E.; Catara, G. ADP-Ribosylation Post-Translational Modification: An Overview with a Focus on RNA Biology and New Pharmacological Perspectives. *Biomolecules* **2022**, *12*, 443. [CrossRef] [PubMed]
22. Chang, P.; Coughlin, M.; Mitchison, T.J. Tankyrase-1 Polymerization of Poly(ADP-Ribose) Is Required for Spindle Structure and Function. *Nat. Cell Biol.* **2005**, *7*, 1133–1139. [CrossRef] [PubMed]

23. Boehler, C.; Gauthier, L.R.; Mortusewicz, O.; Biard, D.S.; Saliou, J.-M.; Bresson, A.; Sanglier-Cianferani, S.; Smith, S.; Schreiber, V.; Boussin, F.; et al. Poly(ADP-Ribose) Polymerase 3 (PARP3), a Newcomer in Cellular Response to DNA Damage and Mitotic Progression. *Proc. Natl. Acad. Sci. USA* **2011**, *108*, 2783–2788. [CrossRef] [PubMed]
24. Yu, M.; Schreek, S.; Cerni, C.; Schamberger, C.; Lesniewicz, K.; Poreba, E.; Vervoorts, J.; Walsemann, G.; Grötzinger, J.; Kremmer, E.; et al. PARP-10, a Novel Myc-Interacting Protein with Poly(ADP-Ribose) Polymerase Activity, Inhibits Transformation. *Oncogene* **2005**, *24*, 1982–1993. [CrossRef] [PubMed]
25. Andrabi, S.A.; Kang, H.C.; Haince, J.-F.; Lee, Y.-I.; Zhang, J.; Chi, Z.; West, A.B.; Koehler, R.C.; Poirier, G.G.; Dawson, T.M.; et al. Iduna Protects the Brain from Glutamate Excitotoxicity and Stroke by Interfering with Poly(ADP-Ribose) Polymer-Induced Cell Death. *Nat. Med.* **2011**, *17*, 692–699. [CrossRef]
26. Bachmann, S.B.; Frommel, S.C.; Camicia, R.; Winkler, H.C.; Santoro, R.; Hassa, P.O. DTX3L and ARTD9 Inhibit IRF1 Expression and Mediate in Cooperation with ARTD8 Survival and Proliferation of Metastatic Prostate Cancer Cells. *Mol. Cancer* **2014**, *13*, 125. [CrossRef]
27. Leung, A.K.L.; Vyas, S.; Rood, J.E.; Bhutkar, A.; Sharp, P.A.; Chang, P. Poly(ADP-Ribose) Regulates Stress Responses and MicroRNA Activity in the Cytoplasm. *Mol. Cell* **2011**, *42*, 489–499. [CrossRef]
28. Catara, G.; Grimaldi, G.; Schembri, L.; Spano, D.; Turacchio, G.; Lo Monte, M.; Beccari, A.R.; Valente, C.; Corda, D. PARP1-Produced Poly-ADP-Ribose Causes the PARP12 Translocation to Stress Granules and Impairment of Golgi Complex Functions. *Sci. Rep.* **2017**, *7*, 14035. [CrossRef]
29. Guo, X.; Ma, J.; Sun, J.; Gao, G. The Zinc-Finger Antiviral Protein Recruits the RNA Processing Exosome to Degrade the Target MRNA. *Proc. Natl. Acad. Sci. USA* **2007**, *104*, 151–156. [CrossRef]
30. Atasheva, S.; Frolova, E.I.; Frolov, I. Interferon-Stimulated Poly(ADP-Ribose) Polymerases Are Potent Inhibitors of Cellular Translation and Virus Replication. *J. Virol.* **2014**, *88*, 2116–2130. [CrossRef]
31. Zhang, Y.; Mao, D.; Roswit, W.T.; Jin, X.; Patel, A.C.; Patel, D.A.; Agapov, E.; Wang, Z.; Tidwell, R.M.; Atkinson, J.J.; et al. PARP9-DTX3L Ubiquitin Ligase Targets Host Histone H2BJ and Viral 3C Protease to Enhance Interferon Signaling and Control Viral Infection. *Nat. Immunol.* **2015**, *16*, 1215–1227. [CrossRef]
32. Kim, C.; Wang, X.-D.; Yu, Y. PARP1 Inhibitors Trigger Innate Immunity via PARP1 Trapping-Induced DNA Damage Response. *eLife* **2020**, *9*, e60637. [CrossRef]
33. Alhammad, Y.M.O.; Fehr, A.R. The Viral Macrodomain Counters Host Antiviral ADP-Ribosylation. *Viruses* **2020**, *12*, 384. [CrossRef]
34. Abd Elmageed, Z.Y.; Naura, A.S.; Errami, Y.; Zerfaoui, M. The Poly(ADP-Ribose) Polymerases (PARPs): New Roles in Intracellular Transport. *Cell. Signal.* **2012**, *24*, 1–8. [CrossRef]
35. Cardamone, M.D.; Gao, Y.; Kwan, J.; Hayashi, V.; Sheeran, M.; Xu, J.; English, J.; Orofino, J.; Emili, A.; Perissi, V. Neuralized-like Protein 4 (NEURL4) Mediates ADP-Ribosylation of Mitochondrial Proteins. *J. Cell Biol.* **2022**, *221*, e202101021. [CrossRef]
36. Jia, A.; Huang, S.; Song, W.; Wang, J.; Meng, Y.; Sun, Y.; Xu, L.; Laessle, H.; Jirschitzka, J.; Hou, J.; et al. TIR-Catalyzed ADP-Ribosylation Reactions Produce Signaling Molecules for Plant Immunity. *Science* **2022**, *377*, eabq8180. [CrossRef]
37. Nordlund, S.; Högbom, M. ADP-Ribosylation, a Mechanism Regulating Nitrogenase Activity. *FEBS J.* **2013**, *280*, 3484–3490. [CrossRef]
38. Jankevicius, G.; Ariza, A.; Ahel, M.; Ahel, I. The Toxin-Antitoxin System DarTG Catalyzes Reversible ADP-Ribosylation of DNA. *Mol. Cell* **2016**, *64*, 1109–1116. [CrossRef]
39. Piscotta, F.J.; Jeffrey, P.D.; Link, A.J. ParST Is a Widespread Toxin-Antitoxin Module That Targets Nucleotide Metabolism. *Proc. Natl. Acad. Sci. USA* **2019**, *116*, 826–834. [CrossRef]
40. Lawarée, E.; Jankevicius, G.; Cooper, C.; Ahel, I.; Uphoff, S.; Tang, C.M. DNA ADP-Ribosylation Stalls Replication and Is Reversed by RecF-Mediated Homologous Recombination and Nucleotide Excision Repair. *Cell Rep.* **2020**, *30*, 1373–1384.e4. [CrossRef]
41. Kong, L.; Feng, B.; Yan, Y.; Zhang, C.; Kim, J.H.; Xu, L.; Rack, J.G.M.; Wang, Y.; Jang, J.-C.; Ahel, I.; et al. Noncanonical Mono(ADP-Ribosyl)Ation of Zinc Finger SZF Proteins Counteracts Ubiquitination for Protein Homeostasis in Plant Immunity. *Mol. Cell* **2021**, *81*, 4591–4604.e8. [CrossRef]
42. Ueda, K.; Hayaishi, O. ADP-Ribosylation. *Annu. Rev. Biochem.* **1985**, *54*, 73–100. [CrossRef] [PubMed]
43. Cohen, M.S.; Chang, P. Insights into the Biogenesis, Function, and Regulation of ADP-Ribosylation. *Nat. Chem. Biol.* **2018**, *14*, 236–243. [CrossRef] [PubMed]
44. Vyas, S.; Matic, I.; Uchima, L.; Rood, J.; Zaja, R.; Hay, R.T.; Ahel, I.; Chang, P. Family-Wide Analysis of Poly(ADP-Ribose) Polymerase Activity. *Nat. Commun.* **2014**, *5*, 4426. [CrossRef] [PubMed]
45. Krueger, K.M.; Barbieri, J.T. The Family of Bacterial ADP-Ribosylating Exotoxins. *Clin. Microbiol. Rev.* **1995**, *8*, 34–47. [CrossRef]
46. Nakano, T.; Takahashi-Nakaguchi, A.; Yamamoto, M.; Watanabe, M. Pierisins and CARP-1: ADP-Ribosylation of DNA by ARTCs in Butterflies and Shellfish. *Curr. Top. Microbiol. Immunol.* **2015**, *384*, 127–149. [CrossRef]
47. Munnur, D.; Ahel, I. Reversible Mono-ADP-Ribosylation of DNA Breaks. *FEBS J.* **2017**, *284*, 4002–4016. [CrossRef]
48. Weixler, L.; Schäringer, K.; Momoh, J.; Lüscher, B.; Feijs, K.L.H.; Žaja, R. ADP-Ribosylation of RNA and DNA: From in Vitro Characterization to in Vivo Function. *Nucleic Acids Res.* **2021**, *49*, 3634–3650. [CrossRef]
49. Groslambert, J.; Prokhorova, E.; Ahel, I. ADP-Ribosylation of DNA and RNA. *DNA Repair* **2021**, *105*, 103144. [CrossRef]
50. Munnur, D.; Bartlett, E.; Mikolčević, P.; Kirby, I.T.; Rack, J.G.M.; Mikoč, A.; Cohen, M.S.; Ahel, I. Reversible ADP-Ribosylation of RNA. *Nucleic Acids Res.* **2019**, *47*, 5658–5669. [CrossRef]

51. Baysarowich, J.; Koteva, K.; Hughes, D.W.; Ejim, L.; Griffiths, E.; Zhang, K.; Junop, M.; Wright, G.D. Rifamycin Antibiotic Resistance by ADP-Ribosylation: Structure and Diversity of Arr. *Proc. Natl. Acad. Sci. USA* **2008**, *105*, 4886–4891. [CrossRef]
52. Hottiger, M.O.; Hassa, P.O.; Lüscher, B.; Schüler, H.; Koch-Nolte, F. Toward a Unified Nomenclature for Mammalian ADP-Ribosyltransferases. *Trends Biochem. Sci.* **2010**, *35*, 208–219. [CrossRef]
53. Rack, J.G.M.; Perina, D.; Ahel, I. Macrodomains: Structure, Function, Evolution, and Catalytic Activities. *Annu. Rev. Biochem.* **2016**, *85*, 431–454. [CrossRef]
54. Rack, J.G.M.; Palazzo, L.; Ahel, I. (ADP-Ribosyl)Hydrolases: Structure, Function, and Biology. *Genes Dev.* **2020**, *34*, 263–284. [CrossRef]
55. Sharifi, R.; Morra, R.; Appel, C.D.; Tallis, M.; Chioza, B.; Jankevicius, G.; Simpson, M.A.; Matic, I.; Ozkan, E.; Golia, B.; et al. Deficiency of Terminal ADP-Ribose Protein Glycohydrolase TARG1/C6orf130 in Neurodegenerative Disease. *EMBO J.* **2013**, *32*, 1225–1237. [CrossRef]
56. Liu, C.; Fang, Y. New Insights of Poly(ADP-Ribosylation) in Neurodegenerative Diseases: A Focus on Protein Phase Separation and Pathologic Aggregation. *Biochem. Pharmacol.* **2019**, *167*, 58–63. [CrossRef]
57. Palazzo, L.; Mikolčević, P.; Mikoč, A.; Ahel, I. ADP-Ribosylation Signalling and Human Disease. *Open Biol.* **2019**, *9*, 190041. [CrossRef]
58. Demény, M.A.; Virág, L. The PARP Enzyme Family and the Hallmarks of Cancer Part 1. Cell Intrinsic Hallmarks. *Cancers* **2021**, *13*, 2042. [CrossRef]
59. Simon, N.C.; Aktories, K.; Barbieri, J.T. Novel Bacterial ADP-Ribosylating Toxins: Structure and Function. *Nat. Rev. Microbiol.* **2014**, *12*, 599–611. [CrossRef]
60. Fehr, A.R.; Channappanavar, R.; Jankevicius, G.; Fett, C.; Zhao, J.; Athmer, J.; Meyerholz, D.K.; Ahel, I.; Perlman, S. The Conserved Coronavirus Macrodomain Promotes Virulence and Suppresses the Innate Immune Response during Severe Acute Respiratory Syndrome Coronavirus Infection. *mBio* **2016**, *7*, e01721-16. [CrossRef] [PubMed]
61. Leung, A.K.L.; McPherson, R.L.; Griffin, D.E. Macrodomain ADP-Ribosylhydrolase and the Pathogenesis of Infectious Diseases. *PLoS Pathog.* **2018**, *14*, e1006864. [CrossRef] [PubMed]
62. Catara, G.; Corteggio, A.; Valente, C.; Grimaldi, G.; Palazzo, L. Targeting ADP-Ribosylation as an Antimicrobial Strategy. *Biochem. Pharmacol.* **2019**, *167*, 13–26. [CrossRef] [PubMed]
63. Fieldhouse, R.J.; Turgeon, Z.; White, D.; Merrill, A.R. Cholera- and Anthrax-like Toxins Are among Several New ADP-Ribosyltransferases. *PLoS Comput. Biol.* **2010**, *6*, e1001029. [CrossRef] [PubMed]
64. Feng, B.; Liu, C.; Shan, L.; He, P. Protein ADP-Ribosylation Takes Control in Plant-Bacterium Interactions. *PLoS Pathog.* **2016**, *12*, e1005941. [CrossRef] [PubMed]
65. Toniti, W.; Yoshida, T.; Tsurumura, T.; Irikura, D.; Monma, C.; Kamata, Y.; Tsuge, H. Crystal Structure and Structure-Based Mutagenesis of Actin-Specific ADP-Ribosylating Toxin CPILE-a as Novel Enterotoxin. *PLoS ONE* **2017**, *12*, e0171278. [CrossRef]
66. Belyy, A.; Lindemann, F.; Roderer, D.; Funk, J.; Bardiaux, B.; Protze, J.; Bieling, P.; Oschkinat, H.; Raunser, S. Mechanism of Threonine ADP-Ribosylation of F-Actin by a Tc Toxin. *Nat. Commun.* **2022**, *13*, 4202. [CrossRef]
67. Rack, J.G.M.; Zorzini, V.; Zhu, Z.; Schuller, M.; Ahel, D.; Ahel, I. Viral Macrodomains: A Structural and Evolutionary Assessment of the Pharmacological Potential. *Open Biol.* **2020**, *10*, 200237. [CrossRef]
68. Russo, L.C.; Tomasin, R.; Matos, I.A.; Manucci, A.C.; Sowa, S.T.; Dale, K.; Caldecott, K.W.; Lehtiö, L.; Schechtman, D.; Meotti, F.C.; et al. The SARS-CoV-2 Nsp3 Macrodomain Reverses PARP9/DTX3L-Dependent ADP-Ribosylation Induced by Interferon Signaling. *J. Biol. Chem.* **2021**, *297*, 101041. [CrossRef]
69. Dasovich, M.; Zhuo, J.; Goodman, J.A.; Thomas, A.; McPherson, R.L.; Jayabalan, A.K.; Busa, V.F.; Cheng, S.-J.; Murphy, B.A.; Redinger, K.R.; et al. High-Throughput Activity Assay for Screening Inhibitors of the SARS-CoV-2 Mac1 Macrodomain. *ACS Chem. Biol.* **2022**, *17*, 17–23. [CrossRef]
70. Leung, A.K.L.; Griffin, D.E.; Bosch, J.; Fehr, A.R. The Conserved Macrodomain Is a Potential Therapeutic Target for Coronaviruses and Alphaviruses. *Pathog. Basel Switz.* **2022**, *11*, 94. [CrossRef]
71. Roy, A.; Alhammad, Y.M.; McDonald, P.; Johnson, D.K.; Zhuo, J.; Wazir, S.; Ferraris, D.; Lehtiö, L.; Leung, A.K.L.; Fehr, A.R. Discovery of Compounds That Inhibit SARS-CoV-2 Mac1-ADP-Ribose Binding by High-Throughput Screening. *Antiviral Res.* **2022**, *203*, 105344. [CrossRef]
72. Zheng, M.; Schultz, M.B.; Sinclair, D.A. NAD+ in COVID-19 and Viral Infections. *Trends Immunol.* **2022**, *43*, 283–295. [CrossRef]
73. Maculins, T.; Fiskin, E.; Bhogaraju, S.; Dikic, I. Bacteria-Host Relationship: Ubiquitin Ligases as Weapons of Invasion. *Cell Res.* **2016**, *26*, 499–510. [CrossRef]
74. Bhogaraju, S.; Dikic, I. Cell Biology: Ubiquitination without E1 and E2 Enzymes. *Nature* **2016**, *533*, 43–44. [CrossRef]
75. Akturk, A.; Wasilko, D.J.; Wu, X.; Liu, Y.; Zhang, Y.; Qiu, J.; Luo, Z.-Q.; Reiter, K.H.; Brzovic, P.S.; Klevit, R.E.; et al. Mechanism of Phosphoribosyl-Ubiquitination Mediated by a Single Legionella Effector. *Nature* **2018**, *557*, 729–733. [CrossRef]
76. Kalayil, S.; Bhogaraju, S.; Bonn, F.; Shin, D.; Liu, Y.; Gan, N.; Basquin, J.; Grumati, P.; Luo, Z.-Q.; Dikic, I. Insights into Catalysis and Function of Phosphoribosyl-Linked Serine Ubiquitination. *Nature* **2018**, *557*, 734–738. [CrossRef]
77. Dikic, I.; Schulman, B.A. An Expanded Lexicon for the Ubiquitin Code. *Nat. Rev. Mol. Cell Biol.* **2022**, *25*, 1–15. [CrossRef]
78. Schuller, M.; Butler, R.E.; Ariza, A.; Tromans-Coia, C.; Jankevicius, G.; Claridge, T.D.W.; Kendall, S.L.; Goh, S.; Stewart, G.R.; Ahel, I. Molecular Basis for DarT ADP-Ribosylation of a DNA Base. *Nature* **2021**, *596*, 597–602. [CrossRef]
79. Rice, L.B. The Clinical Consequences of Antimicrobial Resistance. *Curr. Opin. Microbiol.* **2009**, *12*, 476–481. [CrossRef]

80. Aslam, B.; Khurshid, M.; Arshad, M.I.; Muzammil, S.; Rasool, M.; Yasmeen, N.; Shah, T.; Chaudhry, T.H.; Rasool, M.H.; Shahid, A.; et al. Antibiotic Resistance: One Health One World Outlook. *Front. Cell. Infect. Microbiol.* **2021**, *11*, 771510. [CrossRef]
81. Christaki, E.; Marcou, M.; Tofarides, A. Antimicrobial Resistance in Bacteria: Mechanisms, Evolution, and Persistence. *J. Mol. Evol.* **2020**, *88*, 26–40. [CrossRef]
82. Page, R.; Peti, W. Toxin-Antitoxin Systems in Bacterial Growth Arrest and Persistence. *Nat. Chem. Biol.* **2016**, *12*, 208–214. [CrossRef] [PubMed]
83. Harms, A.; Brodersen, D.E.; Mitarai, N.; Gerdes, K. Toxins, Targets, and Triggers: An Overview of Toxin-Antitoxin Biology. *Mol. Cell* **2018**, *70*, 768–784. [CrossRef] [PubMed]
84. Song, S.; Wood, T.K. A Primary Physiological Role of Toxin/Antitoxin Systems Is Phage Inhibition. *Front. Microbiol.* **2020**, *11*, 1895. [CrossRef] [PubMed]
85. Singh, G.; Yadav, M.; Ghosh, C.; Rathore, J.S. Bacterial Toxin-Antitoxin Modules: Classification, Functions, and Association with Persistence. *Curr. Res. Microb. Sci.* **2021**, *2*, 100047. [CrossRef]
86. Yang, Q.E.; Walsh, T.R. Toxin-Antitoxin Systems and Their Role in Disseminating and Maintaining Antimicrobial Resistance. *FEMS Microbiol. Rev.* **2017**, *41*, 343–353. [CrossRef]
87. Równicki, M.; Lasek, R.; Trylska, J.; Bartosik, D. Targeting Type II Toxin–Antitoxin Systems as Antibacterial Strategies. *Toxins* **2020**, *12*, 568. [CrossRef]
88. Holbourn, K.P.; Shone, C.C.; Acharya, K.R. A Family of Killer Toxins. Exploring the Mechanism of ADP-Ribosylating Toxins. *FEBS J.* **2006**, *273*, 4579–4593. [CrossRef]
89. Rack, J.G.M.; Morra, R.; Barkauskaite, E.; Kraehenbuehl, R.; Ariza, A.; Qu, Y.; Ortmayer, M.; Leidecker, O.; Cameron, D.R.; Matic, I.; et al. Identification of a Class of Protein ADP-Ribosylating Sirtuins in Microbial Pathogens. *Mol. Cell* **2015**, *59*, 309–320. [CrossRef]
90. Bonfiglio, J.J.; Fontana, P.; Zhang, Q.; Colby, T.; Gibbs-Seymour, I.; Atanassov, I.; Bartlett, E.; Zaja, R.; Ahel, I.; Matic, I. Serine ADP-Ribosylation Depends on HPF1. *Mol. Cell* **2017**, *65*, 932–940.e6. [CrossRef]
91. Palazzo, L.; Leidecker, O.; Prokhorova, E.; Dauben, H.; Matic, I.; Ahel, I. Serine Is the Major Residue for ADP-Ribosylation upon DNA Damage. *eLife* **2018**, *7*, e34334. [CrossRef]
92. Suskiewicz, M.J.; Zobel, F.; Ogden, T.E.H.; Fontana, P.; Ariza, A.; Yang, J.-C.; Zhu, K.; Bracken, L.; Hawthorne, W.J.; Ahel, D.; et al. HPF1 Completes the PARP Active Site for DNA Damage-Induced ADP-Ribosylation. *Nature* **2020**, *579*, 598–602. [CrossRef]
93. Carpusca, I.; Jank, T.; Aktories, K. Bacillus Sphaericus Mosquitocidal Toxin (MTX) and Pierisin: The Enigmatic Offspring from the Family of ADP-Ribosyltransferases. *Mol. Microbiol.* **2006**, *62*, 621–630. [CrossRef]
94. Quan, S.; Venter, H.; Dabbs, E.R. Ribosylative Inactivation of Rifampin by Mycobacterium Smegmatis Is a Principal Contributor to Its Low Susceptibility to This Antibiotic. *Antimicrob. Agents Chemother.* **1997**, *41*, 2456–2460. [CrossRef]
95. Ma, Y.; Ludden, P.W. Role of the Dinitrogenase Reductase Arginine 101 Residue in Dinitrogenase Reductase ADP-Ribosyltransferase Binding, NAD Binding, and Cleavage. *J. Bacteriol.* **2001**, *183*, 250–256. [CrossRef]
96. Berthold, C.L.; Wang, H.; Nordlund, S.; Högbom, M. Mechanism of ADP-Ribosylation Removal Revealed by the Structure and Ligand Complexes of the Dimanganese Mono-ADP-Ribosylhydrolase DraG. *Proc. Natl. Acad. Sci. USA* **2009**, *106*, 14247–14252. [CrossRef]
97. Chen, D.; Vollmar, M.; Rossi, M.N.; Phillips, C.; Kraehenbuehl, R.; Slade, D.; Mehrotra, P.V.; von Delft, F.; Crosthwaite, S.K.; Gileadi, O.; et al. Identification of Macrodomain Proteins as Novel O-Acetyl-ADP-Ribose Deacetylases. *J. Biol. Chem.* **2011**, *286*, 13261–13271. [CrossRef]
98. Zhang, W.; Wang, C.; Song, Y.; Shao, C.; Zhang, X.; Zang, J. Structural Insights into the Mechanism of Escherichia Coli YmdB: A 2′-O-Acetyl-ADP-Ribose Deacetylase. *J. Struct. Biol.* **2015**, *192*, 478–486. [CrossRef]
99. García-Saura, A.G.; Zapata-Pérez, R.; Hidalgo, J.F.; Sánchez-Ferrer, Á. Comparative Inhibitory Profile and Distribution of Bacterial PARPs, Using Clostridioides Difficile CD160 PARP as a Model. *Sci. Rep.* **2018**, *8*, 8056. [CrossRef]
100. Cho, C.-C.; Chien, C.-Y.; Chiu, Y.-C.; Lin, M.-H.; Hsu, C.-H. Structural and Biochemical Evidence Supporting Poly ADP-Ribosylation in the Bacterium Deinococcus Radiodurans. *Nat. Commun.* **2019**, *10*, 1491. [CrossRef]
101. Szirák, K.; Keserű, J.; Biró, S.; Schmelczer, I.; Barabás, G.; Penyige, A. Disruption of SCO5461 Gene Coding for a Mono-ADP-Ribosyltransferase Enzyme Produces a Conditional Pleiotropic Phenotype Affecting Morphological Differentiation and Antibiotic Production in Streptomyces Coelicolor. *J. Microbiol.* **2012**, *50*, 409–418. [CrossRef] [PubMed]
102. Lalić, J.; Posavec Marjanović, M.; Palazzo, L.; Perina, D.; Sabljić, I.; Žaja, R.; Colby, T.; Pleše, B.; Halasz, M.; Jankevicius, G.; et al. Disruption of Macrodomain Protein SCO6735 Increases Antibiotic Production in Streptomyces Coelicolor*. *J. Biol. Chem.* **2016**, *291*, 23175–23187. [CrossRef] [PubMed]
103. Moure, V.R.; Costa, F.F.; Cruz, L.M.; Pedrosa, F.O.; Souza, E.M.; Li, X.-D.; Winkler, F.; Huergo, L.F. Regulation of Nitrogenase by Reversible Mono-ADP-Ribosylation. *Curr. Top. Microbiol. Immunol.* **2015**, *384*, 89–106. [CrossRef] [PubMed]
104. Prygiel, M.; Polak, M.; Mosiej, E.; Wdowiak, K.; Formińska, K.; Zasada, A.A. New Corynebacterium Species with the Potential to Produce Diphtheria Toxin. *Pathog. Basel Switz.* **2022**, *11*, 1264. [CrossRef]
105. Ashida, H.; Kim, M.; Sasakawa, C. Exploitation of the Host Ubiquitin System by Human Bacterial Pathogens. *Nat. Rev. Microbiol.* **2014**, *12*, 399–413. [CrossRef]
106. Qiu, J.; Luo, Z.-Q. Hijacking of the Host Ubiquitin Network by Legionella Pneumophila. *Front. Cell. Infect. Microbiol.* **2017**, *7*, 487. [CrossRef]

107. Liu, Y.; Mukherjee, R.; Bonn, F.; Colby, T.; Matic, I.; Glogger, M.; Heilemann, M.; Dikic, I. Serine-Ubiquitination Regulates Golgi Morphology and the Secretory Pathway upon Legionella Infection. *Cell Death Differ.* **2021**, *28*, 2957–2969. [CrossRef]
108. Schuller, M.; Ahel, I. Beyond Protein Modification: The Rise of Non-Canonical ADP-Ribosylation. *Biochem. J.* **2022**, *479*, 463–477. [CrossRef]
109. Watanabe, M.; Takamura-Enya, T.; Kanazawa, T.; Totsuka, Y.; Matsushima-Hibiya, Y.; Koyama, K.; Sugimura, T.; Wakabayashi, K. Mono(ADP-Ribosyl)Ation of DNA by Apoptosis-Inducing Protein, Pierisin. *Nucleic Acids Res. Suppl.* **2002**, *2*, 243–244. [CrossRef]
110. Yamamoto, M.; Nakano, T.; Matsushima-Hibiya, Y.; Totsuka, Y.; Takahashi-Nakaguchi, A.; Matsumoto, Y.; Sugimura, T.; Wakabayashi, K. Molecular Cloning of Apoptosis-Inducing Pierisin-like Proteins, from Two Species of White Butterfly, Pieris Melete and Aporia Crataegi. *Comp. Biochem. Physiol. B Biochem. Mol. Biol.* **2009**, *154*, 326–333. [CrossRef]
111. Lyons, B.; Ravulapalli, R.; Lanoue, J.; Lugo, M.R.; Dutta, D.; Carlin, S.; Merrill, A.R. Scabin, a Novel DNA-Acting ADP-Ribosyltransferase from Streptomyces Scabies*. *J. Biol. Chem.* **2016**, *291*, 11198–11215. [CrossRef]
112. Nakano, T.; Matsushima-Hibiya, Y.; Yamamoto, M.; Takahashi-Nakaguchi, A.; Fukuda, H.; Ono, M.; Takamura-Enya, T.; Kinashi, H.; Totsuka, Y. ADP-Ribosylation of Guanosine by SCO5461 Protein Secreted from Streptomyces Coelicolor. *Toxicon* **2013**, *63*, 55–63. [CrossRef]
113. Langelier, M.-F.; Zandarashvili, L.; Aguiar, P.M.; Black, B.E.; Pascal, J.M. NAD+ Analog Reveals PARP-1 Substrate-Blocking Mechanism and Allosteric Communication from Catalytic Center to DNA-Binding Domains. *Nat. Commun.* **2018**, *9*, 844. [CrossRef]
114. Jørgensen, R.; Wang, Y.; Visschedyk, D.; Merrill, A.R. The nature and character of the transition state for the ADP-ribosyltransferase reaction. *EMBO Rep.* **2008**, *9*, 802–809. [CrossRef]
115. Yoshida, T.; Tsuge, H. Substrate N2 Atom Recognition Mechanism in Pierisin Family DNA-Targeting, Guanine-Specific ADP-Ribosyltransferase ScARP. *J. Biol. Chem.* **2018**, *293*, 13768–13774. [CrossRef]
116. Talhaoui, I.; Lebedeva, N.A.; Zarkovic, G.; Saint-Pierre, C.; Kutuzov, M.M.; Sukhanova, M.V.; Matkarimov, B.T.; Gasparutto, D.; Saparbaev, M.K.; Lavrik, O.I.; et al. Poly(ADP-Ribose) Polymerases Covalently Modify Strand Break Termini in DNA Fragments in Vitro. *Nucleic Acids Res.* **2016**, *44*, 9279–9295. [CrossRef]
117. Zarkovic, G.; Belousova, E.A.; Talhaoui, I.; Saint-Pierre, C.; Kutuzov, M.M.; Matkarimov, B.T.; Biard, D.; Gasparutto, D.; Lavrik, O.I.; Ishchenko, A.A. Characterization of DNA ADP-Ribosyltransferase Activities of PARP2 and PARP3: New Insights into DNA ADP-Ribosylation. *Nucleic Acids Res.* **2018**, *46*, 2417–2431. [CrossRef]
118. Matta, E.; Kiribayeva, A.; Khassenov, B.; Matkarimov, B.T.; Ishchenko, A.A. Insight into DNA Substrate Specificity of PARP1-Catalysed DNA Poly(ADP-Ribosyl)Ation. *Sci. Rep.* **2020**, *10*, 3699. [CrossRef]
119. Hloušek-Kasun, A.; Mikolčević, P.; Rack, J.G.M.; Tromans-Coia, C.; Schuller, M.; Jankevicius, G.; Matković, M.; Bertoša, B.; Ahel, I.; Mikoč, A. Streptomyces Coelicolor Macrodomain Hydrolase SCO6735 Cleaves Thymidine-Linked ADP-Ribosylation of DNA. *Comput. Struct. Biotechnol. J.* **2022**, *20*, 4337–4350. [CrossRef]
120. Ahel, D.; Horejs, Ã.-Z.; Wiechens, N.; Polo, S.; Garcia-Wilson, E.; Ahel, I.; Flynn, H.; Skehel, M.; West, S.; Jackson, S.; et al. Poly(ADP-Ribose)-Dependent Regulation of DNA Repair by the Chromatin Remodeling Enzyme ALC1. *Science* **2009**, *325*, 1240–1243. [CrossRef]
121. Tromans-Coia, C.; Sanchi, A.; Moeller, G.K.; Timinszky, G.; Lopes, M.; Ahel, I. TARG1 Protects against Toxic DNA ADP-Ribosylation. *Nucleic Acids Res.* **2021**, *49*, 10477–10492. [CrossRef] [PubMed]
122. Leplae, R.; Geeraerts, D.; Hallez, R.; Guglielmini, J.; Drèze, P.; Van Melderen, L. Diversity of Bacterial Type II Toxin-Antitoxin Systems: A Comprehensive Search and Functional Analysis of Novel Families. *Nucleic Acids Res.* **2011**, *39*, 5513–5525. [CrossRef] [PubMed]
123. Unterholzner, S.J.; Poppenberger, B.; Rozhon, W. Toxin-Antitoxin Systems: Biology, Identification, and Application. *Mob. Genet. Elem.* **2013**, *3*, e26219. [CrossRef] [PubMed]
124. Srivastava, A.; Pati, S.; Kaushik, H.; Singh, S.; Garg, L.C. Toxin-Antitoxin Systems and Their Medical Applications: Current Status and Future Perspective. *Appl. Microbiol. Biotechnol.* **2021**, *105*, 1803–1821. [CrossRef] [PubMed]
125. Jurėnas, D.; Fraikin, N.; Goormaghtigh, F.; Van Melderen, L. Biology and Evolution of Bacterial Toxin-Antitoxin Systems. *Nat. Rev. Microbiol.* **2022**, *20*, 335–350. [CrossRef]
126. Shao, Y.; Harrison, E.M.; Bi, D.; Tai, C.; He, X.; Ou, H.-Y.; Rajakumar, K.; Deng, Z. TADB: A Web-Based Resource for Type 2 Toxin-Antitoxin Loci in Bacteria and Archaea. *Nucleic Acids Res.* **2011**, *39*, D606–D611. [CrossRef]
127. Akarsu, H.; Bordes, P.; Mansour, M.; Bigot, D.-J.; Genevaux, P.; Falquet, L. TASmania: A Bacterial Toxin-Antitoxin Systems Database. *PLoS Comput. Biol.* **2019**, *15*, e1006946. [CrossRef]
128. Gerdes, K.; Bech, F.W.; Jørgensen, S.T.; Løbner-Olesen, A.; Rasmussen, P.B.; Atlung, T.; Boe, L.; Karlstrom, O.; Molin, S.; von Meyenburg, K. Mechanism of Postsegregational Killing by the Hok Gene Product of the ParB System of Plasmid R1 and Its Homology with the RelF Gene Product of the E. Coli RelB Operon. *EMBO J.* **1986**, *5*, 2023–2029. [CrossRef]
129. Bernard, P.; Couturier, M. Cell Killing by the F Plasmid CcdB Protein Involves Poisoning of DNA-Topoisomerase II Complexes. *J. Mol. Biol.* **1992**, *226*, 735–745. [CrossRef]
130. Castro-Roa, D.; Garcia-Pino, A.; De Gieter, S.; van Nuland, N.A.J.; Loris, R.; Zenkin, N. The Fic Protein Doc Uses an Inverted Substrate to Phosphorylate and Inactivate EF-Tu. *Nat. Chem. Biol.* **2013**, *9*, 811–817. [CrossRef]
131. Fineran, P.C.; Blower, T.R.; Foulds, I.J.; Humphreys, D.P.; Lilley, K.S.; Salmond, G.P.C. The Phage Abortive Infection System, ToxIN, Functions as a Protein-RNA Toxin-Antitoxin Pair. *Proc. Natl. Acad. Sci. USA* **2009**, *106*, 894–899. [CrossRef]

132. Masuda, H.; Tan, Q.; Awano, N.; Wu, K.-P.; Inouye, M. YeeU Enhances the Bundling of Cytoskeletal Polymers of MreB and FtsZ, Antagonizing the CbtA (YeeV) Toxicity in Escherichia Coli. *Mol. Microbiol.* **2012**, *84*, 979–989. [CrossRef]
133. Wang, X.; Lord, D.M.; Cheng, H.-Y.; Osbourne, D.O.; Hong, S.H.; Sanchez-Torres, V.; Quiroga, C.; Zheng, K.; Herrmann, T.; Peti, W.; et al. A New Type V Toxin-Antitoxin System Where MRNA for Toxin GhoT Is Cleaved by Antitoxin GhoS. *Nat. Chem. Biol.* **2012**, *8*, 855–861. [CrossRef]
134. Markovski, M.; Wickner, S. Preventing Bacterial Suicide: A Novel Toxin-Antitoxin Strategy. *Mol. Cell* **2013**, *52*, 611–612. [CrossRef]
135. Marimon, O.; Teixeira, J.M.C.; Cordeiro, T.N.; Soo, V.W.C.; Wood, T.L.; Mayzel, M.; Amata, I.; García, J.; Morera, A.; Gay, M.; et al. An Oxygen-Sensitive Toxin-Antitoxin System. *Nat. Commun.* **2016**, *7*, 13634. [CrossRef]
136. Choi, J.S.; Kim, W.; Suk, S.; Park, H.; Bak, G.; Yoon, J.; Lee, Y. The Small RNA, SdsR, Acts as a Novel Type of Toxin in Escherichia Coli. *RNA Biol.* **2018**, *15*, 1319–1335. [CrossRef]
137. Fraikin, N.; Goormaghtigh, F.; Van Melderen, L. Type II Toxin-Antitoxin Systems: Evolution and Revolutions. *J. Bacteriol.* **2020**, *202*, e00763-19. [CrossRef]
138. Zhang, S.-P.; Feng, H.-Z.; Wang, Q.; Kempher, M.L.; Quan, S.-W.; Tao, X.; Niu, S.; Wang, Y.; Feng, H.-Y.; He, Y.-X. Bacterial Type II Toxin-Antitoxin Systems Acting through Post-Translational Modifications. *Comput. Struct. Biotechnol. J.* **2021**, *19*, 86–93. [CrossRef]
139. LeRoux, M.; Laub, M.T. Toxin-Antitoxin Systems as Phage Defense Elements. *Annu. Rev. Microbiol.* **2022**, *76*, 21–43. [CrossRef]
140. Kaspy, I.; Rotem, E.; Weiss, N.; Ronin, I.; Balaban, N.Q.; Glaser, G. HipA-Mediated Antibiotic Persistence via Phosphorylation of the Glutamyl-TRNA-Synthetase. *Nat. Commun.* **2013**, *4*, 3001. [CrossRef]
141. Germain, E.; Castro-Roa, D.; Zenkin, N.; Gerdes, K. Molecular Mechanism of Bacterial Persistence by HipA. *Mol. Cell* **2013**, *52*, 248–254. [CrossRef] [PubMed]
142. Vang Nielsen, S.; Turnbull, K.J.; Roghanian, M.; Bærentsen, R.; Semanjski, M.; Brodersen, D.E.; Macek, B.; Gerdes, K. Serine-Threonine Kinases Encoded by Split HipA Homologs Inhibit Tryptophanyl-TRNA Synthetase. *mBio* **2019**, *10*, e01138-19. [CrossRef] [PubMed]
143. Harms, A.; Stanger, F.V.; Scheu, P.D.; de Jong, I.G.; Goepfert, A.; Glatter, T.; Gerdes, K.; Schirmer, T.; Dehio, C. Adenylylation of Gyrase and Topo IV by FicT Toxins Disrupts Bacterial DNA Topology. *Cell Rep.* **2015**, *12*, 1497–1507. [CrossRef]
144. Lu, C.; Nakayasu, E.S.; Zhang, L.-Q.; Luo, Z.-Q. Identification of Fic-1 as an Enzyme That Inhibits Bacterial DNA Replication by AMPylating GyrB, Promoting Filament Formation. *Sci. Signal.* **2016**, *9*, ra11. [CrossRef] [PubMed]
145. Harms, A.; Liesch, M.; Körner, J.; Québatte, M.; Engel, P.; Dehio, C. A Bacterial Toxin-Antitoxin Module Is the Origin of Inter-Bacterial and Inter-Kingdom Effectors of Bartonella. *PLoS Genet.* **2017**, *13*, e1007077. [CrossRef]
146. LeRoux, M.; Srikant, S.; Teodoro, G.I.C.; Zhang, T.; Littlehale, M.L.; Doron, S.; Badiee, M.; Leung, A.K.L.; Sorek, R.; Laub, M.T. The DarTG Toxin-Antitoxin System Provides Phage Defence by ADP-Ribosylating Viral DNA. *Nat. Microbiol.* **2022**, *7*, 1028–1040. [CrossRef]
147. Ting, S.-Y.; Bosch, D.E.; Mangiameli, S.M.; Radey, M.C.; Huang, S.; Park, Y.-J.; Kelly, K.A.; Filip, S.K.; Goo, Y.A.; Eng, J.K.; et al. Bifunctional Immunity Proteins Protect Bacteria against FtsZ-Targeting ADP-Ribosylating Toxins. *Cell* **2018**, *175*, 1380–1392.e14. [CrossRef]
148. Freire, D.M.; Gutierrez, C.; Garza-Garcia, A.; Grabowska, A.D.; Sala, A.J.; Ariyachaokun, K.; Panikova, T.; Beckham, K.S.H.; Colom, A.; Pogenberg, V.; et al. An NAD+ Phosphorylase Toxin Triggers Mycobacterium Tuberculosis Cell Death. *Mol. Cell* **2019**, *73*, 1282–1291.e8. [CrossRef]
149. Cheverton, A.M.; Gollan, B.; Przydacz, M.; Wong, C.T.; Mylona, A.; Hare, S.A.; Helaine, S. A Salmonella Toxin Promotes Persister Formation through Acetylation of TRNA. *Mol. Cell* **2016**, *63*, 86–96. [CrossRef]
150. Jurėnas, D.; Garcia-Pino, A.; Van Melderen, L. Novel Toxins from Type II Toxin-Antitoxin Systems with Acetyltransferase Activity. *Plasmid* **2017**, *93*, 30–35. [CrossRef]
151. Qian, H.; Yu, H.; Li, P.; Zhu, E.; Yao, Q.; Tai, C.; Deng, Z.; Gerdes, K.; He, X.; Gan, J.; et al. Toxin-Antitoxin Operon KacAT of Klebsiella Pneumoniae Is Regulated by Conditional Cooperativity via a W-Shaped KacA-KacT Complex. *Nucleic Acids Res.* **2019**, *47*, 7690–7702. [CrossRef]
152. Wilcox, B.; Osterman, I.; Serebryakova, M.; Lukyanov, D.; Komarova, E.; Gollan, B.; Morozova, N.; Wolf, Y.I.; Makarova, K.S.; Helaine, S.; et al. Escherichia Coli ItaT Is a Type II Toxin That Inhibits Translation by Acetylating Isoleucyl-TRNAIle. *Nucleic Acids Res.* **2018**, *46*, 7873–7885. [CrossRef]
153. Zaveri, A.; Wang, R.; Botella, L.; Sharma, R.; Zhu, L.; Wallach, J.B.; Song, N.; Jansen, R.S.; Rhee, K.Y.; Ehrt, S.; et al. Depletion of the DarG Antitoxin in Mycobacterium Tuberculosis Triggers the DNA-Damage Response and Leads to Cell Death. *Mol. Microbiol.* **2020**, *114*, 641–652. [CrossRef]
154. Magnuson, R.D. Hypothetical Functions of Toxin-Antitoxin Systems. *J. Bacteriol.* **2007**, *189*, 6089–6092. [CrossRef]
155. Korch, S.B.; Hill, T.M. Ectopic Overexpression of Wild-Type and Mutant HipA Genes in Escherichia Coli: Effects on Macromolecular Synthesis and Persister Formation. *J. Bacteriol.* **2006**, *188*, 3826–3836. [CrossRef]
156. Gerdes, K.; Christensen, S.K.; Løbner-Olesen, A. Prokaryotic Toxin-Antitoxin Stress Response Loci. *Nat. Rev. Microbiol.* **2005**, *3*, 371–382. [CrossRef]
157. Wang, X.; Wood, T.K. Toxin-Antitoxin Systems Influence Biofilm and Persister Cell Formation and the General Stress Response. *Appl. Environ. Microbiol.* **2011**, *77*, 5577–5583. [CrossRef]

158. Wen, Y.; Behiels, E.; Devreese, B. Toxin-Antitoxin Systems: Their Role in Persistence, Biofilm Formation, and Pathogenicity. *Pathog. Dis.* **2014**, *70*, 240–249. [CrossRef]
159. Qiu, J.; Zhai, Y.; Wei, M.; Zheng, C.; Jiao, X. Toxin-Antitoxin Systems: Classification, Biological Roles, and Applications. *Microbiol. Res.* **2022**, *264*, 127159. [CrossRef]
160. Tal, N.; Sorek, R. SnapShot: Bacterial Immunity. *Cell* **2022**, *185*, 578–578.e1. [CrossRef]
161. Tesson, F.; Hervé, A.; Mordret, E.; Touchon, M.; d'Humières, C.; Cury, J.; Bernheim, A. Systematic and Quantitative View of the Antiviral Arsenal of Prokaryotes. *Nat. Commun.* **2022**, *13*, 2561. [CrossRef] [PubMed]
162. Koga, M.; Otsuka, Y.; Lemire, S.; Yonesaki, T. Escherichia Coli RnlA and RnlB Compose a Novel Toxin-Antitoxin System. *Genetics* **2011**, *187*, 123–130. [CrossRef] [PubMed]
163. Samson, J.E.; Spinelli, S.; Cambillau, C.; Moineau, S. Structure and Activity of AbiQ, a Lactococcal Endoribonuclease Belonging to the Type III Toxin-Antitoxin System. *Mol. Microbiol.* **2013**, *87*, 756–768. [CrossRef] [PubMed]
164. Dy, R.L.; Przybilski, R.; Semeijn, K.; Salmond, G.P.C.; Fineran, P.C. A Widespread Bacteriophage Abortive Infection System Functions through a Type IV Toxin-Antitoxin Mechanism. *Nucleic Acids Res.* **2014**, *42*, 4590–4605. [CrossRef]
165. Lopatina, A.; Tal, N.; Sorek, R. Abortive Infection: Bacterial Suicide as an Antiviral Immune Strategy. *Annu. Rev. Virol.* **2020**, *7*, 371–384. [CrossRef]
166. Saha, D.; Mukherjee, R. Ameliorating the Antimicrobial Resistance Crisis: Phage Therapy. *IUBMB Life* **2019**, *71*, 781–790. [CrossRef]
167. Xu, H.-M.; Xu, W.-M.; Zhang, L. Current Status of Phage Therapy against Infectious Diseases and Potential Application beyond Infectious Diseases. *Int. J. Clin. Pract.* **2022**, *2022*, 4913146. [CrossRef]
168. Årdal, C.; Balasegaram, M.; Laxminarayan, R.; McAdams, D.; Outterson, K.; Rex, J.H.; Sumpradit, N. Antibiotic Development—Economic, Regulatory and Societal Challenges. *Nat. Rev. Microbiol.* **2020**, *18*, 267–274. [CrossRef]
169. Machowska, A.; Stålsby Lundborg, C. Drivers of Irrational Use of Antibiotics in Europe. *Int. J. Environ. Res. Public. Health* **2018**, *16*, 27. [CrossRef]
170. Lewis, K. The Science of Antibiotic Discovery. *Cell* **2020**, *181*, 29–45. [CrossRef]
171. Czaplewski, L.; Bax, R.; Clokie, M.; Dawson, M.; Fairhead, H.; Fischetti, V.A.; Foster, S.; Gilmore, B.F.; Hancock, R.E.W.; Harper, D.; et al. Alternatives to Antibiotics—A Pipeline Portfolio Review. *Lancet Infect. Dis.* **2016**, *16*, 239–251. [CrossRef]
172. Roussin, M.; Salcedo, S.P. NAD+-Targeting by Bacteria: An Emerging Weapon in Pathogenesis. *FEMS Microbiol. Rev.* **2021**, *45*, fuab037. [CrossRef]
173. Mesquita, I.; Varela, P.; Belinha, A.; Gaifem, J.; Laforge, M.; Vergnes, B.; Estaquier, J.; Silvestre, R. Exploring NAD+ Metabolism in Host-Pathogen Interactions. *Cell. Mol. Life Sci. CMLS* **2016**, *73*, 1225–1236. [CrossRef]
174. van Doorn, C.L.R.; Steenbergen, S.A.M.; Walburg, K.V.; Ottenhoff, T.H.M. Pharmacological Poly (ADP-Ribose) Polymerase Inhibitors Decrease Mycobacterium Tuberculosis Survival in Human Macrophages. *Front. Immunol.* **2021**, *12*, 712021. [CrossRef]
175. Lugo, M.R.; Merrill, A.R. Development of Anti-Virulence Therapeutics against Mono-ADP-Ribosyltransferase Toxins. *Toxins* **2020**, *13*, 16. [CrossRef]
176. Turgeon, Z.; Jørgensen, R.; Visschedyk, D.; Edwards, P.R.; Legree, S.; McGregor, C.; Fieldhouse, R.J.; Mangroo, D.; Schapira, M.; Merrill, A.R. Newly Discovered and Characterized Antivirulence Compounds Inhibit Bacterial Mono-ADP-Ribosyltransferase Toxins. *Antimicrob. Agents Chemother.* **2011**, *55*, 983–991. [CrossRef]
177. Cherubin, P.; Garcia, M.C.; Curtis, D.; Britt, C.B.T.; Craft, J.W.; Burress, H.; Berndt, C.; Reddy, S.; Guyette, J.; Zheng, T.; et al. Inhibition of Cholera Toxin and Other AB Toxins by Polyphenolic Compounds. *PLoS ONE* **2016**, *11*, e0166477. [CrossRef]
178. Yates, S.P.; Taylor, P.L.; Jørgensen, R.; Ferraris, D.; Zhang, J.; Andersen, G.R.; Merrill, A.R. Structure–Function Analysis of Water-Soluble Inhibitors of the Catalytic Domain of Exotoxin A from Pseudomonas Aeruginosa. *Biochem. J.* **2005**, *385*, 667–675. [CrossRef]
179. Pinto, A.F.; Ebrahimi, M.; Saleeb, M.; Forsberg, Å.; Elofsson, M.; Schüler, H. Identification of Inhibitors of Pseudomonas Aeruginosa Exotoxin-S ADP-Ribosyltransferase Activity. *SLAS Discov.* **2016**, *21*, 590–595. [CrossRef]
180. Saleeb, M.; Sundin, C.; Aglar, Ö.; Pinto, A.F.; Ebrahimi, M.; Forsberg, Å.; Schüler, H.; Elofsson, M. Structure–Activity Relationships for Inhibitors of Pseudomonas Aeruginosa Exoenzyme S ADP-Ribosyltransferase Activity. *Eur. J. Med. Chem.* **2018**, *143*, 568–576. [CrossRef]
181. Wu, S.-C.; Liu, F.; Zhu, K.; Shen, J.-Z. Natural Products That Target Virulence Factors in Antibiotic-Resistant *Staphylococcus Aureus*. *J. Agric. Food Chem.* **2019**, *67*, 13195–13211. [CrossRef]
182. Abdelhamid, A.G.; El-Dougdoug, N.K. Controlling Foodborne Pathogens with Natural Antimicrobials by Biological Control and Antivirulence Strategies. *Heliyon* **2020**, *6*, e05020. [CrossRef] [PubMed]
183. Bogan, A.A.; Thorn, K.S. Anatomy of Hot Spots in Protein Interfaces. *J. Mol. Biol.* **1998**, *280*, 1–9. [CrossRef]
184. Modell, A.E.; Blosser, S.L.; Arora, P.S. Systematic Targeting of Protein-Protein Interactions. *Trends Pharmacol. Sci.* **2016**, *37*, 702–713. [CrossRef]
185. Scott, D.E.; Bayly, A.R.; Abell, C.; Skidmore, J. Small Molecules, Big Targets: Drug Discovery Faces the Protein-Protein Interaction Challenge. *Nat. Rev. Drug Discov.* **2016**, *15*, 533–550. [CrossRef]
186. Cossar, P.J.; Lewis, P.J.; McCluskey, A. Protein-Protein Interactions as Antibiotic Targets: A Medicinal Chemistry Perspective. *Med. Res. Rev.* **2020**, *40*, 469–494. [CrossRef]

187. Carro, L. Protein-Protein Interactions in Bacteria: A Promising and Challenging Avenue towards the Discovery of New Antibiotics. *Beilstein J. Org. Chem.* **2018**, *14*, 2881–2896. [CrossRef]
188. Kramer, V.G.; Schader, S.M.; Oliveira, M.; Colby-Germinario, S.P.; Donahue, D.A.; Singhroy, D.N.; Tressler, R.; Sloan, R.D.; Wainberg, M.A. Maraviroc and Other HIV-1 Entry Inhibitors Exhibit a Class-Specific Redistribution Effect That Results in Increased Extracellular Viral Load. *Antimicrob. Agents Chemother.* **2012**, *56*, 4154–4160. [CrossRef]
189. Kahan, R.; Worm, D.J.; de Castro, G.V.; Ng, S.; Barnard, A. Modulators of Protein-Protein Interactions as Antimicrobial Agents. *RSC Chem. Biol.* **2021**, *2*, 387–409. [CrossRef]
190. Gómez Borrego, J.; Torrent Burgas, M. Analysis of Host–Bacteria Protein Interactions Reveals Conserved Domains and Motifs That Mediate Fundamental Infection Pathways. *Int. J. Mol. Sci.* **2022**, *23*, 11489. [CrossRef]
191. Park, S.J.; Son, W.S.; Lee, B.-J. Structural Overview of Toxin-Antitoxin Systems in Infectious Bacteria: A Target for Developing Antimicrobial Agents. *Biochim. Biophys. Acta* **2013**, *1834*, 1155–1167. [CrossRef]
192. Belousova, E.A.; Ishchenko, A.A.; Lavrik, O.I. Dna Is a New Target of Parp3. *Sci. Rep.* **2018**, *8*, 4176. [CrossRef]
193. Musheev, M.U.; Schomacher, L.; Basu, A.; Han, D.; Krebs, L.; Scholz, C.; Niehrs, C. Mammalian N1-Adenosine PARylation Is a Reversible DNA Modification. *Nat. Commun.* **2022**, *13*, 6138. [CrossRef]
194. Fehr, A.R.; Singh, S.A.; Kerr, C.M.; Mukai, S.; Higashi, H.; Aikawa, M. The Impact of PARPs and ADP-Ribosylation on Inflammation and Host–Pathogen Interactions. *Genes Dev.* **2020**, *34*, 341–359. [CrossRef]
195. Eckei, L.; Krieg, S.; Bütepage, M.; Lehmann, A.; Gross, A.; Lippok, B.; Grimm, A.R.; Kümmerer, B.M.; Rossetti, G.; Lüscher, B.; et al. The Conserved Macrodomains of the Non-Structural Proteins of Chikungunya Virus and Other Pathogenic Positive Strand RNA Viruses Function as Mono-ADP-Ribosylhydrolases. *Sci. Rep.* **2017**, *7*, 41746. [CrossRef]
196. Fehr, A.R.; Jankevicius, G.; Ahel, I.; Perlman, S. Viral Macrodomains: Unique Mediators of Viral Replication and Pathogenesis. *Trends Microbiol.* **2018**, *26*, 598–610. [CrossRef]
197. Abraham, R.; Hauer, D.; McPherson, R.L.; Utt, A.; Kirby, I.T.; Cohen, M.S.; Merits, A.; Leung, A.K.L.; Griffin, D.E. ADP-Ribosyl-Binding and Hydrolase Activities of the Alphavirus NsP3 Macrodomain Are Critical for Initiation of Virus Replication. *Proc. Natl. Acad. Sci. USA* **2018**, *115*, E10457–E10466. [CrossRef]
198. Grunewald, M.E.; Chen, Y.; Kuny, C.; Maejima, T.; Lease, R.; Ferraris, D.; Aikawa, M.; Sullivan, C.S.; Perlman, S.; Fehr, A.R. The Coronavirus Macrodomain Is Required to Prevent PARP-Mediated Inhibition of Virus Replication and Enhancement of IFN Expression. *PLoS Pathog.* **2019**, *15*, e1007756. [CrossRef]
199. Schuller, M.; Correy, G.J.; Gahbauer, S.; Fearon, D.; Wu, T.; Díaz, R.E.; Young, I.D.; Carvalho Martins, L.; Smith, D.H.; Schulze-Gahmen, U.; et al. Fragment Binding to the Nsp3 Macrodomain of SARS-CoV-2 Identified through Crystallographic Screening and Computational Docking. *Sci. Adv.* **2021**, *7*, eabf8711. [CrossRef]
200. Gahbauer, S.; Correy, G.J.; Schuller, M.; Ferla, M.P.; Doruk, Y.U.; Rachman, M.; Wu, T.; Diolaiti, M.; Wang, S.; Neitz, R.J.; et al. Iterative Computational Design and Crystallographic Screening Identifies Potent Inhibitors Targeting the Nsp3 Macrodomain of SARS-CoV-2. *BioRxiv Prepr. Serv. Biol.* **2022**, *120*, e2212931120. [CrossRef] [PubMed]
201. Sherrill, L.M.; Joya, E.E.; Walker, A.; Roy, A.; Alhammad, Y.M.; Atobatele, M.; Wazir, S.; Abbas, G.; Keane, P.; Zhuo, J.; et al. Design, Synthesis and Evaluation of Inhibitors of the SARS-CoV-2 Nsp3 Macrodomain. *Bioorg. Med. Chem.* **2022**, *67*, 116788. [CrossRef] [PubMed]
202. Bernheim, A.; Sorek, R. The Pan-Immune System of Bacteria: Antiviral Defence as a Community Resource. *Nat. Rev. Microbiol.* **2020**, *18*, 113–119. [CrossRef] [PubMed]
203. Millman, A.; Bernheim, A.; Stokar-Avihail, A.; Fedorenko, T.; Voichek, M.; Leavitt, A.; Oppenheimer-Shaanan, Y.; Sorek, R. Bacterial Retrons Function In Anti-Phage Defense. *Cell* **2020**, *183*, 1551–1561.e12. [CrossRef]
204. González-Delgado, A.; Mestre, M.R.; Martínez-Abarca, F.; Toro, N. Prokaryotic Reverse Transcriptases: From Retroelements to Specialized Defense Systems. *FEMS Microbiol. Rev.* **2021**, *45*, fuab025. [CrossRef]
205. Bobonis, J.; Mitosch, K.; Mateus, A.; Karcher, N.; Kritikos, G.; Selkrig, J.; Zietek, M.; Monzon, V.; Pfalz, B.; Garcia-Santamarina, S.; et al. Bacterial Retrons Encode Phage-Defending Tripartite Toxin–Antitoxin Systems. *Nature* **2022**, *609*, 144–150. [CrossRef]
206. Lüscher, B.; Bütepage, M.; Eckei, L.; Krieg, S.; Verheugd, P.; Shilton, B.H. ADP-Ribosylation, a Multifaceted Posttranslational Modification Involved in the Control of Cell Physiology in Health and Disease. *Chem. Rev.* **2018**, *118*, 1092–1136. [CrossRef]
207. Iyer, L.M.; Burroughs, A.M.; Anantharaman, V.; Aravind, L. Apprehending the NAD+-ADPr-Dependent Systems in the Virus World. *Viruses* **2022**, *14*, 1977. [CrossRef]

pathogens MDPI

Article

14-3-3 Activated Bacterial Exotoxins AexT and ExoT Share Actin and the SH2 Domains of CRK Proteins as Targets for ADP-Ribosylation

Carmen Ebenwaldner [1], Peter Hornyak [2,†], Antonio Ginés García-Saura [1], Archimede Torretta [1], Saber Anoosheh [3], Anders Hofer [3] and Herwig Schüler [1,2,*]

[1] Center for Molecular Protein Science (CMPS), Department of Chemistry, Lund University, 22100 Lund, Sweden
[2] Department of Biosciences, Karolinska Institutet, 14157 Huddinge, Sweden
[3] Department of Medical Biochemistry and Biophysics, Umeå University, 90736 Umeå, Sweden
* Correspondence: herwig.schuler@biochemistry.lu.se
† Current address: Vascular Research Group, 1083 Budapest, Hungary.

Abstract: Bacterial exotoxins with ADP-ribosyltransferase activity can be divided into distinct clades based on their domain organization. Exotoxins from several clades are known to modify actin at Arg177; but of the 14-3-3 dependent exotoxins only *Aeromonas salmonicida* exoenzyme T (AexT) has been reported to ADP-ribosylate actin. Given the extensive similarity among the 14-3-3 dependent exotoxins, we initiated a structural and biochemical comparison of these proteins. Structural modeling of AexT indicated a target binding site that shared homology with *Pseudomonas aeruginosa* Exoenzyme T (ExoT) but not with Exoenzyme S (ExoS). Biochemical analyses confirmed that the catalytic activities of both exotoxins were stimulated by agmatine, indicating that they ADP-ribosylate arginine residues in their targets. Side-by-side comparison of target protein modification showed that AexT had activity toward the SH2 domain of the Crk-like protein (CRKL), a known target for ExoT. We found that both AexT and ExoT ADP-ribosylated actin and in both cases, the modification compromised actin polymerization. Our results indicate that AexT and ExoT are functional homologs that affect cytoskeletal integrity via actin and signaling pathways to the cytoskeleton.

Keywords: 14-3-3 activated bacterial exotoxins; actin; ADP-ribosylation; *Aeromonas salmonicid*; protein refolding; type III secretion system; *Pseudomonas aeruginosa*

Citation: Ebenwaldner, C.; Hornyak, P.; García-Saura, A.G.; Torretta, A.; Anoosheh, S.; Hofer, A.; Schüler, H. 14-3-3 Activated Bacterial Exotoxins AexT and ExoT Share Actin and the SH2 Domains of CRK Proteins as Targets for ADP-Ribosylation. *Pathogens* **2022**, *11*, 1497. https://doi.org/10.3390/pathogens11121497

Academic Editors: Anthony K L Leung, Anthony Fehr and Rachy Abraham

Received: 10 November 2022
Accepted: 6 December 2022
Published: 8 December 2022

1. Introduction

Aeromonas and *Pseudomonas* are ubiquitous Gram-negative bacteria that can cause infections in a range of tissues. The planktonic *Aeromonas* species are opportunistic pathogens of diverse aquatic animals; they can display resistance to multiple antibiotics and can cause serious economic damage to fisheries [1]. The most common clinical manifestation of aeromonads is seafoodborne gastroenteritis caused by *Aeromonas salmonicida*. Both aeromonads and pseudomonads cause difficult to treat wound and airway infections [2–4], with pseudomonads being among the principal causative agents of nosocomial infections [5].

Contact with eukaryotic target cells triggers signalling cascades in *Pseudomonas aeruginosa* leading to the expression of the transcription factor ExsA, which in turn regulates the expression of all type III secretion system (T3SS) genes [6]. These genes encode the structural components of the T3SS needle apparatus as well as four exotoxins [7], and their expression is the major determinant of virulence in *Pseudomonas* [8]. Expression of the ADP-ribosyltransferase (ART) toxin exotoxin-S (exoenzyme-S; ExoS) and the phospholipase ExoU is mutually exclusive in nearly all *Pseudomonas* isolates examined [8,9]. A second ART exotoxin, ExoT, is expressed in a majority of clinical isolates [9]. Presence of either ExoS or ExoT has been shown to be sufficient to induce morphological changes in infected cultured cell lines [10,11]. In addition, ExoT stimulates apoptosis in infected host cells [12]. In *A. salmonicida* AexT, an ortholog of ExoS and ExoT, is also injected via a T3SS; AexT appears to be the major determinant of virulence in *Aeromonas* strains [13]. Fish gonad RTG-2 cells infected with *A. salmonicida* undergo cell rounding whereas infection with a mutant strain devoid of AexT does not affect cell morphology [13].

Generally, ADP-ribosyl transferases catalyze the formation of a posttranslational modification that affects or determines the activities and the fates of a multitude of proteins within human cells [14]. Diverse pathogenic bacteria have evolved mechanisms to employ ADP-ribosylation in generating a replication competent niche for themselves. ADP-ribosylating toxins that target eEF-2, trimeric G-proteins, as well as actin and other cytoskeletal proteins and their regulators including small GTPases have been identified [15–17]. A distinct clade of ADP-ribosylating bacterial toxins, with *P. aeruginosa* ExoS as their founding member, get activated by binding to host-derived 14-3-3 proteins upon their T3SS-dependent delivery into the host cytosol [18]. Most members of this clade, including *Pseudomonas* ExoT [19] and *Aeromonas* AexT [20], combine a GTPase activating protein (GAP) domain with an ART domain [21]. Whereas the GAP domains of ExoS, ExoT and AexT appear to target the same host GTPases, the ART domain specificities differ between the three exotoxins [7,19,20,22–24]. ExoS has been shown to affect host cell morphology by ADP-ribosylation of Ras homologs, vimentin, and the microfilament linked ezrin/radixin/moesin (ERM) proteins [7,25]. ExoT is known to inactivate Ras-mediated phagocytosis through ADP-ribosylation of the SH2 domain of the ubiquitous upstream regulatory factors CT10 regulator of kinase protein (CRK) and the CRK-like protein (CRKL) [24]. Although affecting the host cell cytoskeleton by various means, neither ExoS nor ExoT have been shown to ADP-ribosylate actin [7,26]. Therefore, it was unexpected that AexT from *A. salmonicida* was reported to modify actin [20], which is a common target protein for toxins of other clades [16,17,21,27].

Given the implications of these pathogens it is of continued interest to characterize their T3SS effectors. The recent determination of the crystal structures of *Pseudomonas* ExoS and ExoT ART domains [28] provided a new opportunity to address this aim. Here, we modelled the structure of the ART domain of AexT and predicted its target protein binding surface to be more similar to that of ExoT. This observation prompted us to purify full length AexT and ExoT proteins and to analyze their ADP-ribosyltransferase activities toward different target proteins. We present evidence supporting previous findings that both exotoxins catalyze arginine-linked ADP-ribosylation. We found that both AexT and ExoT modified actin, previously the only known target for AexT. Additionally, both AexT and ExoT modified the SH2 domain of CRKL, previously the only known target for ExoT. Thus, a common in vitro target specificity could be inferred from conserved surface features of the two exotoxins. Together, these results expand the known repertoire of virulence mechanisms in *P. aeruginosa* and *A. salmonicida* that lead to disruption of the cytoskeleton in infected host cells.

2. Results

2.1. Structural Model of AexT

Phylogenetic analyses indicated a close relationship between AexT and the more well-characterized 14-3-3 dependent ART toxins [21]. Alignment of the C-terminal sequences of AexT, ExoS and ExoT illustrates the near-perfect conservation of the 14-3-3 interaction sites (Figure 1a). These consist of (i) a recently discovered extensive hydrophobic surface patch with contributions of residues V414-LLSA($\times$8)RV-L429 (AexT sequence numbering); and (ii) the two "LDLA" boxes that interact with the phosphopeptide binding grooves in each monomer of the 14-3-3 dimer [28]. This sequence conservation confirms that AexT is a member of the 14-3-3 dependent ART toxins. The AexT GAP domain shares lower sequence homology with ExoS and ExoT than the latter share with one another. Although all three exotoxins share more than 80% homology within their ART domains, the AexT ART domain is more similar to ExoT than to ExoS (Figure 1b).

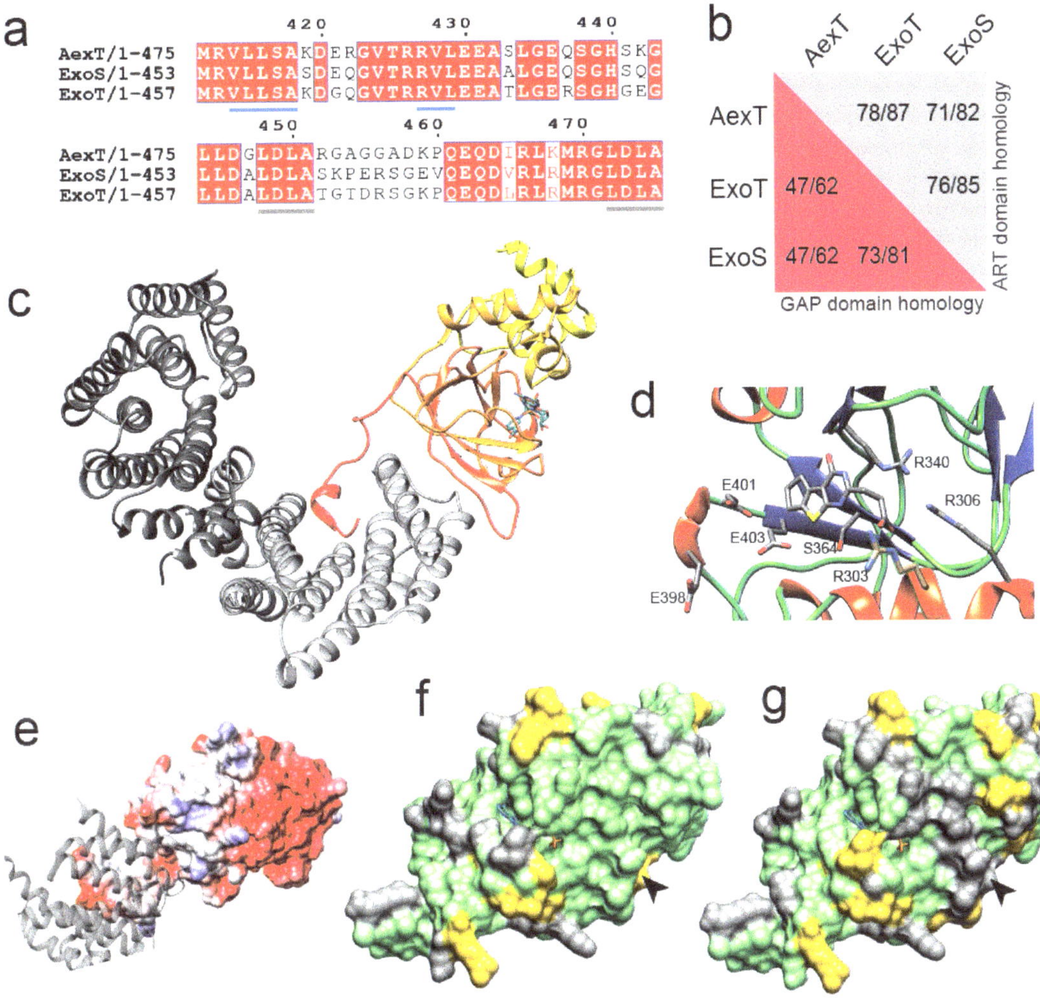

Figure 1. Homology model of the ART domain of *Aeromonas salmonicida* AexT. (**a**) Alignment of the C-terminal sequences of AexT and *Pseudomonas aeruginosa* ExoS and ExoT. The hydrophobic interaction site with 14-3-3 is underlined in blue and the LDLA-boxes are underlined in grey. (**b**) Extent of sequence homology (expressed as identity/similarity) between the GAP domains (red background) and ART domains (grey background) of the three exotoxins. (**c**) Homology model of the ART domain of AexT (color gradient from yellow (N-terminus) to red (C-terminus)), based on the ExoT structure (PDB: 6GNN). A carbaNAD+ molecule (sticks) was modeled based on the ExoS structure (PDB: 6GNK) to indicate the location of the active site. The 14-3-3β homodimer is shown in grey (different shades for each monomer). (**d**) Close-up of the active site containing ExoT inhibitor STO1101, with key side chains shown as sticks. (**e**) Electrostatic surface model of the AexT ART domain, showing the hydrophobic 14-3-3 binding site in light grey and a 14-3-3 monomer as grey ribbon. (**f**,**g**) Surface representation of the AexT ART domain colored for sequence conservation with either ExoT (**f**) or ExoS (**g**). Arrowheads indicate the location of the target binding sites (see text). Green, identity; yellow, similarity; grey, divergence.

To explore the phylogenetic relationship within this clade of ART toxins for new insights into AexT functions, we constructed a homology model of the ART domain of

AexT based on our recent crystal structure of the 14-3-3β:ExoT ART domain complex [28]. The quality of the model is good, with 2% of side chains in disallowed regions of the Ramachandran plot and an overall quality score of 0.68 determined by the QMEAN server [29] (Figure S1a in the Supplementary document). The model aligns with a root mean square deviation of 1.4 Å over all Cα-atom pairs with a model produced by AlphaFold [30] (Figure S1b). A heterotrimer model of AexT with the 14-3-3β homodimer is illustrated in Figure 1c. The signature motif of toxins in this clade, the conserved R-S-E triad [21] within the dinucleotide binding site, is in the expected position in our model (residues R306, S364, E403 of AexT; Figure 1d). Additionally, the surface of the AexT model shares the hydrophobic surface features in the 14-3-3 binding site with ExoT and ExoS (Figure 1e) [28].

Analysis of the AexT surface residues that are conserved with either ExoT or ExoS provided an important clue to shared functions among the three toxins: The overall number of conserved residues is similar in both pairwise comparisons; however, there is a prominent ridge of surface residues near the NAD+ binding site that is strictly conserved between AexT and ExoT, but divergent between AexT and ExoS (Figure 1f,g). This surface region coincides well with the surface that has been implicated in target protein recognition [19,31]. Furthermore, we noted that the canonical ExE motif, involved in catalysis in most ADP-ribosylating toxins, appears to be extended to 398ExxExE403 (AexT numbering) in AexT and ExoT exclusively. Fehr and co-workers found that the first and second, rather than second and third, glutamate of the motif were essential for catalysis in AexT (residues E401 and E403; Figure 1d) [20]. The conservation of these extended motifs among AexT and ExoT may be related to a shared target spectrum. In summary, structural modeling suggested a near homology of target recognition sites in AexT and ExoT, which was not anticipated previously. This prompted us to compare target protein ADP-ribosylation catalyzed by recombinant AexT and ExoT.

Table 1. Multimeric states of exotoxin-chaperone complexes in solution.

		AexT-ART: 14-3-3β	FL-AexT: 14-3-3β	ExoT-ART: 14-3-3β	FL-ExoT: 14-3-3β	14-3-3β
Theoretical MW (kDa)	Exotoxin	27.503	52.658	26.689	51.067	
	14-3-3β *	28.534	29.356	28.534	29.356	29.356
	Complex	84.571 **	111.370 **	83.757 **	109.779 **	58.712 ***
Measured MW (kDa) ****	Mass Photometry	85 ± 19	106 ± 23	81 ± 18.8	107 ± 69	61 ± 12.8
	SEC-MALS	82.323	ND	82.352	ND	57.506

* With or without affinity tag, depending on protocol (see Materials and Methods); ** Assuming heterotrimer of two 14-3-3β molecules and one toxin molecule; *** Assuming homodimer; **** Means ± S.D. ND, not determined.

2.2. Recombinant Production of Full-Length Exotoxins

Exotoxin target protein modification has previously been investigated for full-length recombinant AexT [20] and both the recombinant ExoT ART domain in isolation and the full-length toxin secreted into mammalian cell cultures. [24,26,32] To achieve a comparative analysis, we determined that it was necessary to establish protocols for the recombinant production of the full-length and ART domain constructs of both AexT and ExoT. While the ART domains could be produced to satisfactory quantity and purity by purification from the soluble fractions of *E. coli* lysates, the full-length toxins were purified by refolding the proteins from the insoluble fractions (Figure S2). Purification of functional ExoT, as judged by its NAD+ glycohydrolase activity, was strongly aided by addition of purified 14-3-3β protein in excess (Figure S2b) and therefore, 14-3-3β protein was added also during the refolding of AexT.

Recombinant ExoT and ExoS ART domains in complex with 14-3-3β were previously shown to form both heterotrimers and heterotetramers. [28] Here, we employed mass photometry to analyze the solution properties of the purified proteins. The results indicated that both ART domains and full-length exotoxins formed mainly heterotrimers with 14-3-3β (Figure 2a–d and Table 1). Heterotrimer formation was confirmed by analytical size exclusion chromatography in case of the ART domains (Figure 2f,g and Table 1). Full-length exotoxin complexes could not be assessed by this method, and exotoxin constructs in the absence of 14-3-3 protein could not be assessed by either of the two methods, due to extensive protein aggregation during the experiment. Using mass photometry at ~15 nM protein concentration, heterotrimers were apparently unstable, as indicated by the presence of the 57-kDa 14-3-3β homodimer peaks (Figure 2a–e). In contrast, in analytical size exclusion chromatography, the peak concentration was estimated to ~6 μM. Under these conditions, single peaks were observed indicating stable complexes. Together, these results suggest that K_D values describing the affinities of AexT and ExoT for 14-3-3β homodimers would lie in the mid-nanomolar range. For comparison, the affinity of the homologous ExoS-ART domain for 14-3-3β homodimers has been determined to a K_D of roughly 40 nM [28].

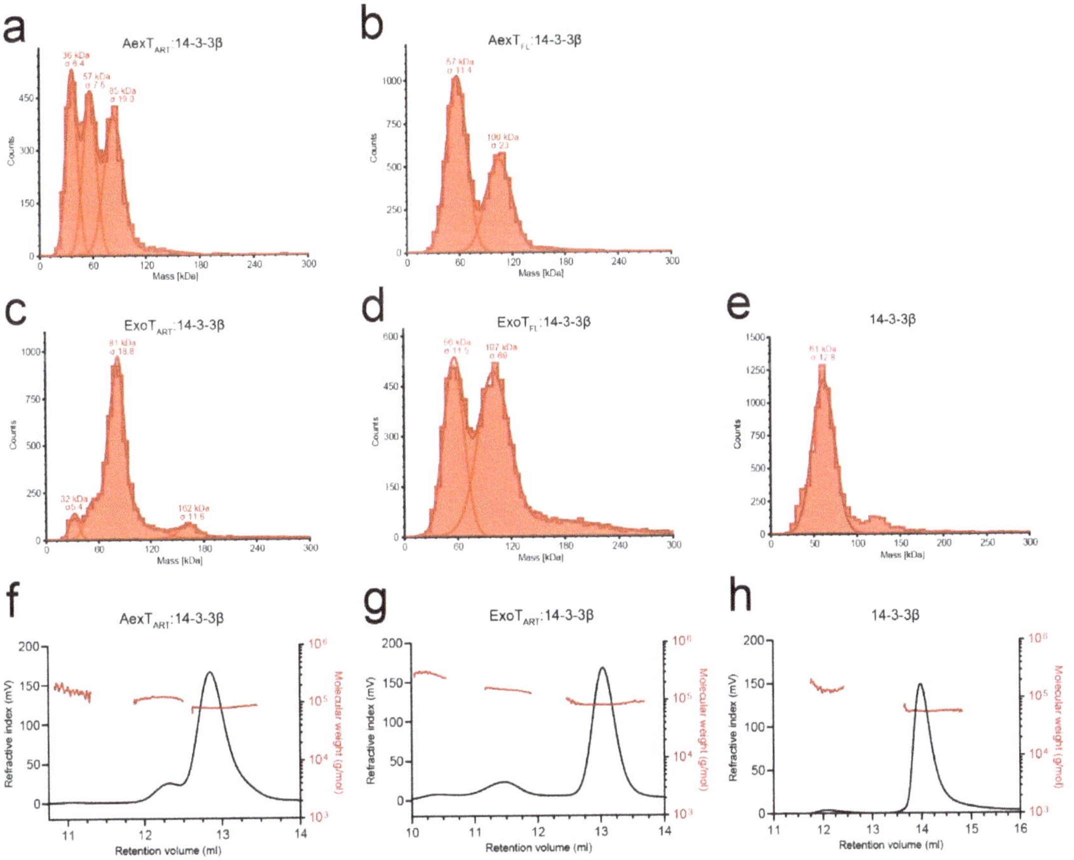

Figure 2. Solution properties of the recombinant exotoxins. (**a**–**e**) Size distributions of the 14-3-3β complexes of AexT-ART (**a**) and full-length (**b**) as well as ExoT-ART (**c**) and full-length proteins (**d**) measured using mass photometry. (**f**–**h**) Size distributions of the 14-3-3β complexes of AexT-ART (**f**) and ExoT-ART (**g**) assessed by analytical size exclusion chromatography. Size determinations of 14-3-3β homodimers by each method are shown for comparison (**e**,**h**). ART, ADP-ribosyltransferase domain; FL, full-length protein.

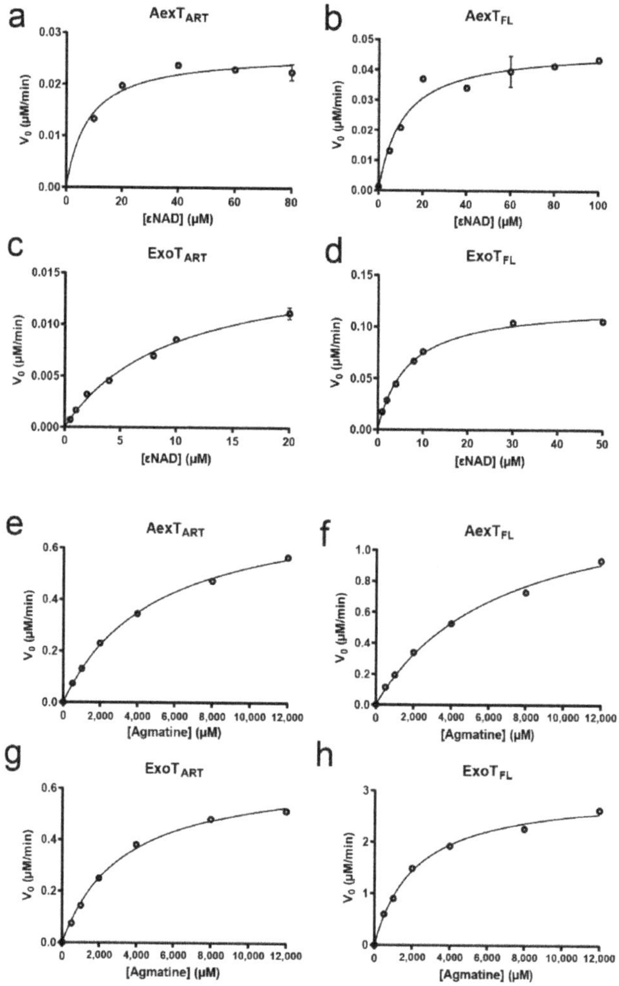

Figure 3. Steady-state kinetic parameters of the recombinant exotoxins. (**a**–**d**) Co-substrate dependent N^6-etheno-NAD+-glycohydrolase activities of the indicated exotoxins, in presence of 14-3-3β. (**e**–**h**) Agmatine dependent N^6-etheno-NAD+-glycohydrolase activities of the indicated exotoxins, in presence of 14-3-3β. The enzymatic parameters derived from these experiments are given in Table 2. Representative experiments are shown (n = 3).

Table 2. Steady-state kinetic parameters of the recombinant exotoxins.

Type of Reaction	NAD+Glycohydrolysis/Automodification		+Agmatine Modification
Exotoxin Construct [1]	K_M (NAD+) [2]	k_{cat} [3]	k_{cat} [3]
AexT-ART	8.1 ± 1.5	0.13	3.96
AexT-FL	10.9 ± 2.0	0.24	7.05
ExoT-ART	10.1 ± 1.0	0.09	3.37
ExoT-FL	6.6 ± 0.2	0.62	15.0

[1] ART, ADP-ribosyltransferase domain; FL, full-length protein; both as 14-3-3β complexes. [2] in µM. [3] in min^{-1}. Original data shown in Figure 3.

2.3. Enzymatic Properties of AexT and ExoT

To establish basic enzymatic properties of our AexT and ExoT constructs, we employed an assay of NAD+ glycohydrolase activity in which the dinucleotide analog εNAD+ was used as co-substrate, resulting in a fluorescence increase upon its hydrolysis to εADP-ribose [33]. The result indicated that, in the absence of target protein (i.e., when εNAD+ was either hydrolyzed or processed in an auto-ADP-ribosylation reaction), both AexT and ExoT processed εNAD+ with a K_M value in the low micromolar range (Figure 3 and Table 2). This was independent of whether the full-length proteins or the isolated catalytic domains were analyzed and indicated a dinucleotide affinity in the range of published values for other ADP-ribosylating bacterial toxins using the same assay (Table 2) [34]. Next, we analyzed AexT and ExoT activities in presence of agmatine, an analog for arginine [35]. Addition of agmatine led to a 20- to 30-fold increase in k_{cat}. These results confirm previous analyses [20,36] showing that both AexT and ExoT can modify arginine residues (Figure 3 and Table 2).

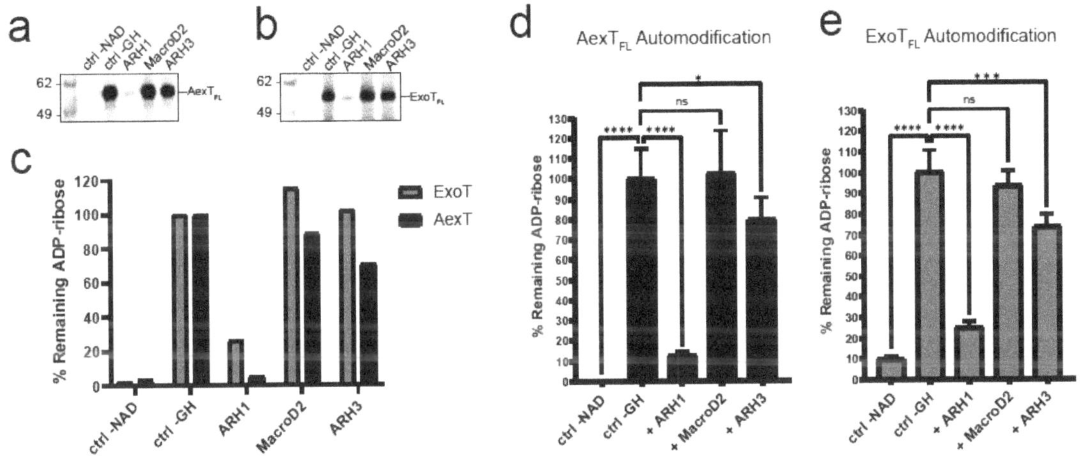

Figure 4. Exotoxin ADP-ribosylation and its removal by ADP-ribosylhydrolases. (**a**,**b**) Exotoxins were incubated with biotin-NAD+ and subsequently with the ADP-ribosylhydrolases indicated. Membranes were probed with streptavidin-HRP. (**a**) AexT; (**b**) ExoT. Uncropped images and loading controls are shown in Figure S3. (**c**) Quantification of exotoxin ADP-ribosylation from the membranes shown above. (**d**,**e**) Plate reader-based overlay assay using MacroGreen (see Materials and Methods). Shown are quantification of GFP signal indicating ADP-ribosylation after incubation with the ADP-ribosylhydrolases indicated. Significance levels: $p \geq 0.05$ (ns), $p < 0.05$ (*), $p < 0.001$ (***), $p < 0.0001$ (****); n = 8. ART, ADP-ribosyltransferase domain; FL, full-length protein.

We wanted to further validate these findings by analyzing whether arginines were the only target residues for AexT and ExoT. To do this, we employed the residue specific ADP-ribosylhydrolases ARH1, ARH3, and MacroD2. We set up reactions that enable NAD+glycohydrolase and automodification activities of full-length exotoxins. We included 10% N^6-biotin-NAD+ in the reaction buffer to analyze the results by Western blotting and detection using streptavidin-HRP. We analyzed exotoxins treated either with the arginine-ADP-ribosyl specific ADP-ribosylhydrolase ARH1 [37] or with the carboxyl-ADP-ribosyl specific MacroD2, [38] or with the serine-ADP-ribosyl specific ARH3 [39]. This revealed that AexT was capable of efficiently modifying itself, which has not been documented before (Figure 4a). Furthermore, ARH1 efficiently removed ADP-ribosyl from both AexT and ExoT, suggesting that automodification in both exotoxins occurred primarily at arginine residues (Figure 4). Neither ARH3 nor MacroD2 activities resulted in major reduction of ADP-ribosylation levels. However, ARH3 treatment gave a weak but significant reduction in ADP-ribosyl levels on both exotoxins. These results were confirmed by the MacroGreen assay (Figure 4d,e) that measures binding of an engineered Af1521 macrodomain [40], fused to GFP [41], to ADP-ribosylated protein immobilized on a microtiter plate. We conclude that both AexT and ExoT ADP-ribosylate themselves primarily on arginine residues.

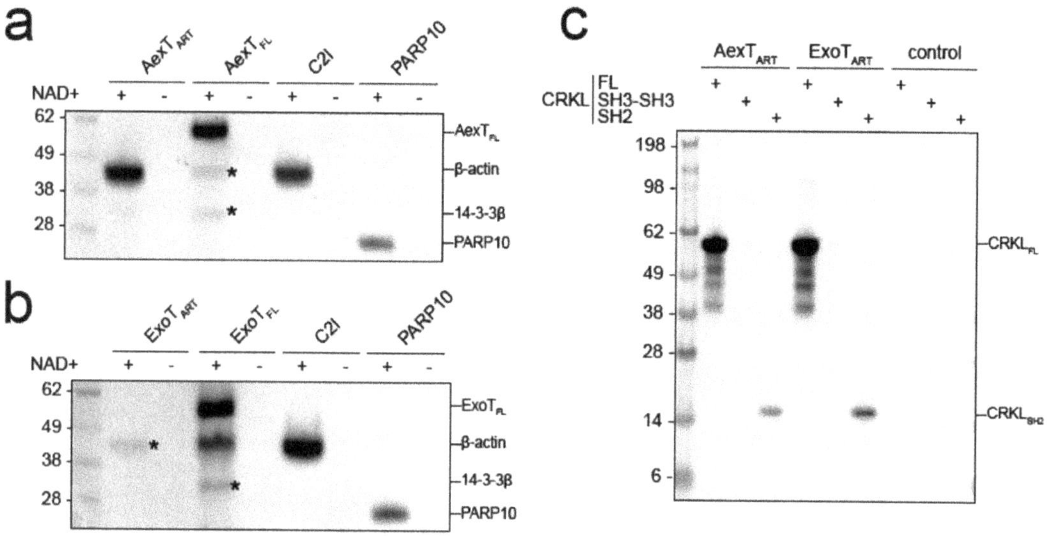

Figure 5. ADP-ribosylation of canonical AexT and ExoT targets by both exotoxins. (**a**,**b**) Exotoxin constructs were incubated with biotin-NAD+ and purified β-actin. Membranes were probed with streptavidin-HRP. C2I toxin, a known actin specific ADP-ribosyltransferase, and PARP10 ART domain, a strongly auto-ADP-ribosylating enzyme, were included for comparison. (**c**) ADP-ribosylation of CRKL constructs by exotoxin ART domains. Reactions were processed as in the previous panels. Uncropped images and loading controls are shown in Figure S4. ART, ADP-ribosyltransferase domain; FL, full-length protein.

Since AexT and ExoT have been reported to modify different target proteins, we sought to confirm those observations by analyzing ADP-ribosylation of purified AexT and ExoT target proteins by Western blotting. Both AexT (Figure 5a) and ExoT (Figure 5b) ADP-ribosylated actin, albeit with lower efficiency than the *Clostridium* C2I subunit. We observed that the isolated ART domain of AexT was more efficient at modifying actin than the full-length toxin, while the relationship was reversed for ExoT. We cannot exclude that low levels of actin modification by full-length AexT were caused by misfolding of the re-folded toxin; but this appears as an unlikely explanation in the light of the efficient automodification of AexT (Figures 4a and 5a). Actin modification might still be significant in infected cells, where the ratio of actin over AexT is likely much higher than under our experimental conditions. The experiments shown in Figure 5a,b also reveal that AexT and ExoT automodification occurs only in the full-length exotoxins. This may explain the differences in NAD+ glycohydrolase rates observed between the ART domain and full-length constructs (Table 2): Whereas the ART domains catalyze NAD+ glycohydrolysis only, the full-length exotoxins catalyze both the glycohydrolysis reaction and their automodification. Interestingly, *Pseudomonas* ExoS automodification occurs in its GAP domain and suppresses the activity of that domain [42]. Our results imply that AexT and ExoT might employ the same mechanism of autoregulation.

AexT has been suggested to target actin exclusively [20] and be unable to modify the CRK proteins, which are known ExoT targets. However, we show here that both toxins were able to modify the SH2 domain of the CRK-like protein (CRKL) and the modification appeared to be more efficient when the SH2 domain was presented in the context of the full-length protein (Figure 5c). This experiment was carried out with the ART domain constructs in order to avoid overlap of automodified full-length toxins with the CRKL constructs, which have similar mobility in SDS-PAGE (see loading controls in Supplementary materials Figure S4).

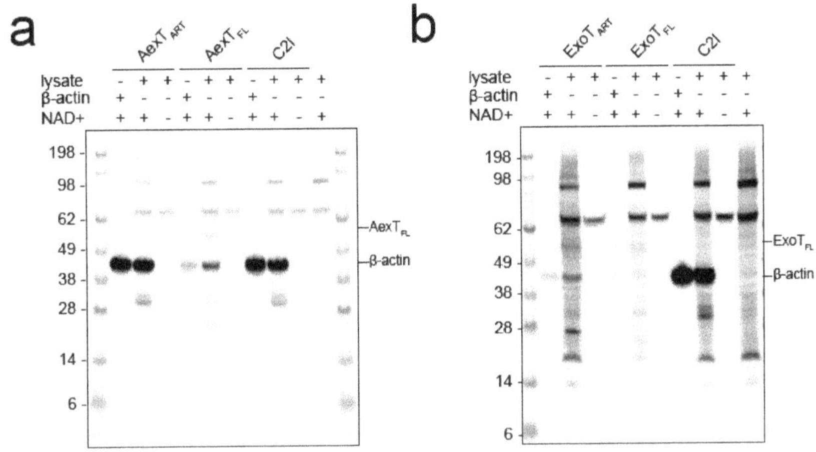

Figure 6. ADP-ribosylation of actin in THP-1 cell lysates. Exotoxin:14-3-3 complex or C2I and excess biotin-NAD+ were added to cell lysates. The reactions were processed by Western blotting followed by detection using streptavidin-HRP. (**a**) AexT constructs; (**b**) ExoT constructs. Uncropped images, alternative exposures and loading controls are shown in Figure S5. ART, ADP-ribosyltransferase domain; FL, full-length protein.

Modification of actin has not been reported for ExoT, to the best of our knowledge. Therefore, we explored this observation further. When incubated with cell lysates, ExoT as well as AexT and C2I modified a protein with a mobility in SDS-PAGE indicative of actin (Figure 6). All ADP-ribosylating toxins analyzed so far modify actin at residue R177 [43] and AexT is no exception [20]. ADP-ribosylation at R177 is known to block actin polymerization [44]. To test the effect of AexT and ExoT on the extent of actin polymerization, we conducted a sedimentation assay that is illustrated in Figure 7a. While actin ADP-ribosylation by C2I abolished actin polymerization, ADP-ribosylation by both AexT and ExoT led to partial inhibition of polymerization (Figure 7b). AexT and ExoT ART domain constructs were more efficient than the respective full-length proteins in preventing actin polymerization. This can likely be explained by the fact that actin ADP-ribosylation competed with automodification only in the case of the full-length exotoxins (Figure 5a,b). Nevertheless, these results show that AexT mediated ADP-ribosylation of actin compromises actin polymer formation. Furthermore, they suggest that ExoT can ADP-ribosylate actin on the same target residue as AexT and toxins from other clades.

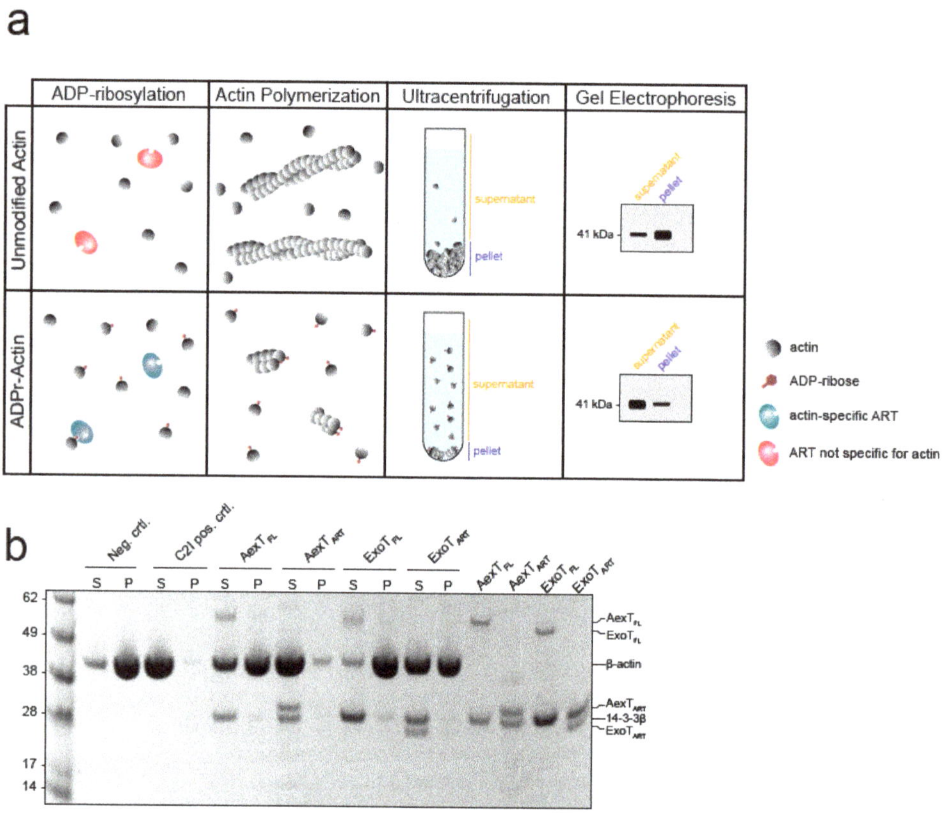

Figure 7. Exotoxin mediated modification compromises actin polymerization. (**a**) Schematic representation of the F-actin sedimentation assay employed. Exotoxin:14-3-3 complex and NAD+ were added to monomeric actin. After incubation to allow actin ADP-ribosylation, actin polymerization was induced and allowed to proceed to steady-state before actin polymers were sedimented. (**b**) The Coomassie stained SDS-PAGE gel shows equal amounts of pellet and supernatant fractions containing actin polymers and monomers, respectively. ART, ADP-ribosyltransferase domain; FL, full-length protein.

3. Conclusions

We show that *Pseudomonas* ExoT and *Aeromonas* AexT are homologous toxins that share actin and CRK-proteins as targets *in vitro*. The existence of common targets for these two exotoxins was suggested by homology modeling (Figure 1) that placed conserved AexT residues in the experimentally determined target recognition site of ExoT [31]. Our data suggest a new mechanism for the ADP-ribosyltransferase activitiy of each toxin: ExoT might modify cellular actin to prevent polymerization; and AexT might affect the integrity of the microfilament system via CRK proteins and possibly contribute to induction of apoptosis. Therefore, it will be important to re-examine the ADP-ribosyltransferase activities of ExoT and AexT to test these findings in models of *Pseudomonas* and *Aeromonas* infection. If inhibitors of the exotoxin:14-3-3 interaction can be developed, they will be helpful tools to test exotoxin effects on host cell invasion and spreading of the pathogens within infected tissue. This would contribute to clarification whether these exotoxins are feasible targets toward development of therapeutic agents.

4. Materials and Methods

4.1. Materials

Fine chemicals and growth media were purchased from Merck Life Science AB, Solna, Sweden. 1,N^6-biotin-NAD+ (biotin-NAD+) was obtained from BPS Bioscience, San Diego, Ca, USA, and 1,N^6-etheno-AMP (εAMP) and -NAD+ (εNAD+) were obtained from Merck Life Science AB, Solna, Swedenand Jena Bioscience, Jena, Germany, respectively.

4.2. Structural Homology Model of AexT

The AexT, ExoT and ExoS sequences were downloaded from Uniprot (accession numbers Q93Q17, Q9I788, and G3XDA1, respectively) and aligned in Jalview [45] with the ClustalO algorithm. Given the sequence similarities within the ART domains, we used the structure of ExoT-ART:14-3-3β (PDB: 6GNN) [28] as a template and modelled the AexT-ART domain structure using SWISS-MODEL [46]. Model quality was assessed using Molprobity [47], which resulted in a score of 2.08, a clash score of 2.02, and 4.08% outliers in the Ramachandran plot. The model was superimposed with the ExoT-ART:14-3-3β structure using Chimera 1.16 [48].

4.3. Molecular Cloning

The cDNA fragment encoding the ART domain of AexT (AexT-ART; residues 252–475), codon-optimized for expression in *Escherichia coli* strains, was obtained from GeneArt (Thermo Fisher Scientific, Gothenburg, Sweden). For co-expression with 14-3-3β, the fragment was sub-cloned into the NcoI and EcoRI sites of a modified pET vector to obtain an N-terminal hexahistidine fusion. The ExoT ART domain (ExoT-ART; residues 235–457) sub-cloned in the corresponding vector was described earlier. [28] The hexahistidine tag on the ART domains enabled co-purification of un-tagged 14-3-3β in the toxin chaperone complex. The cDNAs encoding full-length AexT and ExoT were amplified by PCR from whole cell lysates of *Aeromonas salmonicida* strain A449 and *Pseudomonas aeruginosa* strain PAK, respectively. These cDNAs were inserted into pNIC28 (GenBank: EF198106) using ligase independent cloning. A cDNA encoding *Clostridium botulinum* C2I toxin subunit (residues 6–428), codon-optimized for expression in *E. coli* strains, was obtained from GeneArt (Thermo Fisher Scientific, Gothenburg, Sweden) and sub-cloned into the NcoI and BamHI sites of pET28 to obtain an N-terminal hexahistidine fusion. Human CRKL GST fusion plasmids (pGEX-2T) coding for either the SH2-SH3-SH3 domains, residues 7–303 or the SH3-SH3 domains, residues 115–303 were obtained from Addgene (#36400 and #36403). [49,50] Expression vector pNIC-H102 encoding the SH2 domain (residues 1–98) with an N-terminal decahistidine tag were obtained from Open Biosystems, Huntsville, Ga, USA (OHS4902). The expression plasmids for human 14-3-3β, ARH1, ARH3, MacroD2 and PARP10 have been described [28,41].

4.4. Protein Expression and Purification

Bovine cytosolic actin (β- and γ-isoform mixture) was purified from calf thymus and the isoforms were separated by chromatography on hydroxyapatite as described [51]. All other proteins were expressed in either *E. coli* BL21(DE3)T1R (Merck Life Science AB, Solna, Sweden) carrying the pRARE2 plasmid (Karolinska Institutet Protein Science Facility) or BL21(DE3) (New England Biolabs, C2530H). Expression cultures were set up in 2 L Schott flasks containing 1.5 L Terrific Broth supplemented with the appropriate antibiotics. Cells were grown at 37 °C in a LEX Bioreactor (Epiphyte3) until OD_{600} reached 2. The temperature was reduced to 18 °C and protein overexpression was induced with 0.5 mM IPTG for 16 h. Cells were pelleted by centrifugation at 4600× *g* at 4 °C and resuspended in lysis buffer (2.5 mL per gram of wet cell weight; 50 mM HEPES-NaOH pH 7.5, 500 mM NaCl, 10% glycerol, 1 mM TCEP supplemented with one tablet protease inhibitor cocktail (cat. no. S8830; Merck Life Science AB, Solna, Sweden,) and 5 μL benzonase (cat. no. E1014; Merck Life Science AB, Solna, Sweden). The suspension was sonicated on ice (Sonics VibraCell; 100 W with a 0.5-inch tip) for 4 min in 10 s intervals and subsequently centrifuged at

22,000× *g* for 25 min at 4 °C. The supernatant was clarified using a 0.45 μm filter prior to chromatography.

Immobilized metal affinity chromatography (IMAC) was carried out using 5 mL HiTrap TALON columns (Cytiva, Uppsala, Sweden) on an ÄKTA Pure system. The bound proteins were eluted with 50 mM HEPES-NaOH pH 7.5, 500 mM NaCl, 300 mM imidazole, 10% glycerol, and 1 mM TCEP over a 10-column volume gradient. Peak fractions were analyzed by SDS-PAGE (NuPage 4–12% Bis-Tris gels; Invitrogen, Thermo Fisher Scientific, Gothenburg, Sweden) and Coomassie staining [52] before subsequent purification steps.

AexT-ART:14-3-3β complex was further purified by cation exchange chromatography. Pooled IMAC fractions were desalted and injected in a 5 mL HiTrap Heparin HP column (Cytiva, Uppsala, Sweden) equilibrated in 20 mM MES pH 6.0, 1 mM $MgCl_2$, 1 mM TCEP. The bound proteins were eluted with a 10-column volume gradient of the same buffer but containing 1 M NaCl.

For all proteins (except for full-length exotoxins; see 4.5), size exclusion chromatography (SEC) was carried out using a HiLoad 16/600 Superdex 75 pg column (Cytiva, Uppsala, Sweden) equilibrated in 50 mM HEPES-NaOH pH 7.5, 300 mM NaCl, 10% glycerol, 0.5 mM TCEP. Individually expressed 14-3-3β was further purified by hydrophobic interaction chromatography as described [28]. Protein purity was verified by SDS-PAGE and pure fractions were pooled and concentrated with Vivaspin centrifugal concentrators (cat. no. Z614025 or similar Merck Life Science AB, Solna, Sweden). Samples were frozen in aliquots and stored at –80 °C.

4.5. Protein Recovery from Inclusion Bodies

For full-length AexT and ExoT, protein overexpression and bacterial cell lysis were performed as described above. The supernatant insoluble fraction containing the inclusion bodies was resuspended in wash buffer A (100 mM Tris-HCl pH 7.5, 500 mM NaCl, 2 M Urea, 10 mM EDTA, 5 mM TCEP, 2% Triton X-100; 5 mL per gram of wet cell weight) using a glass-glass tissue homogenizer. The suspension was centrifuged at 15,000× *g* for 15 min at 4 °C to pellet inclusion bodies. This washing step was performed three times. Then, the pellet was resuspended in wash buffer B (100 mM Tris-HCl pH 7.5, 10 mM EDTA; 5 mL per gram of wet cell weight). The suspension was centrifuged at 15,000× *g* for 10 min at 4 °C, and the supernatant was discarded. The pellet was resuspended in solubilization buffer [6.6 M guanidine hydrochloride (GdnHCl), 50 mM Tris-HCl pH 8; 1 mL per 20–40 mg inclusion bodies], using a tissue homogenizer. The suspension was incubated for 2–3 h at room temperature under constant stirring. Cell debris and non-solubilized material were removed by centrifugation at 50,000× *g* for 20 min. For ExoT, the supernatant was incubated with Ni-NTA agarose (Qiagen, Kista, Sweden) at room temperature overnight under gentle shaking. The beads were collected by low-speed centrifugation and washed twice with 6 M GdnHCl, 50 mM Tris-HCl pH 8 and once with 6 M GdnHCl, 50 mM Tris-HCl pH 8.0, 12 mM imidazole. ExoT was eluted in 6 M GdnHCl, 50 mM Tris-HCl pH 8.0, 400 mM imidazole and the Ni-NTA bead eluate was diluted to a final concentration of 0.1 mg/mL with 6 M GdnHCl, 50 mM Tris-HCl pH 8.0. For the AexT-containing supernatant, the protein concentration was kept at 2 mg/mL. Both proteins were dialyzed at room temperature under gentle stirring, using high retention seamless cellulose tubings (14 kDa cut-off; cat. no. D0405; Merck Life Science AB, Solna, Sweden). Each dialysis step was performed in approximately 80 mL buffer per ml of protein solution for at least 2h, using the following buffers with decreasing GdnHCl concentration: Buffer A (4 M GdnHCl, 50 mM Tris-HCl pH 7.5), buffer B (2 M GdnHCl, 50 mM Tris-HCl pH 7.5), buffer C (0.5 M GdnHCl, 50 mM Tris-HCl pH 7.5), buffer D (50 mM Tris-HCl pH 7.5, 300 mM NaCl, 10% glycerol, 0.5 mM TCEP). Before starting the final dialysis step, purified 14-3-3β was added in approximately equimolar concentrations to the exotoxins to assist refolding. The refolded proteins were separated from aggregates by centrifugation at 4000× *g* for 5 min and SEC was performed as above but sing a HiLoad 16/600 Superdex 200 pg column (Cytiva). Protein purity was verified by SDS-PAGE and pure fractions were pooled and concentrated with Vivaspin

20 centrifugal concentrators (30 kDa MWCO, cat. no. GE28–9323–61, Merck Life Science AB, Solna, Sweden). The proteins were frozen in aliquots and the degree of refolding was evaluated by measuring enzymatic activity.

4.6. Analysis of Protein Solution Properties

The oligomeric states of the exotoxin:14-3-3β complexes were analyzed using mass photometry on a REFEYN TwoMP mass photometer. Microscopy slides (24 × 50 mm) were rinsed consecutively with milli-Q water, isopropanol, and milli-Q water and dried under a clean nitrogen stream. A row of 2 × 3 silicone gasket wells (Grace Biolabs, Merck Life Science AB, Solna, Sweden) was placed onto a slide. Protein batches were thawed and diluted to approximately 100 nM in buffer containing 20 mM HEPES-NaOH pH 7.5, 50 mM NaCl, 4 mM $MgCl_2$, 0.5 mM TCEP. The focus was adjusted to a well containing 15 μL buffer before 3 μL of protein solution was added, yielding a final protein concentration of approximately 16 nM. The landing events within the recorded frame (900 × 354 pixels) were measured for 60 s at a frame rate of 496 Hz and at least 4900 events were recorded per acquisition. Data were analyzed using DiscoverMP software (REFEYN, Oxford, UK). For calibration, a native PAGE marker (Invitrogen, Thermo Fisher Scientific, Gothenburg, Sweden) was used. Oligomeric states were also characterized by analytical size exclusion chromatography (OMNISEC; Malvern Panalytical, Uppsala, Sweden) equipped with a Superdex 200 Increase 10/300 GL column (Cytiva, Uppsala, Sweden) equilibrated in 50 mM HEPES-NaOH pH 7.5, 300 mM NaCl and 1 mM TCEP. The detectors were calibrated with bovine serum albumin. Protein aliquots were filtered with 0.2 μm centrifugal filters and kept at 4 °C until injected into the column. Each sample was injected in duplicates of 120 μL, corresponding to 50–60 μg of protein per injection. Proteins were eluted at 0.5 mL/min. Data were analyzed using OMNISEC v11.32 software (Malvern Panalytical). One of two reproducible runs was chosen for analysis. The refractive index and measured molecular weight parameters were re-plotted in Prism (GraphPad Software LLC, DanDiego, CA, USA).

4.7. Enzymatic Analyses

In general, exotoxin assays contained 200 nM exotoxins and 2–2.5-fold molar excess of 14.3.3β in 20 mM HEPES-NaOH pH 7.5, 50 mM NaCl, 4 mM $MgCl_2$, 0.5 mM TCEP. The source of 14-3-3β depended on the production protocol (see Sections 4.3–4.5). In reactions containing agmatine (Merck Life Science AB, Solna, Sweden), εNAD+ was kept at 25 μM. For negative control reactions, εNAD+ or toxin was omitted. Measurements were carried out in black flat-bottom half-area non-binding 96-well plates (Corning CLS3993; Merck Life Science AB, Solna, Sweden) in final volumes of 50 μL. Enzymatic reactions, at room temperature (ca. 22 °C), were started by the addition of εNAD+, and fluorescence was followed over time in a CLARIOstar multimode plate reader (BMG Labtech, Ortenberg, Germany) using λ_{ex} = 302/20 nm (filter) and λ_{em} = 410/20 nm (monochromator). All kinetic parameters were determined based on fluorescence increase in the linear time range. Fluorescence intensities were related to concentrations of the fluorescent product using dilution series of εAMP. Kinetic parameters were calculated using Prism (GraphPad Software LLC, DanDiego, Ca, USA).

For detection of ADP-ribosylation by Western blotting, enzymatic reactions were performed with 100 μM NAD+ (Figure 5a,b and Figure 6) or 50 μM NAD+ (Figure 5c) containing 10% biotin-NAD+. Enzyme concentrations were 1 μM 14-3-3 dependent exotoxins with 2–2.5-fold molar excess of 14-3-3β, 20 nM *Clostridium botulinum* C2I toxin subunit, and 2 μM PARP10 catalytic domain. For the experiments depicted in Figures 5c and 6, the concentration of exotoxins was reduced to 0.5 μM. Target protein concentrations were 10 μM purified protein or 0.5 mg/mL total protein for reactions containing cell lysate. Reactions (20–40 μL in 20 mM HEPES-NaOH 7.5, 50 mM NaCl, 4 mM $MgCl_2$, 0.5 mM TCEP) were started by the addition of either NAD+ or toxin. Reactions were incubated for 1 h at room temperature and stopped by addition of Laemmli buffer and heating at

95 °C for 4 min. Automodification reactions prior to ADP-ribosylhydrolase treatment were conducted essentially as above, with 1 µM 14-3-3 dependent exotoxins, 2–2.5-fold molar excess of 14-3-3 β and 5 µM NAD+. After 1 h of incubation, the reactions were stopped by addition of 100 µM inhibitor STO1101 [53]. After 20 min of incubation at room temperature, (ADP) ribosylhydrolases ARH1, MACROD2 and ARH3 were added at 10 µM, followed by incubation for 2 h. Reactions were stopped with Laemmli buffer and heating at 95 °C for 4 min. Proteins were separated on 4–12% Bis-Tris polyacrylamide gels (Invitrogen, Thermo Fisher Scientific, Gothenburg, Sweden) and transferred onto a PVDF membrane in transfer buffer (3% (m/v) Trizma, 14.4% (m/v) glycine and 10% (v/v) methanol) for 1h at 160 mA. Ponceau-S staining of the membrane served as a control for loaded protein and transfer efficiency. Membranes were de-stained, blocked for 1 h in 1% BSA in TBST, and incubated for 1 h in 0.5 µg/mL HRP-linked Streptavidin (cat. no. 21126; Thermo Fisher Scientific, Gothenburg, Sweden) in 1% BSA in TBST. Proteins modified with biotin-NAD+ as co-substrate were visualized with SuperSignal™ West Pico PLUS Chemiluminescent Substrate (Thermo Fisher Scientific, Gothenburg, Sweden). Images were obtained using a ChemiDoc imaging system (Bio-Rad, Solna, Sweden) and densitometric analysis was performed in Image Lab software (Bio-Rad, Solna, Sweden).

Target residue specificity of the exotoxins was also determined using a macrodomain-GFP fusion overlay assay [41]. Auto-ADP-ribosylation reactions of 250 nM exotoxin and 2-2.5-fold molar excess of 14-3-3β were carried out in reaction buffer (25 mM HEPES-NaOH pH 7.5, 100 mM NaCl, 0.2 mM TCEP, 4 mM $MgCl_2$) in the presence of 1 mM NAD+ at room temperature for 30 min. The reactions were transferred into a 96-well high-binding plate (Nunc MaxiSorp™; Thermo Fisher Scientific, Gothenburg, Sweden). Proteins were allowed to bind to the plate for 30 min. To remove NAD+, the wells were rinsed three times with reaction buffer and then blocked for 5 min with 1% BSA in reaction buffer. The wells were rinsed twice before ARH1, MacroD2, or ARH3 were added at 1.5 µM in reaction buffer. After 1h of incubation, the wells were rinsed twice with reaction buffer and then washed with 1 mM ADP-ribose in reaction buffer for 10 min. The wells were rinsed three times with TBST buffer, blocked again with 1% BSA in TBST buffer and rinsed twice. 1 µM of eAF1521-GFP [41] was added for detection of remaining ADP-ribose and three rinsing steps in TBST buffer were performed to remove unbound eAF1521-GFP. 150 µL TBST buffer was added to the wells for measurement. GFP fluorescence was measured on a CLARIOstar multimode reader (BMG Labtech, Ortenberg, Germany) using a filter set of λ_{ex} = 470/15 nm and λ_{em} = 515/20 nm. The data were normalized to the background signal (blank wells) and the percentage of remaining ADP-ribose was determined by comparison to auto-modification control reactions with no ADP-ribosylhydrolases added. The normalized data were plotted in Prism (GraphPad Software) and the significance was determined with a Welch ANOVA test, in which the mean of each column was compared to the mean of the auto-modification reaction.

4.8. Preparation of Cell Lysates

THP-1 human monocytic leukemia cells were grown in RPMI medium supplemented with 10% fetal bovine serum and 1% penicillin/streptomycin. 300×10^6 cells were harvested and washed in 2 mL ice-cold D-PBS and subsequently lysed in 800 µL hypotonic buffer (50 mM Tris-HCl pH 7.5) supplemented with protease and phosphatase inhibitor cocktails (Complete, 11504400 and PhosSTOP EASYpack, 04906845001, respectively; Roche, Basel, Switzerland). The cells were resuspended by vortexing and pipetting thoroughly during 20 min incubation on ice. Protein concentration was determined using Bradford reagent (Bio-Rad). Before freezing, 80µL of 200 mM HEPES-NaOH pH 7.5, 500 mM NaCl, 40 mM $MgCl_2$, 5 mM TCEP and 0.1% benzonase were added.

4.9. Actin Polymerization Assay

Bovine cytosolic β-actin in G-actin buffer (20 mM HEPES-NaOH pH 7.5, 0.2 mM $CaCl_2$, 0.2 mM ATP, 0.5 mM TCEP) was thawed and diluted 1:1 in fresh G-actin buffer. The sample was centrifuged for 15 min at 21,000× *g*, 4 °C, and the supernatant was transferred to a new tube. 200 μM NAD+, 10 μM β-actin, and either 800 nM exotoxin:14-3-3β complex or 20 nM Clostridium botulinum C2I toxin subunit in G-actin buffer were added in 100 μL reactions and the samples were incubated for 30 min at room temperature. Actin polymerization was triggered by the addition of 100 mM KCl and 2 mM $MgCl_2$. The samples were transferred into airfuge tubes (cat. no. 342630; Beckman Coulter, Bromma, Sweden). After 45 min at room temperature, actin polymers were sedimented by ultracentrifugation in a Beckman Coulter Airfuge (A-100/18 rotor) for 15 min at 30 psi (corresponding to roughly149,000× *g*). The supernatant was separated from the pellet and the pellet was resuspended in 100 μL G-actin buffer. Equal volumes of supernatant and pellet fractions were loaded on a 4–12% NuPage Bis-Tris gel (Invitrogen, Thermo Fisher Scientific, Gothenburg, Sweden) and subjected to gel electrophoresis. The gel was stained with Coomassie.

Supplementary Materials: The following supporting information can be downloaded at: https://www.mdpi.com/article/10.3390/pathogens11121497/s1, Supplementary Information document containing Figures S1–S5.

Author Contributions: Conceptualization, C.E., P.H. and H.S.; investigation, C.E., P.H., A.G.G.-S., A.T., S.A. and H.S.; formal analysis, C.E., P.H. and H.S.; writing—original draft preparation, H.S.; writing—review and editing, C.E. and H.S.; visualization, C.E. and A.T.; supervision, A.H. and H.S.; funding acquisition, A.H. and H.S. All authors have read and agreed to the published version of the manuscript.

Funding: This research was funded by the Swedish Research Council, grant numbers 2019-4871 (to HS) and 2019-1242 (to AH), and by the Crafoord Foundation, grant number 2021–0673 (to HS). CE was recipient of an ERASMUS+ scholarship and support from the section "Wirtschaft für Bildung" of the Federation of Austrian Industries Kärnten.

Institutional Review Board Statement: Not applicable.

Informed Consent Statement: Not applicable.

Data Availability Statement: Data is contained within the article or Supplementary Materials.

Acknowledgments: We thank Janos Padra and Sara Lindén (University of Gothenburg) for *Aeromonas salmonicida* whole-cell extracts; Johannes Rack and Ivan Ahel (Oxford University) for the ARH1 expression plasmid; and Tomas Nyman (Karolinska Institutet, Protein Science Facility) for the CRKL SH2 domain expression plasmid. We also thank Céleste Sele (Lund University Protein Production Platform) for conducting OmniSec experiments.

Conflicts of Interest: The authors declare no conflict of interest. The funders had no role in the design of the study; in the collection, analyses, or interpretation of data; in the writing of the manuscript; or in the decision to publish the results.

References

1. Zdanowicz, M.; Mudryk, Z.J.; Perlinski, P. Abundance and antibiotic resistance of Aeromonas isolated from the water of three carp ponds. *Vet. Res. Commun.* **2020**, *44*, 9–18. [CrossRef] [PubMed]
2. Parker, J.L.; Shaw, J.G. Aeromonas spp. clinical microbiology and disease. *J. Infect.* **2011**, *62*, 109–118. [CrossRef] [PubMed]
3. Menanteau-Ledouble, S.; Kumar, G.; Saleh, M.; El-Matbouli, M. Aeromonas salmonicida: Updates on an old acquaintance. *Dis. Aquat. Org.* **2016**, *120*, 49–68. [CrossRef]
4. Janda, J.M.; Abbott, S.L. The genus Aeromonas: Taxonomy, pathogenicity, and infection. *Clin. Microbiol. Rev.* **2010**, *23*, 35–73. [CrossRef]
5. Collaborators, A.R. Global burden of bacterial antimicrobial resistance in 2019: A systematic analysis. *Lancet* **2022**, *399*, 629–655. [CrossRef]
6. Yahr, T.L.; Wolfgang, M.C. Transcriptional regulation of the Pseudomonas aeruginosa type III secretion system. *Mol. Microbiol.* **2006**, *62*, 631–640. [CrossRef]
7. Hauser, A.R. The type III secretion system of Pseudomonas aeruginosa: Infection by injection. *Nat. Rev. Microbiol.* **2009**, *7*, 654–665. [CrossRef] [PubMed]
8. Klockgether, J.; Tummler, B. Recent advances in understanding Pseudomonas aeruginosa as a pathogen. *F1000Res* **2017**, *6*, 1261. [CrossRef]
9. Horna, G.; Ruiz, J. Type 3 secretion system of Pseudomonas aeruginosa. *Microbiol. Res.* **2021**, *246*, 126719. [CrossRef]

10. Vallis, A.J.; Finck-Barbancon, V.; Yahr, T.L.; Frank, D.W. Biological effects of Pseudomonas aeruginosa type III-secreted proteins on CHO cells. *Infect. Immun.* **1999**, *67*, 2040–2044. [CrossRef]
11. Garrity-Ryan, L.; Shafikhani, S.; Balachandran, P.; Nguyen, L.; Oza, J.; Jakobsen, T.; Sargent, J.; Fang, X.; Cordwell, S.; Matthay, M.A.; et al. The ADP ribosyltransferase domain of Pseudomonas aeruginosa ExoT contributes to its biological activities. *Infect. Immun.* **2004**, *72*, 546–558. [CrossRef]
12. Wood, S.; Goldufsky, J.; Shafikhani, S.H. Pseudomonas aeruginosa ExoT Induces Atypical Anoikis Apoptosis in Target Host Cells by Transforming Crk Adaptor Protein into a Cytotoxin. *PLoS Pathog.* **2015**, *11*, e1004934. [CrossRef]
13. Braun, M.; Stuber, K.; Schlatter, Y.; Wahli, T.; Kuhnert, P.; Frey, J. Characterization of an ADP-ribosyltransferase toxin (AexT) from Aeromonas salmonicida subsp. salmonicida. *J. Bacteriol.* **2002**, *184*, 1851–1858. [CrossRef]
14. Boehi, F.; Manetsch, P.; Hottiger, M.O. Interplay between ADP-ribosyltransferases and essential cell signaling pathways controls cellular responses. *Cell Discov.* **2021**, *7*, 104. [CrossRef]
15. Schwan, C.; Lang, A.E.; Schlosser, A.; Fujita-Becker, S.; AlHaj, A.; Schroder, R.R.; Faix, J.; Aktories, K.; Mannherz, H.G. Inhibition of Arp2/3 Complex after ADP-Ribosylation of Arp2 by Binary Clostridioides Toxins. *Cells* **2022**, *11*, 3661. [CrossRef] [PubMed]
16. Deng, Q.; Barbieri, J.T. Molecular mechanisms of the cytotoxicity of ADP-ribosylating toxins. *Annu. Rev. Microbiol.* **2008**, *62*, 271–288. [CrossRef] [PubMed]
17. Simon, N.C.; Aktories, K.; Barbieri, J.T. Novel bacterial ADP-ribosylating toxins: Structure and function. *Nat. Rev. Microbiol.* **2014**, *12*, 599–611. [CrossRef] [PubMed]
18. Fu, H.; Coburn, J.; Collier, R.J. The eukaryotic host factor that activates exoenzyme S of Pseudomonas aeruginosa is a member of the 14-3-3 protein family. *Proc. Natl. Acad. Sci. USA* **1993**, *90*, 2320–2324. [CrossRef]
19. Barbieri, J.T.; Sun, J. Pseudomonas aeruginosa ExoS and ExoT. *Rev. Physiol. Biochem. Pharm.* **2004**, *152*, 79–92. [CrossRef]
20. Fehr, D.; Burr, S.E.; Gibert, M.; d'Alayer, J.; Frey, J.; Popoff, M.R. Aeromonas exoenzyme T of Aeromonas salmonicida is a bifunctional protein that targets the host cytoskeleton. *J. Biol. Chem.* **2007**, *282*, 28843–28852. [CrossRef] [PubMed]
21. Fieldhouse, R.J.; Turgeon, Z.; White, D.; Merrill, A.R. Cholera- and anthrax-like toxins are among several new ADP-ribosyltransferases. *PLoS Comput. Biol.* **2010**, *6*, e1001029. [CrossRef] [PubMed]
22. Krall, R.; Schmidt, G.; Aktories, K.; Barbieri, J.T. Pseudomonas aeruginosa ExoT is a Rho GTPase-activating protein. *Infect. Immun.* **2000**, *68*, 6066–6068. [CrossRef] [PubMed]
23. Kazmierczak, B.I.; Engel, J.N. Pseudomonas aeruginosa ExoT acts in vivo as a GTPase-activating protein for RhoA, Rac1, and Cdc42. *Infect. Immun.* **2002**, *70*, 2198–2205. [CrossRef]
24. Sun, J.; Barbieri, J.T. Pseudomonas aeruginosa ExoT ADP-ribosylates CT10 regulator of kinase (Crk) proteins. *J. Biol. Chem.* **2003**, *278*, 32794–32800. [CrossRef] [PubMed]
25. Fehon, R.G.; McClatchey, A.I.; Bretscher, A. Organizing the cell cortex: The role of ERM proteins. *Nat. Rev. Mol. Cell Biol.* **2010**, *11*, 276–287. [CrossRef]
26. Sundin, C.; Henriksson, M.L.; Hallberg, B.; Forsberg, A.; Frithz-Lindsten, E. Exoenzyme T of Pseudomonas aeruginosa elicits cytotoxicity without interfering with Ras signal transduction. *Cell. Microbiol.* **2001**, *3*, 237–246. [CrossRef]
27. Otto, H.; Reche, P.A.; Bazan, F.; Dittmar, K.; Haag, F.; Koch-Nolte, F. In silico characterization of the family of PARP-like poly(ADP-ribosyl)transferases (pARTs). *BMC Genom.* **2005**, *6*, 139. [CrossRef]
28. Karlberg, T.; Hornyak, P.; Pinto, A.F.; Milanova, S.; Ebrahimi, M.; Lindberg, M.; Püllen, N.; Nordström, A.; Löverli, E.; Caraballo, R.; et al. 14-3-3 proteins activate Pseudomonas exotoxins-S and -T by chaperoning a hydrophobic surface. *Nat. Commun.* **2018**, *9*, 3785. [CrossRef]
29. Studer, G.; Rempfer, C.; Waterhouse, A.M.; Gumienny, R.; Haas, J.; Schwede, T. QMEANDisCo-distance constraints applied on model quality estimation. *Bioinformatics* **2020**, *36*, 2647. [CrossRef] [PubMed]
30. Jumper, J.; Evans, R.; Pritzel, A.; Green, T.; Figurnov, M.; Ronneberger, O.; Tunyasuvunakool, K.; Bates, R.; Zidek, A.; Potapenko, A.; et al. Highly accurate protein structure prediction with AlphaFold. *Nature* **2021**, *596*, 583–589. [CrossRef]
31. Sun, J.; Maresso, A.W.; Kim, J.J.; Barbieri, J.T. How bacterial ADP-ribosylating toxins recognize substrates. *Nat. Struct. Mol. Biol.* **2004**, *11*, 868–876. [CrossRef] [PubMed]
32. Shafikhani, S.H.; Engel, J. Pseudomonas aeruginosa type III-secreted toxin ExoT inhibits host-cell division by targeting cytokinesis at multiple steps. *Proc. Natl. Acad. Sci. USA* **2006**, *103*, 15605–15610. [CrossRef] [PubMed]
33. Armstrong, S.; Merrill, A.R. Application of a fluorometric assay for characterization of the catalytic competency of a domain III fragment of Pseudomonas aeruginosa exotoxin A. *Anal. Biochem.* **2001**, *292*, 26–33. [CrossRef] [PubMed]
34. Icenogle, L.M.; Hengel, S.M.; Coye, L.H.; Streifel, A.; Collins, C.M.; Goodlett, D.R.; Moseley, S.L. Molecular and biological characterization of Streptococcal SpyA-mediated ADP-ribosylation of intermediate filament protein vimentin. *J. Biol. Chem.* **2012**, *287*, 21481–21491. [CrossRef]
35. Ravulapalli, R.; Lugo, M.R.; Pfoh, R.; Visschedyk, D.; Poole, A.; Fieldhouse, R.J.; Pai, E.F.; Merrill, A.R. Characterization of Vis Toxin, a Novel ADP-Ribosyltransferase from Vibrio splendidus. *Biochemistry* **2015**, *54*, 5920–5936. [CrossRef]
36. Deng, Q.; Sun, J.; Barbieri, J.T. Uncoupling Crk signal transduction by Pseudomonas exoenzyme T. *J. Biol. Chem.* **2005**, *280*, 35953–35960. [CrossRef]
37. Rack, J.G.M.; Ariza, A.; Drown, B.S.; Henfrey, C.; Bartlett, E.; Shirai, T.; Hergenrother, P.J.; Ahel, I. (ADP-ribosyl)hydrolases: Structural Basis for Differential Substrate Recognition and Inhibition. *Cell Chem. Biol.* **2018**, *25*, 1533–1546.e12. [CrossRef]

38. Rosenthal, F.; Feijs, K.L.; Frugier, E.; Bonalli, M.; Forst, A.H.; Imhof, R.; Winkler, H.C.; Fischer, D.; Caflisch, A.; Hassa, P.O.; et al. Macrodomain-containing proteins are new mono-ADP-ribosylhydrolases. *Nat. Struct. Mol. Biol.* **2013**, *20*, 502–507. [CrossRef]
39. Fontana, P.; Bonfiglio, J.J.; Palazzo, L.; Bartlett, E.; Matic, I.; Ahel, I. Serine ADP-ribosylation reversal by the hydrolase ARH3. *Elife* **2017**, *6*. [CrossRef]
40. Nowak, K.; Rosenthal, F.; Karlberg, T.; Butepage, M.; Thorsell, A.G.; Dreier, B.; Grossmann, J.; Sobek, J.; Imhof, R.; Luscher, B.; et al. Engineering Af1521 improves ADP-ribose binding and identification of ADP-ribosylated proteins. *Nat. Commun.* **2020**, *11*, 5199. [CrossRef]
41. Garcia-Saura, A.G.; Herzog, L.K.; Dantuma, N.P.; Schuler, H. MacroGreen, a simple tool for detection of ADP-ribosylated proteins. *Commun. Biol.* **2021**, *4*, 919. [CrossRef] [PubMed]
42. Riese, M.J.; Goehring, U.M.; Ehrmantraut, M.E.; Moss, J.; Barbieri, J.T.; Aktories, K.; Schmidt, G. Auto-ADP-ribosylation of Pseudomonas aeruginosa ExoS. *J. Biol. Chem.* **2002**, *277*, 12082–12088. [CrossRef] [PubMed]
43. Aktories, K.; Lang, A.E.; Schwan, C.; Mannherz, H.G. Actin as target for modification by bacterial protein toxins. *FEBS J.* **2011**, *278*, 4526–4543. [CrossRef]
44. Wegner, A.; Aktories, K. ADP-ribosylated actin caps the barbed ends of actin filaments. *J. Biol. Chem.* **1988**, *263*, 13739–13742. [CrossRef] [PubMed]
45. Waterhouse, A.M.; Procter, J.B.; Martin, D.M.; Clamp, M.; Barton, G.J. Jalview Version 2–a multiple sequence alignment editor and analysis workbench. *Bioinformatics* **2009**, *25*, 1189–1191. [CrossRef]
46. Waterhouse, A.; Bertoni, M.; Bienert, S.; Studer, G.; Tauriello, G.; Gumienny, R.; Heer, F.T.; de Beer, T.A.P.; Rempfer, C.; Bordoli, L.; et al. SWISS-MODEL: Homology modelling of protein structures and complexes. *Nucleic. Acids Res.* **2018**, *46*, W296–W303. [CrossRef]
47. Williams, C.J.; Headd, J.J.; Moriarty, N.W.; Prisant, M.G.; Videau, L.L.; Deis, L.N.; Verma, V.; Keedy, D.A.; Hintze, B.J.; Chen, V.B.; et al. MolProbity: More and better reference data for improved all-atom structure validation. *Protein Sci.* **2018**, *27*, 293–315. [CrossRef]
48. Pettersen, E.F.; Goddard, T.D.; Huang, C.C.; Couch, G.S.; Greenblatt, D.M.; Meng, E.C.; Ferrin, T.E. UCSF Chimera–a visualization system for exploratory research and analysis. *J. Comput. Chem.* **2004**, *25*, 1605–1612. [CrossRef]
49. de Jong, R.; ten Hoeve, J.; Heisterkamp, N.; Groffen, J. Crkl is complexed with tyrosine-phosphorylated Cbl in Ph-positive leukemia. *J. Biol. Chem.* **1995**, *270*, 21468–21471. [CrossRef]
50. ten Hoeve, J.; Kaartinen, V.; Fioretos, T.; Haataja, L.; Voncken, J.W.; Heisterkamp, N.; Groffen, J. Cellular interactions of CRKL, and SH2-SH3 adaptor protein. *Cancer Res.* **1994**, *54*, 2563–2567.
51. Schüler, H.; Karlsson, R.; Lindberg, U. Purification of nonmuscle actin. In *Cell Biology: A Laboratory Handbook*; Celis, J.E., Ed.; Elsevier Academic Press: Cambridge, MA, USA, 2006; Volume 1, pp. 165–171.
52. Yasumitsu, H.; Ozeki, Y.; Kawsar, S.M.; Toda, T.; Kanaly, R. CGP stain: An inexpensive, odorless, rapid, sensitive, and in principle in vitro methylation-free Coomassie Brilliant Blue stain. *Anal. Biochem.* **2010**, *406*, 86–88. [CrossRef] [PubMed]
53. Pinto, A.F.; Ebrahimi, M.; Saleeb, M.; Forsberg, A.; Elofsson, M.; Schuler, H. Identification of Inhibitors of Pseudomonas aeruginosa Exotoxin-S ADP-Ribosyltransferase Activity. *J. Biomol. Screen.* **2016**, *21*, 590–595. [CrossRef] [PubMed]

pathogens

MDPI

Review

An Update on the Current State of SARS-CoV-2 Mac1 Inhibitors

Joseph J. O'Connor [1], Dana Ferraris [2] and Anthony R. Fehr [1,*]

[1] Department of Molecular Biosciences, University of Kansas, Lawrence, KS 66045, USA; joseph.oconnor@ku.edu
[2] Department of Chemistry, McDaniel College, 2 College Hill, Westminster, MD 21157, USA; dferraris@mcdaniel.edu
* Correspondence: arfehr@ku.edu; Tel.: +1-(785)-864-6626

Abstract: Non-structural protein 3 (nsp3) from all coronaviruses (CoVs) contains a conserved macrodomain, known as Mac1, that has been proposed as a potential therapeutic target for CoVs due to its critical role in viral pathogenesis. Mac1 is an ADP-ribose binding protein and ADP-ribosylhydrolase that promotes replication and blocks IFN responses, though the precise mechanisms it uses to carry out these functions remain unknown. Over the past 3 years following the onset of COVID-19, several groups have used high-throughput screening with multiple assays and chemical modifications to create unique chemical inhibitors of the SARS-CoV-2 Mac1 protein. Here, we summarize the current efforts to identify selective and potent inhibitors of SARS-CoV-2 Mac1.

Keywords: coronavirus; SARS-CoV-2; macrodomain; Mac1; ADP-ribosylation; inhibitors; ADP-ribosylhydrolase

1. Introduction

The advent of COVID-19 highlighted both the importance of rapid responses to viral outbreaks and the necessity of a diverse toolset in implementing those strategies. Though the historic rollout of the COVID-19 vaccines was instrumental to global recovery, there remains a need for ready-made, fast-acting antivirals that can be rapidly deployed when needed. While a vaccine's development timeline may begin in the early stages of a future pandemic, a plethora of drug-based solutions may be developed in the interim. One promising antiviral drug target is the highly conserved coronavirus macrodomain, known as Mac1.

Macrodomains are small globular protein domains that display a mixed α-β-α sandwich fold structure that is highly conserved across all forms of life [1]. Previous studies have established ADP-ribose as their target ligand [2,3]. ADP-ribosylation is the modification of proteins and nucleic acids via the addition of ADP-ribose subunits by ADP-ribosyltransferases (ARTs) [4]. This occurs with either single subunits or polymers of ADP-ribose, known as mono-ADP-ribosylation (MARylation) or poly-ADP-ribosylation (PARylation). ADP-ribosylation regulates intracellular processes, including DNA damage repair, protein degradation, stress responses, and immune signaling [4]. Furthermore, macrodomains are found in several families of positive-sense RNA viruses and one family of DNA viruses and promote their replication and pathogenesis [5].

SARS-CoV-2 Mac1 is contained within non-structural protein 3 (nsp3) and belongs to the MacroD-type class, characterized via hydrolysis activity against MARylation [6] and also includes human macrodomains MacroD1, MacroD2, PARG, and the PARP9 and PARP14 N-terminal macrodomains. It shares three major regions of homology with related coronaviruses, namely, NAAN and GGG motifs in loop 1 between β3 and α2, a GIF motif in loop 2 between β6 and α5, and a VGP motif between β5 and α4 [7]. Previous biochemical characterizations of ligand binding within the cleft have implicated several stabilizing interactions with ADP-ribose. A conserved interaction between the D22 residue

Citation: O'Connor, J.J.; Ferraris, D.; Fehr, A.R. An Update on the Current State of SARS-CoV-2 Mac1 Inhibitors. *Pathogens* **2023**, *12*, 1221. https://doi.org/10.3390/pathogens12101221

Academic Editor: Theresa Chang

Received: 7 September 2023
Revised: 29 September 2023
Accepted: 4 October 2023
Published: 7 October 2023

and the N6 of the adenine is critical for substrate binding, while hydrogen bonds between N40 and distal ribose are conserved in catalytically active macrodomains [7,8]. Comparisons between the SARS-CoV-2 Mac1 and human MacroD-type macrodomains reveal a sequence similarity at or below 38%, suggesting a potential for the specific inhibition of viral macrodomains [7], though it should be noted that human macrodomains also encode the previously highlighted structural motifs.

Mac1 is a critical factor for the replication and/or pathogenesis of all CoVs where it has been tested, including MHV, MERS-CoV, and both SARS-CoV and SARS-CoV-2 [9–11]. Point mutations in the Mac1-binding pocket sensitize CoVs to ART activity and IFN signaling [10,12]. More recently, we demonstrated that a complete deletion of Mac1 in MHV and MERS-CoV was not recoverable. However, a Mac1 deletion in SARS-CoV-2 was easily recovered, indicating that it is not essential for replication. However, the Mac1 deletion virus was highly sensitive to IFN-γ signaling and replicated poorly and caused no disease in mice [11]. Recombinant alphaviruses and HEV with macrodomain mutations also result in severe defects in virus replication, indicating that macrodomains clearly play important roles during infection. However, the precise function of viral macrodomains in the viral lifecycle remains unclear, though mutational analysis indicates that it targets multiple stages of the lifecycle [13–15]. Together, these studies comprise a framework that has established viral macrodomains as druggable targets. Mac1 inhibitors could be used as probes to better understand the function of Mac1 during infection or could be used as antiviral therapeutics. Here, we will comprehensively review the collective efforts among multiple groups to identify and develop SARS-CoV-2 Mac1 inhibitors with the aim of highlighting emerging techniques and chemotypes.

2. Virtual and Crystallography Screens Identify Several Mac1 Interacting Compounds

Several groups initially focused on determining the structure and biochemical activities of Mac1 and related macrodomains [1,6,8,16]. Dozens of different Mac1 structures in complex with a variety of ligands are available in the PDB. These structures have included not only the main substrate of Mac1 and ADP-ribose (Figure 1A,B), but also several nucleotides and compounds, including HEPES, MES, GMP, cAMP, and ADP-ribose-2″-phosphate [17,18]. These structural analyses have revealed three pockets near the active site of the enzyme. The two pockets that comprise the adenosine binding site have the highest solvent exposure and together are the preeminent site of interest for ligand design. One especially interesting ligand that has been crystallized with Mac1 is GS441524 (**1**), the active metabolite of remdesivir, an approved anti-CoV drug that is well known for its ability to inhibit CoV polymerase [18]. Compound **1** ($K_D = 10.8 \pm 1.8$ μM) and the phosphonated version, **2**, demonstrated low micromolar binding affinity against Mac1 ($K_D = 8.6 \pm 2.5$ μM) (Figure 1C). Tsika et al. found that, when used in 500-fold excess, **1** can inhibit Mac1 ADP-ribosylhydrolase activity, and this activity was specific for the SARS-CoV-2 Mac1, as other viral or human macrodomains were not similarly affected [19]. However, isothermal titration calorimetry (ITC) data demonstrated that the compound bound to SARS- and MERS-Mac1 proteins, but with reduced affinity. These results raise the possibility that remdesivir's Mac1 inhibitory activity may play a small role in its antiviral activity against SARS-CoV-2. However, this hypothesis has not been formally tested. Also, due to its inhibitory activity against the CoV polymerase, remdesivir would not make for a useful Mac1-specific inhibitor, though it may indicate a potential role as a dual polymerase/Mac1 inhibitor.

Concurrent with these structural efforts, several groups have performed virtual screens and biochemical binding assays to identify potential Mac1 inhibitors. Babar et al., Pandey et al., Singh et al., Selvaraj et al., and Russo et al. each used virtual screening to identify several compounds that had docking scores as good or better than ADP-ribose [20–23]. However, none of these compounds have yet to be confirmed to interact or inhibit Mac1 in biochemical assays. Russo et al. tested 69 different virtual screen hits in a cell-based de-MARylation assay, but none of them demonstrated any inhibition of enzyme activity.

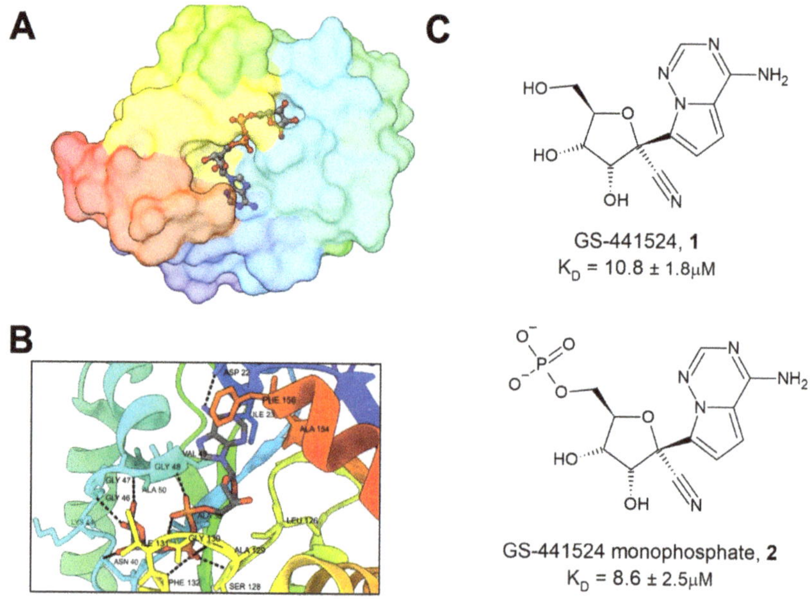

Figure 1. The structure of SARS-CoV-2 Mac1. (**A**) SARS-CoV-2 Mac1 complexed with ADP-ribose within the active site (PDB 6WOJ). (**B**) Coordination of ADP-ribose with the binding pocket of Mac1 is reliant on a network of coordinated hydrogen bonds, critically with residues D22, N40, and a GIF motif at 130–132. (**C**) GS-441524, the active metabolite of remdesivir, with its phosphonated version.

Knowing that poly(ADP-ribose) glycohydrolases (PARGs) are structurally related to Mac1, and that there are several PARG inhibitors, Brosey et al. utilized virtual screening and scaffold optimization to identify PARG-inhibitor-related morpholine-based structures that might also bind to Mac1 [24]. From this work, the group discovered two compounds, PARG-345 (**3**) and PARG-329 (**4**), that interacted with Mac1 via crystallography. Both compounds anchor the methylxanthine head in the adenine binding site. The morpholine group on PARG-345 interacts with N40 while the sulfonyl linker makes backbone contacts with the G130 and F132 residues of Loop 2 (Figure 2A). Conversely, the thio-urea of PARG-329 interacts with Mac1 largely through water-mediated contacts. The morpholine group extends further into the active site but loses the hydrogen bonding with N40 (Figure 2B). While these compounds interact with Mac1, demonstrating K_D values of between 16.6 and 32 μM, no further reports have been published demonstrating any inhibitory activity of these compounds.

To streamline fragment screens, Bajusz et al. developed a new strategy of assembling fragment libraries that utilize confirmed binding pharmacophores [25]. Their pilot library, termed SpotXplorer0, utilized ~3000 experimental protein–fragment structures from the PDB which were then used to generate pharmacophore sets via the FTMap protein-mapping algorithm. The validated workflow, when applied to the SARS-CoV-2 Mac1, identified five small fragments that crystallized with Mac1: SX003, SX005, SX048, SX051, and SX054 (**5–9**) (Figure 3). Compounds **5**, **6**, and **9** bind in the adenine binding site of Mac1 and all target the aromatic sidechain of F156, utilizing π-π stacking interactions, while **7** and **8** occupy the proximal ribose site. None of these molecules were tested for Mac1 inhibition, but they could offer viable starting points for further compound iteration.

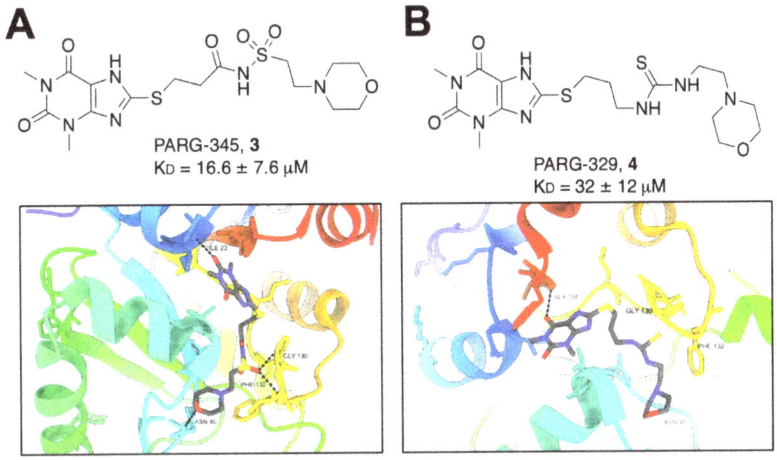

Figure 2. Morpholine-based compounds PARG-345 and PARG-329 bind the Mac1 active site. (**A**) Crystallography shows that PARG-345 (PBD 7LG7) makes several hydrogen bonds within the Mac1-binding pocket, with the methylxanthine group binding at the adenine binding site. (**B**) PARG-329 (PDB 7KXB) binds in the same orientation; however, it interacts with Mac1 primarily through water-mediated contacts (waters not shown for clarity) and takes on a more strained conformation [24].

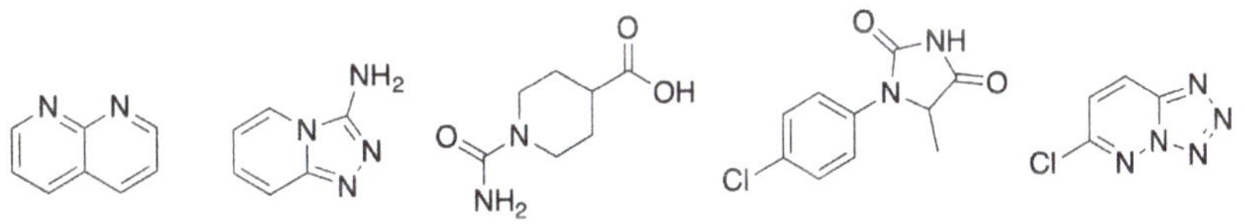

SX003, **5** SX005, **6** SX048, **7** SX051, **8** SX054, **9**

Figure 3. Small binding fragments were identified using SpotXplorer0-generated pharmacophore sets that were mapped to Mac1 via the FTMap algorithm. All interact with the F156 side chain through π-π stacking [25].

Schuller et al. took crystallographic screening to the next level by performing a crystallography screen of >2500 diverse fragments, which identified 214 Mac1-binding fragments [26]. In addition, they also used computational docking of 20 million fragments, from which they tested 60 fragments and confirmed 20, for a total of 234 confirmed Mac1 binders. Of these hits, about 80% occupied the active site of the enzyme, while the remaining hits were scattered across the surface of the protein and in a conserved pocket near K90. The most common and effective structure amongst these fragments was a 7H-pyrrolo[2,3-d]pyrimidine, a component of compounds **10** and **11** (Figure 4), which use hydrogen bonds with D22, I23, and π interactions with F156 to anchor fragments in the adenine-binding pocket. Many compounds also included carboxylic acid (e.g., **10**), which formed hydrogen bonds with the backbone amides of F156 and D157 in the "oxyanion" subsite (Figure 4). Combined with Bajusz et al.'s screen, a notable hallmark of these fragment screens was the identification of fused heterocyclic rings that anchor the fragments to D22, I23, and/or F156. In addition to crystallography, several fragments also showed a thermal shift with Mac1 in a differential scanning fluorimetry (DSF) assay, had measurable binding via ITC, and showed the inhibition of Mac1 activity in a peptide displacement assay using a homogenous time-resolved fluorescence assay (HTRF) similar to the previously developed AlphaScreen assay (**10–13**) [27]. The IC_{50} values for inhibition ranged from 180 μM (**10**) to more than 2 mM (**13**), which are reasonable values for fragments. These fragments would prove to serve as excellent starting points for further inhibitor derivatization (see below).

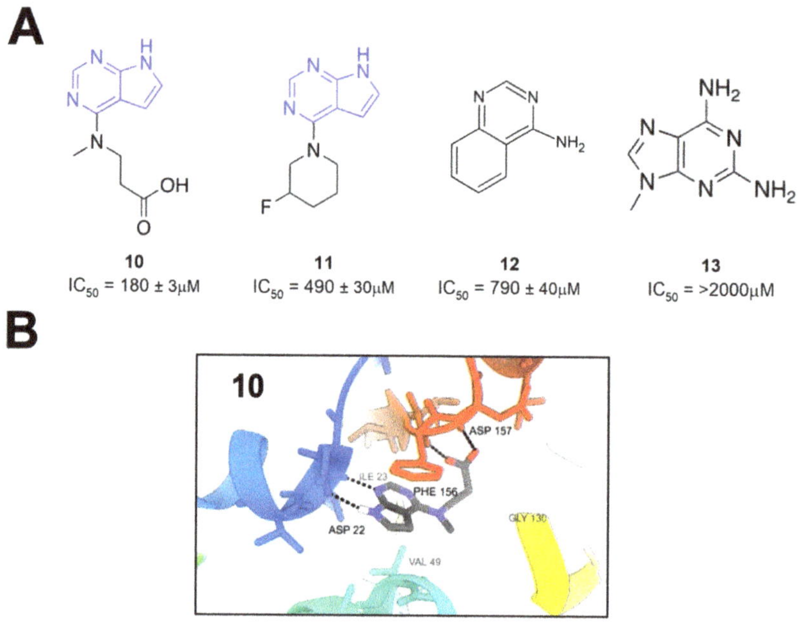

Figure 4. A large crystallography screen identifies several small-molecule Mac1 binders. (**A**) Fragments identified from crystallography screen for Mac1 binders. (**B**) A 7H-pyrrolo[2,3-d]pyrimidine moiety anchors fragment **10** within the adenine binding site (PDB 5RSG) [26].

The same group later performed another molecular docking screen followed by crystallography screens of the top hits [28]. After docking approximately 400 million compounds, 124 were selected for further testing via crystallography and HTRF. Of these 124 compounds, many either bound to Mac1 via crystallography (47 compounds) or demonstrated the inhibition of binding in the HTRF assay with IC_{50} values ranging from 42 to 504 μM (**14–17**) (Figure 5A). These hits consistently targeted the adenosine binding pocket, with those compounds that showed activity in the HTRF assay typically consisting of a pyrrolopyrimidine or pyrrolopyridine head group (in blue, Figure 5A) that bound in the adenine binding pocket and stack with F156 and used acidic moieties to interact with the oxyanion subsite, similar to the results obtained from their fragment screen. However, two compounds, Z7873 (**16**) and Z6923 (**17**), instead utilized a tricyclic pyrimidoindole scaffold (in red, Figure 5A) that is closely related to the pyrrolopyrimidine head group. This series was the starting point for optimization towards low-micromolar Mac1 inhibitors by this same group, as outlined below. The top compound from these screening efforts was Z3122 (**18**), with an IC_{50} of 2.5 μM (Figure 5B). This molecule was identified by screening for molecules that docked in the "open" state of the phosphate-binding region, though via crystallography it was shown to bind in the "closed" state. This molecule uses the commonly identified pyrrolopyrimidine to interact with the adenine binding site, a carboxylic acid to make contact in the oxyanion subsite, but notably has a 4-bromobenzyl ring that extends further into the phosphate-binding region than other compounds (Figure 5C).

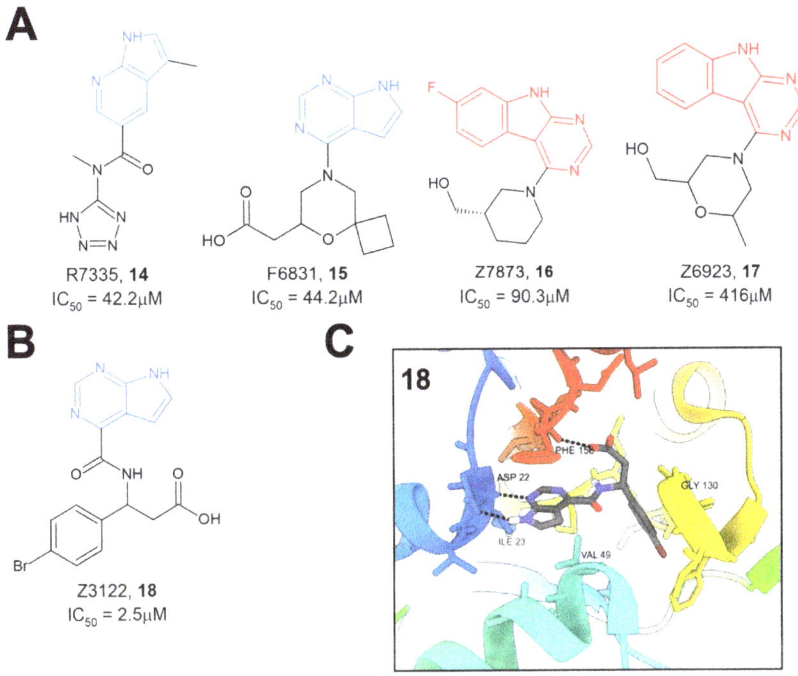

Figure 5. Additional crystallography screening hit compounds. (**A**) Binding to Mac1 by compounds **14–17** is facilitated by either pyrrolopyrimidine (blue) or tricyclic pyrimidoindole (red) head group. (**B**,**C**) The structure of the top compound from this crystallography screen, **18** (**B**), and its binding to Mac1 (PDB 5SS9) (**C**) [28].

3. High-Throughput Screening Using Both Binding and Enzyme Assays Have Identified Several Mac1 Inhibitors

Most assays that have been used to identify Mac1 inhibitors are based on inhibiting Mac1 binding to ADP-ribosylated peptides. However, a significant aspect of its function is its ability to remove ADP-ribose from proteins through hydrolysis. Dasovich et al. developed a novel technique for the evaluation of ADP-ribosylhydrolase activity, the luminescence-based ADPR-Glo assay, to serve as a high-throughput alternative to gel-based measures of Mac1 enzyme activity and inhibition [29]. This approach utilizes co-incubation of an ADP-ribosylated substrate, Mac1, and the NudF phosphodiesterase. Hydrolyzed ADP-ribose is cleaved by NudF into AMP, the levels of which can be quantified as luminescence with the AMP-Glo kit. Dasovich et al. applied this technique to over 3000 compounds from the Selleck-FDA and LOPAC libraries and identified dasatinib (**19**) as a compound with the specific, dose-dependent inhibition of Mac1 [29]. This compound had an IC_{50} between 37.5 and 57.5 μM (Figure 6A). Surface plasmon resonance analysis confirmed specificity for SARS-CoV-2 Mac1 over the human MacroD2, and molecular docking revealed that only 15 out of 25 contacting residues were conserved between the two enzymes, potentially explaining the selectivity. Dasatinib extended into multiple pockets within the active site of Mac1 (Figure 6B), but biochemical data indicate that it will require further optimization to be a potent Mac1 inhibitor.

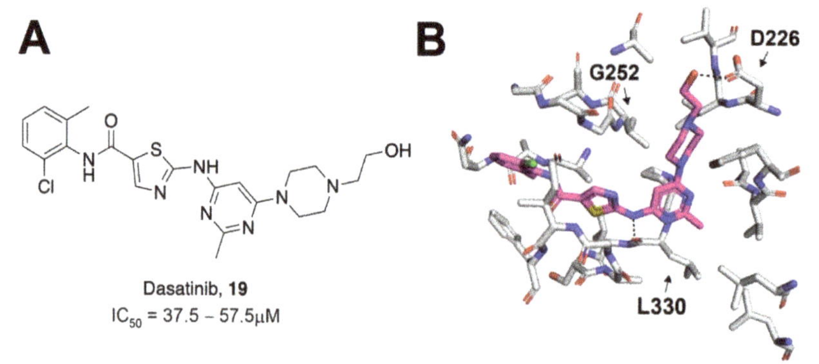

Figure 6. Dasatinib was identified as a Mac1 inhibitor using the ADPr-Glo assay. (**A**) Chemical structure of dasatinib and its measured IC_{50} value. (**B**) Molecular modeling of dasatinib into the Mac1-binding pocket [29].

A novel high-throughput ADP-ribose binding assay developed by Sowa et al. is based on the known cysteine ADP-ribosylation of the α_i subunits of heterotrimeric G proteins ($G\alpha_i$) via pertussis toxin (PtxS1) [30]. Because the ADP-ribose is attached to a cysteine, it is unlikely to be hydrolyzed by macrodomains, which are only known to hydrolyze ADP-ribose from proteins at acidic residues. In this assay, a YFP-fused $G\alpha_i$ subunit is co-incubated with a CFP-fused ADP-ribose binding protein of interest. Close proximity between the YFP and CFP fluorophores produces a fluorescence energy transfer (FRET) signal that correlates with the substrate binding affinity. Applying this technique, Wazir et al. identified three derivatives of a 2-amine-3-methylester thiophene scaffold with high affinity and specificity for SARS-CoV-2 Mac1 [31]. The initial top hit, **20**, contains a seven-membered aliphatic ring fused with a heteroaromatic thiophene ring and a cyclohexenyl group fused to a central amine linkage (Figure 7). This compound had an IC_{50} of 14 μM for SARS-CoV-2 Mac1 and showed greater levels of inhibition of Mac1 at 100 μM (~80% inhibition at 100 μM) than alternative binding partners PARP9 MD1 (~70% inhibition at 100 μM) or ALC1 (~60% inhibition at 100 μM). The crystal structures of this compound complexed with SARS-CoV-2 Mac1 show that the fused aliphatic and thiophene ring structure forms hydrophobic interactions with F156, while the carbonyl oxygen of the methyl ester group forms hydrogen bonds with the I23 backbone. Iterating on the basic structure of **20** led to the development of derivative **21**, which substitutes the cyclohexenyl unit from **20** with a saturated cyclohexanyl unit in an *(R,R)-trans* configuration. Further iteration on this structure produced **22**, which substitutes an eight- rather than a seven-membered aliphatic ring fused to the thiophene ring. Both **21** and **22** show improvements in IC_{50} (2.7 μM and 2.1 μM, respectively). Crystal structures of **22** complexed with SARS-CoV-2 Mac1 reveal similar interactions as in **20**, except for a substituted hydrogen bonding interaction between the carboxylate and I32 [31]. Notably, hydrophobic interactions with F156 are enhanced due to the larger eight-membered aliphatic ring and a slight change in F156 orientation. This compound demonstrated improvements in selectivity against ALC1 and also inhibited macrodomains from SARS-CoV (64% inhibition at 100 μM) and MERS-CoV (43% inhibition at 100 μM), suggesting the suitability of this scaffold as a selective inhibitor of viral macrodomains.

Roy et al. utilized a previously published AlphaScreen assay and, as an orthogonal assay, a novel fluorescence polarization (FP) assay to screen ~38,000 compounds from the Analyticon, 3D Biodiversity, and Peptidomimetics libraries [32]. Following HTS, the selected compounds were further tested for Mac1 binding affinity using thermal shift assays and for their inhibition of hydrolysis activity using the ADP-Glo assay developed by Dasovich et al. [29]. Of those screened, the highest-performing compounds included compounds FS2MD-1 (**23**) and FS2MD-6 (**24**) with IC_{50} values of 6.1 and 8.5 μM, respectively

(Figure 8). Compound **23** contained thienopyrimidine and resembled previously identified compounds from Ekblad and Schuller [26,27]. Compound **24**, which is comprised of a β-alanine core, an N-chlorobenzyl group, a methoxy benzoyl group, and a piperazine amide, was the highest-performing compound in the study. In addition to its low IC_{50}, the compound demonstrated highest ΔT_m values at 1.67 $\pm$ 0.21 °C and showed substantial enzyme inhibition in both the ADP-Glo and gel-based assays.

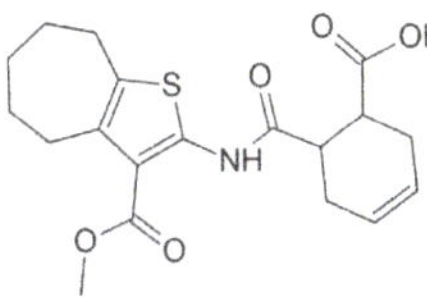

20
IC_{50} = 14μM

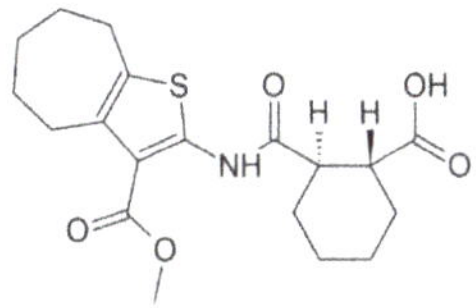

21
IC_{50} = 2.7μM

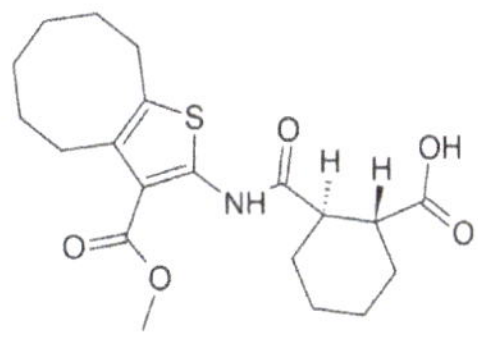

22
IC_{50} = 2.1μM

Figure 7. Mac1 binding compounds featuring fused aliphatic and thiophene ring structures [31].

A

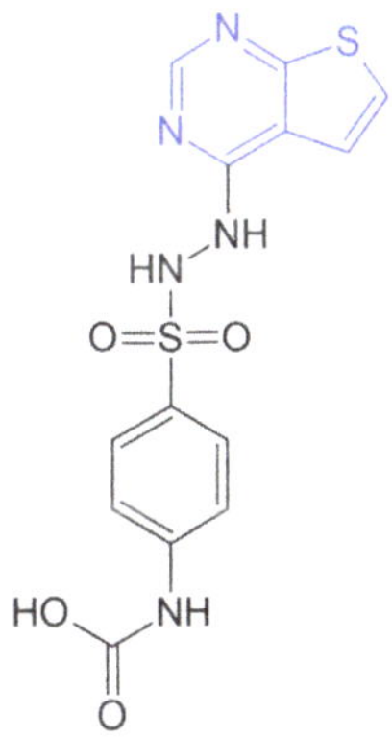

FS2MD-1, **23**
IC_{50} = 6.1μM

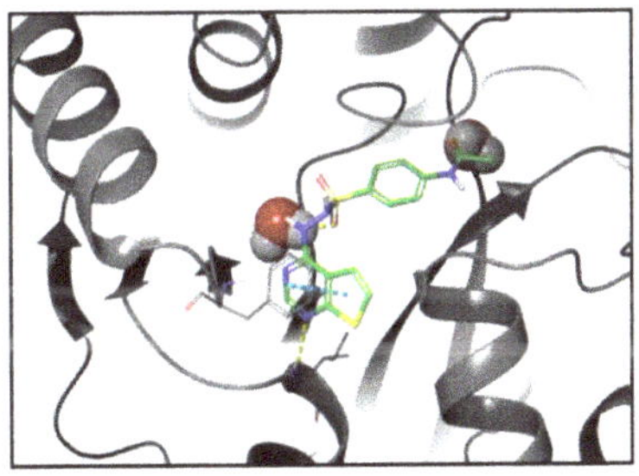

B

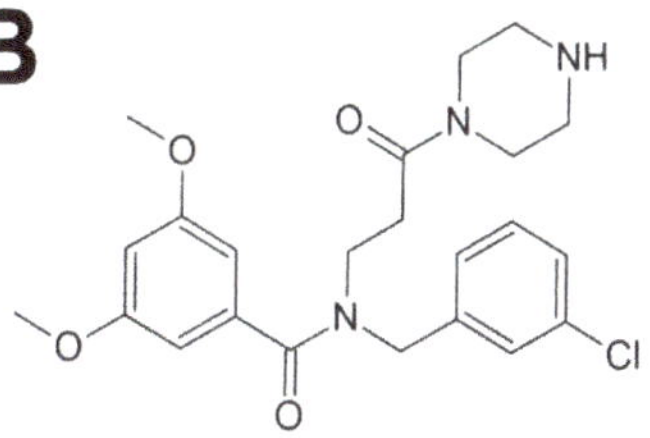

FS2MD-6, **24**
IC_{50} = 8.5μM

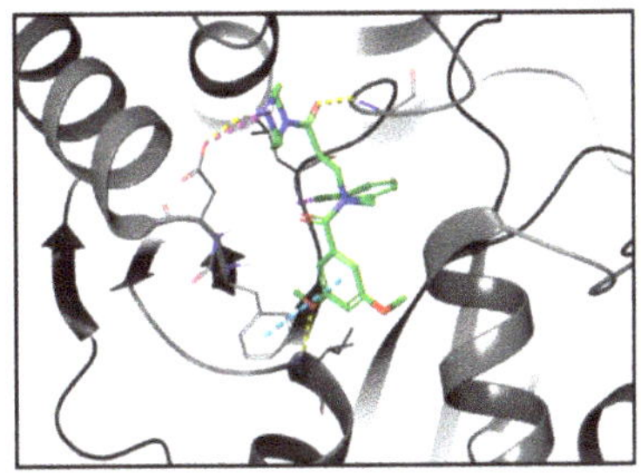

Figure 8. Top hit compounds were identified via high-throughput screening using an AlphaScreen assay. (**A**) Compound **23** contains a thienopyrimidine head group (blue) resembling the pyrrolopyrimidine head groups in compounds identified by Schuller et al. (**B**) Compound **24** contains a β-alanine core, a chlorobenzyl group, a methoxy benzoyl group, and a piperazine amide [32].

Schuller et al. utilized both virtual and high-throughput screening using the HTRF assay described previously to identify novel Mac1 inhibitors [33]. The virtual screening of 125,000 compounds from the BioAscent library in addition to HTRF screening of 1786 compounds identified two molecules, IAL-MD0305 (**25**) and IAL-MD0306 (**26**), with IC_{50}'s of 28 and 18 µM, respectively (Figure 9A). Next, the screening of 10,100 small molecules from the Manchester Institute Diversity Set (MIDAS), as well as an array of FDA-approved compounds, identified four primary scaffold types, three of which contained a pyrrolopyridine core (Figure 9B). Preliminary SAR was able to be gathered due to the availability of several derivatives from each of these scaffold types. Scaffold type I, characterized by a furanyl-pyrrolo[2,3-*b*]pyridine structure, yielded six compounds with IC_{50} values below 9 µM and one at 25.3 µM. The top compound from this set, IAL-MD0131, **27**, contains a morpholine amide and had an IC_{50} of 4.9 µM. Co-crystallization of **27** with SARS-CoV-2 Mac1 showed ligand binding at the adenosine binding site of the ADP-ribose binding pocket (Figure 9C). This was mediated via interactions between the scaffold backbone and the residues I23 and F156, resembling those established by ADP-ribose. Scaffold type II was characterized by a pyridinyl-pyrrolo[2,3-*b*]pyridine, with the top-performing compound from this scaffold being IAL-MD0128, **28**, with an IC_{50} of 3.1 µM. Scaffold III was characterized by a thiophenyl-pyrazolo[3,4-b]pyridine core with the most potent compound identified also containing a morpholino amide IAL-MD0051, **29**, with an IC_{50} of 13 µM. Scaffold type IV was defined by phenylquinoline-4-carboxylic acid, and the most potent derivative identified in this series was IAL-MD0031, **30**, with an IC_{50} of 19 µM.

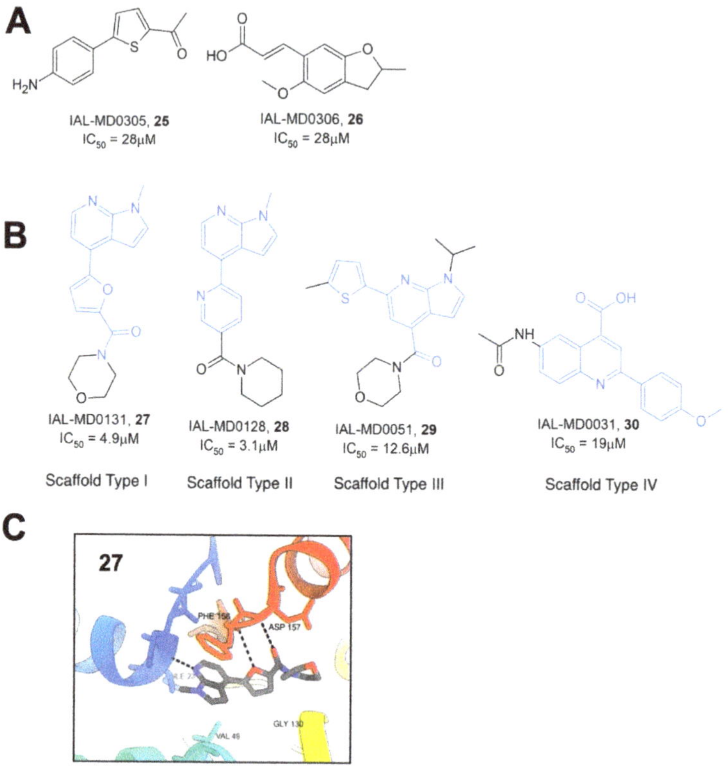

Figure 9. Mac1 inhibitors identified through HTS with the HTRF assay. (**A**) Hits from the BioAscent library. (**B**) Example small-molecule hits generated from the Manchester Institute Diversity Set with primary scaffold types highlighted in blue. (**C**) Crystal structure of **27** within the Mac1-binding pocket (PDB 8C19) [33].

Additionally, an in vitro screen of FDA-approved compounds revealed several antibiotics with inhibitory activity against SARS-CoV-2 Mac1. Aztreonam, **31**, a monocyclic beta-lactam, demonstrated an IC_{50} of 29.3 μM and was similarly co-crystallized with SARS-CoV-2 nsp3 (Figure 10). These structures revealed that, rather than binding within the ADP-ribose binding pocket, the compound occupies an adjacent groove coordinated by hydrogen bonding of the carboxylic acid group to the oxyanion subsite of Mac1 by coordinating over a water molecule to the backbone amides of A154 and P125. This observation, combined with the frequency of carboxylic acid groups in a number of already identified Mac1 inhibitors, caused researchers to speculate that this functional group may play similar roles in other compounds. Interestingly, both **27** from the MIDAS library and **31** show π-π stacking with F156 similar to ADP-ribose and many other compounds identified by multiple groups.

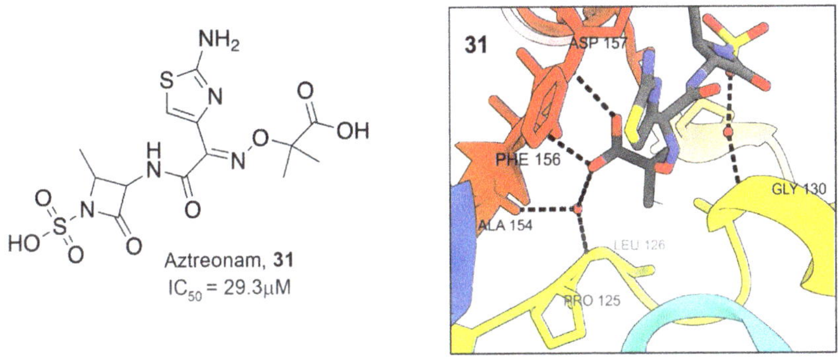

Figure 10. Beta-lactam antibiotic Aztreonam binds Mac1 [33].

4. Early Efforts in the Optimization of Fragment Hits to Potent Mac1 Inhibitors In Vitro

Sherrill et al. synthesized and assayed primary and secondary amino acid derivatives that served as iterations on the high-performing 7H-pyrrolo[2,3-d]pyrimidine scaffolds previously identified by Schuller et al. [26,34]. From this optimization, the three highest-performing amino acid derivatives—two secondary and one primary—were demonstrated to have IC_{50} values less than 24 μM (Figure 11), a nearly 10-fold improvement over the original fragments. The two secondary amino acid derivatives, **32** and **33**, incorporated piperidine rings with a carboxylic acid attached at the three- or four-position. As expected, molecular modeling revealed a common feature of hydrogen bonding between the pyrrolopyrimidine core and the D22 and I23 residues of Mac1, while the carboxylic acid moiety from the six-membered piperidine makes hydrogen bonds with the oxyanion subsite. Several primary amino acid derivates also demonstrated significant Mac1 inhibitory activity, including valinate and β-amino acid derivates. Of the primary amino acid derivates, the addition of tryptophanate was the most potent, resulting in compound **34**, which had an IC_{50} of 6.1 μM. This result was confirmed in a FRET-based assay, which also showed that **33** had a high level of specificity for coronavirus macrodomains. Finally, the differential scanning fluorimetry of **34** resulted in thermal shift values of >4 °C, which was within 0.2 °C of ADP-ribose. Molecular modeling indicates that the tryptophan moiety extends deep into the phosphate-binding pocket, likely accounting for the increased potency (Figure 11). Nearly all ester derivatives of each tested compound were less potent inhibitors of Mac1-ADP-ribose binding when compared to their carboxylic acid counterparts, suggesting that the carboxylic acid group is important to efficient Mac1 binding. However, several methyl- and ethyl-esters had only mildly reduced activity, indicating the feasibility of creating neutral Mac1 inhibitors using this backbone.

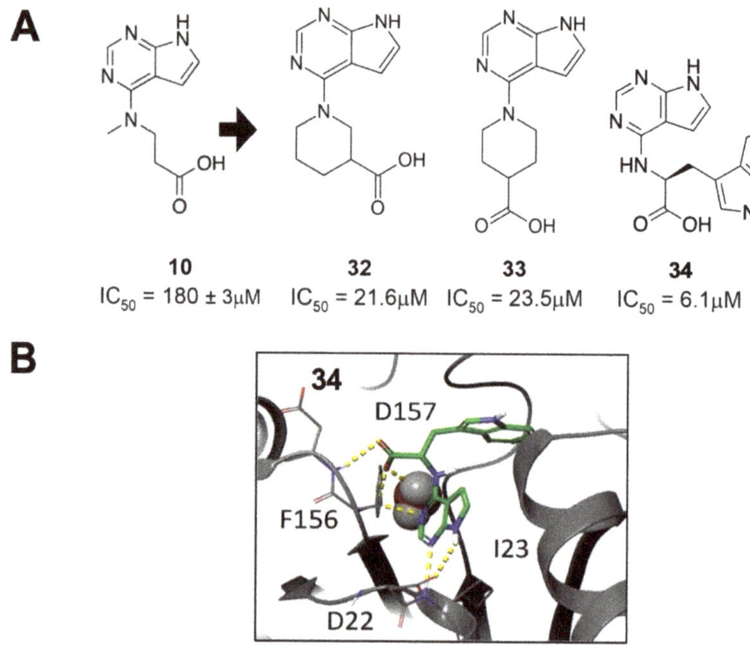

Figure 11. Primary and secondary amino acid derivatives of **10** showed improved Mac1 binding. (**A**) Compounds **32–34** improve the IC_{50} of **10** 20- to 30-fold. (**B**) Molecular model of **34** with Mac1 demonstrates that the tryptophan extends deep into the phosphate-binding pocket [34].

Gahbauer et al. utilized fragment merging to create the most potent Mac1 inhibitors yet to date. Using an automated fragment-linking approach, termed Fragmenstein, they virtually merged fragment hits and modeled them into the Mac1-binding pocket [28]. These merged molecules were then used as templates to search for available chemical matter in the Enamine REAL database. They ultimately tested 13 molecules for interactions with Mac1 via crystallography and thermal shift analysis. Eight of these hits bound to Mac1 via crystallography, while seven of them induced a thermal shift. Additional analogs of the merger between ZINC922 (**35**) and ZINC337835 (**36**) were purchased, allowing for some initial SAR (Figure 12A). The most potent molecules from these mergers were Z8515 (**38**) and Z8539 (**39**), with IC_{50} values of 10.3 and 1.1 μM, respectively. These compounds share a phenylurea group that stacks with F156 but differ in that Z8539 has a cyclopropyl group that extends further into the adenine binding pocket. Furthermore, upon chiral separation, it was found that the (*R*,*R*) steroisomer, Z8601 (**40**), had an even better IC_{50} value of 0.5 μM. The improved efficacy was explained through crystallography, showing that the indane group of the (*R*,*R*) stereoisomer flipped into the phosphate-binding domain and hydrogen bonds with the backbone oxygen of L126 (Figure 11B). Importantly, these highly potent inhibitors had minimal to no effect against human macrodomains. Next, further modifications to these fragments were created at the cyclopropyl moiety, the phenylurea, and the acid-carrying indane group. From here, several compounds were created with IC_{50} values ranging from 0.4 to 83.8 μM, including at least five with sub-micromolar IC_{50}'s. Z8539_0023 (**41**), where the cyclopropyl group was replaced by a phenyl group ($IC_{50.}$ = 0.5 μM, thermal shift of 9 °C), and Z8539_0072 (**42**), where an alcohol group was added to the central benzene (IC_{50} = 0.4 μM) (Figure 12B), along with **40**, which are the most potent SARS-CoV-2 Mac1 inhibitors derived to date.

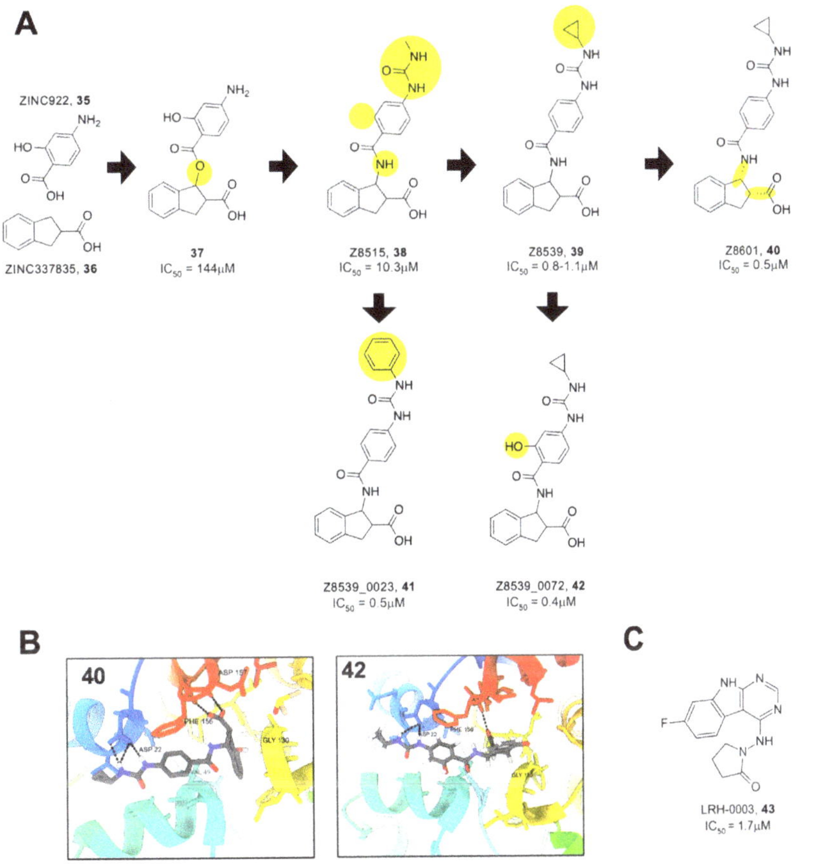

Figure 12. Compounds identified via computational fragment-linking. (**A**) Progressive modifications of **37** led to compounds with sub-μM IC_{50} values (modifications highlighted in yellow). (**B**) Structures of highly potent compounds **40** and **42** (PDB 5SQJ, 5SQW). (**C**) Fragment-linking identified neutral compound with potent inhibitor activity [28].

As mentioned above, many of the most potent inhibitors use carboxylic acids to interact with the oxyanion subsite. To address this issue, Gahbauer et al. identified analogs linking the pyrrolopyrimidine or pyrimidoindole to small functional moieties containing hydrogen bond donors or acceptors (e.g., sulfones, hydroxyls, pyridines, or ketones) [28]. Out of 124 possible compounds, 20 were ultimately tested, and 14 were found to bind to the SARS-CoV-2 Mac1 protein via crystallography and 4 via HTRF. The most promising molecule, LRH-003 (**43**), had an IC_{50} of 1.7 μM, not far from the most potent acidic compounds (Figure 12C).

5. Discussion/Conclusions

The advancements in applied modeling and biochemical techniques have greatly expanded the repertoire of candidate SARS-CoV-2 Mac1 inhibitors within the last few years. There remain, however, significant challenges which must be addressed before this objective can be fully realized. The presence of acidic side chains in several of the identified compounds has raised potential concerns regarding membrane permeability. However, the potency of several neutral modifications indicates that developing potent cell permeable inhibitors is highly feasible [28,34].

Another major challenge will be to move inhibitor testing from in vitro assays to demonstrating efficacy in cell-based activity and virus replication assays. Russo et al. recently developed an immunofluorescent-based assay for testing inhibitor efficacy in cell culture [23]. This assay is based on the observation that MARylation is drastically increased in cells following polyI:C or IFN-γ treatment. This assay, and potentially others, should allow for cell-based testing, though the dynamic range of these assays may not be very large. Alhammad et al. recently demonstrated that a recombinant SARS-CoV-2 bearing a full Mac1 deletion had only minor defects in virus replication, unless cells were pretreated with IFN-γ, though the virus was extremely attenuated in mice [11]. These results indicate that testing inhibitors for SARS-CoV-2 inhibition in cell culture will be difficult. However, MHV and MERS-CoV Mac1-deleted viruses were unrecoverable, indicating that Mac1 is critical for their ability to replicate. Thus MHV/MERS-CoV may be excellent viruses to test Mac1 inhibitors. In fact, Wazir et al. recently showed that **22** was not toxic to cells and inhibited MHV replication in a dose-dependent manner at mid–high micromolar concentrations, as might be expected considering it is unknown what the IC_{50} of **22** would be for MHV Mac1. Identifying Mac1 targets and biomarkers during infection could substantially aid in demonstrating inhibitor efficacy, making the pivot from in vitro to in situ testing a critical next step for identified compounds.

However, while the SARS-CoV-2 Mac1 has a highly conserved structure and function, variations between coronavirus macrodomains demonstrate the difficulty of applying identified compounds as broad-spectrum antivirals. Commonly identified features such as π-π stacking between F156 and heterocycles, such as pyrrolopyrimidine groups or hydrogen bonding to D22, feature in some of the strongest candidates. However, both SARS- and MERS-CoV encode asparagine in place of F156 [7]. The reliance of many compounds on interacting with the F156 residue creates the possibility for a resistant mutant to arise, though it is currently unclear how such a mutation would affect overall SARS-CoV-2 fitness. This illustrates the inherent difficulty in designing drugs that can inhibit multiple viral macrodomains, as many within the same family contain slight variations in their overall binding scheme that undercut the potency of the established inhibitors. In cases where SARS-CoV-2 Mac1 inhibitors were tested against related macrodomains from SARS- and MERS-CoV, they demonstrated some activity, but it was generally much less than their activity against the SARS-CoV-2 Mac1 [19,31,32,34]. However, as inhibitor development becomes more advanced, there should be opportunities to develop more broad-spectrum inhibitors, or at least SAR should indicate minor changes that could quickly change the specificity of a particular compound from one macrodomain to another.

Despite these challenges, the future of macrodomain inhibitor development is bright, as more compounds with low-to-sub-micromolar activity continue to be identified. With several groups currently working to develop Mac1 inhibitors, it is likely that there will soon be compounds with mid- to even low-nanomolar affinity activity and have appropriate ADME and PK properties for cell and animal testing. With any luck, these efforts will soon fully unlock the potential of utilizing Mac1 inhibitors as research tools and even antiviral therapeutics.

Author Contributions: Conceptualization, J.J.O., D.F. and A.R.F.; methodology, J.J.O., D.F. and A.R.F.; validation, J.J.O., D.F. and A.R.F.; formal analysis, J.J.O., D.F. and A.R.F.; investigation, J.J.O., D.F. and A.R.F.; resources, J.J.O., D.F. and A.R.F.; data curation, J.J.O., D.F. and A.R.F.; writing—original draft preparation, J.J.O., D.F. and A.R.F.; writing—review and editing, J.J.O., D.F. and A.R.F.; visualization, J.J.O., D.F. and A.R.F.; supervision, A.R.F.; project administration, A.R.F.; funding acquisition, A.R.F. All authors have read and agreed to the published version of the manuscript.

Funding: This research was funded by NIH grants R35GM138029 (A.R.F.) and P20GM113117 (A.R.F.).

Acknowledgments: Structures were created using ChimeraX Modeling Software version 1.6.1. Compound structures were created with ChemSketch and ChemDraw. We thank Lari Lehtio for critical reading of this manuscript.

Conflicts of Interest: The authors declare no conflict of interest.

References

1. Rack, J.G.M.; Perina, D.; Ahel, I. Macrodomains: Structure, Function, Evolution, and Catalytic Activities. *Annu. Rev. Biochem.* **2016**, *85*, 431–454. [CrossRef] [PubMed]
2. Allen, M.D.; Buckle, A.M.; Cordell, S.C.; Löwe, J.; Bycroft, M. The Crystal Structure of AF1521 a Protein from Archaeoglobus Fulgidus with Homology to the Non-Histone Domain of MacroH2A. *J. Mol. Biol.* **2003**, *330*, 503–511. [CrossRef] [PubMed]
3. Karras, G.I.; Kustatscher, G.; Buhecha, H.R.; Allen, M.D.; Pugieux, C.; Sait, F.; Bycroft, M.; Ladurner, A.G. The Macro Domain Is an ADP-Ribose Binding Module. *EMBO J.* **2005**, *24*, 1911–1920. [CrossRef] [PubMed]
4. Lüscher, B.; Ahel, I.; Altmeyer, M.; Ashworth, A.; Bai, P.; Chang, P.; Cohen, M.; Corda, D.; Dantzer, F.; Daugherty, M.D.; et al. ADP-Ribosyltransferases, an Update on Function and Nomenclature. *FEBS J.* **2021**, *289*, 7399–7410. [CrossRef]
5. Leung, A.K.L.; McPherson, R.L.; Griffin, D.E. Macrodomain ADP-Ribosylhydrolase and the Pathogenesis of Infectious Diseases. *PLoS Pathog.* **2018**, *14*, e1006864. [CrossRef]
6. Jankevicius, G.; Hassler, M.; Golia, B.; Rybin, V.; Zacharias, M.; Timinszky, G.; Ladurner, A.G. A Family of Macrodomain Proteins Reverses Cellular Mono-ADP-Ribosylation. *Nat. Struct. Mol. Biol.* **2013**, *20*, 508–514. [CrossRef]
7. Rack, J.G.M.; Zorzini, V.; Zhu, Z.; Schuller, M.; Ahel, D.; Ahel, I. Viral Macrodomains: A Structural and Evolutionary Assessment of the Pharmacological Potential. *Open Biol.* **2020**, *10*, 200237. [CrossRef]
8. Correy, G.J.; Kneller, D.W.; Phillips, G.; Pant, S.; Russi, S.; Cohen, A.E.; Meigs, G.; Holton, J.M.; Gahbauer, S.; Thompson, M.C.; et al. The Mechanisms of Catalysis and Ligand Binding for the SARS-CoV-2 NSP3 Macrodomain from Neutron and X-ray Diffraction at Room Temperature. *Sci. Adv.* **2022**, *8*, eabo5083. [CrossRef]
9. Fehr, A.R.; Channappanavar, R.; Jankevicius, G.; Fett, C.; Zhao, J.; Athmer, J.; Meyerholz, D.K.; Ahel, I.; Perlman, S. The Conserved Coronavirus Macrodomain Promotes Virulence and Suppresses the Innate Immune Response during Severe Acute Respiratory Syndrome Coronavirus Infection. *mBio* **2016**, *7*, e01721-16. [CrossRef]
10. Fehr, A.R.; Athmer, J.; Channappanavar, R.; Phillips, J.M.; Meyerholz, D.K.; Perlman, S. The Nsp3 Macrodomain Promotes Virulence in Mice with Coronavirus-Induced Encephalitis. *J. Virol.* **2015**, *89*, 1523–1536. [CrossRef]
11. Alhammad, Y.M.; Parthasarathy, S.; Ghimire, R.; Kerr, C.M.; O'Connor, J.J.; Pfannenstiel, J.J.; Chanda, D.; Miller, C.A.; Baumlin, N.; Salathe, M.; et al. SARS-CoV-2 Mac1 Is Required for IFN Antagonism and Efficient Virus Replication in Cell Culture and in Mice. *Proc. Natl. Acad. Sci. USA* **2023**, *120*, e2302083120. [CrossRef] [PubMed]
12. Grunewald, M.E.; Chen, Y.; Kuny, C.; Maejima, T.; Lease, R.; Ferraris, D.; Aikawa, M.; Sullivan, C.S.; Perlman, S.; Fehr, A.R. The Coronavirus Macrodomain Is Required to Prevent PARP-Mediated Inhibition of Virus Replication and Enhancement of IFN Expression. *PLoS Pathog.* **2019**, *15*, e1007756. [CrossRef] [PubMed]
13. McPherson, R.L.; Abraham, R.; Sreekumar, E.; Ong, S.E.; Cheng, S.J.; Baxter, V.K.; Kistemaker, H.A.V.; Filippov, D.V.; Griffin, D.E.; Leung, A.K.L. ADP-Ribosylhydrolase Activity of Chikungunya Virus Macrodomain Is Critical for Virus Replication and Virulence. *Proc. Natl. Acad. Sci. USA* **2017**, *114*, 1666–1671. [CrossRef]
14. Abraham, R.; Hauer, D.; McPherson, R.L.; Utt, A.; Kirby, I.T.; Cohen, M.S.; Merits, A.; Leung, A.K.L.; Griffin, D.E. ADP-Ribosyl-Binding and Hydrolase Activities of the Alphavirus NsP3 Macrodomain Are Critical for Initiation of Virus Replication. *Proc. Natl. Acad. Sci. USA* **2018**, *115*, E10457–E10466. [CrossRef] [PubMed]
15. Parvez, M.K. The Hepatitis E Virus ORF1 "X-Domain" Residues Form a Putative Macrodomain Protein/Appr-1″-Pase Catalytic-Site, Critical for Viral RNA Replication. *Gene* **2015**, *566*, 47–53. [CrossRef] [PubMed]
16. Eckei, L.; Krieg, S.; Bütepage, M.; Lehmann, A.; Gross, A.; Lippok, B.; Grimm, A.R.; Kümmerer, B.M.; Rossetti, G.; Lüscher, B.; et al. The Conserved Macrodomains of the Non-Structural Proteins of Chikungunya Virus and Other Pathogenic Positive Strand RNA Viruses Function as Mono-ADP-Ribosylhydrolases. *Sci. Rep.* **2017**, *7*, 41746. [CrossRef] [PubMed]
17. Virdi, R.S.; Bavisotto, R.V.; Hopper, N.C.; Vuksanovic, N.; Melkonian, T.R.; Silvaggi, N.R.; Frick, D.N. Discovery of Drug-Like Ligands for the Mac1 Domain of SARS-CoV-2 Nsp3. *SLAS Discov.* **2020**, *25*, 1162–1170. [CrossRef]
18. Ni, X.; Schröder, M.; Olieric, V.; Sharpe, M.E.; Hernandez-Olmos, V.; Proschak, E.; Merk, D.; Knapp, S.; Chaikuad, A. Structural Insights into Plasticity and Discovery of Remdesivir Metabolite GS-441524 Binding in SARS-CoV-2 Macrodomain. *ACS Med. Chem. Lett.* **2021**, *12*, 603–609. [CrossRef]
19. Tsika, A.C.; Gallo, A.; Fourkiotis, N.K.; Argyriou, A.I.; Sreeramulu, S.; Löhr, F.; Rogov, V.V.; Richter, C.; Linhard, V.; Gande, S.L.; et al. Binding Adaptation of GS-441524 Diversifies Macro Domains and Downregulates SARS-CoV-2 de-MARylation Capacity. *J. Mol. Biol.* **2022**, *434*, 167720. [CrossRef]
20. Singh, A.K.; Kushwaha, P.P.; Prajapati, K.S.; Shuaib, M.; Gupta, S.; Kumar, S. Identification of FDA Approved Drugs and Nucleoside Analogues as Potential SARS-CoV-2 A1pp Domain Inhibitor: An in Silico Study. *Comput. Biol. Med.* **2021**, *130*, 104185. [CrossRef]
21. Selvaraj, C.; Dinesh, D.C.; Panwar, U.; Boura, E.; Singh, S.K. High-Throughput Screening and Quantum Mechanics for Identifying Potent Inhibitors against Mac1 Domain of SARS-CoV-2 Nsp3. *IEEE/ACM Trans. Comput. Biol. Bioinform.* **2021**, *18*, 1262–1270. [CrossRef] [PubMed]
22. Babar, Z.; Khan, M.; Zahra, M.; Anwar, M.; Noor, K.; Hashmi, H.F.; Suleman, M.; Waseem, M.; Shah, A.; Ali, S.; et al. Drug Similarity and Structure-Based Screening of Medicinal Compounds to Target Macrodomain-I from SARS-CoV-2 to Rescue the Host Immune System: A Molecular Dynamics Study. *J. Biomol. Struct. Dyn.* **2022**, *40*, 523–537. [CrossRef] [PubMed]

23. Russo, L.C.; Tomasin, R.; Matos, I.A.; Manucci, A.C.; Sowa, S.T.; Dale, K.; Caldecott, K.W.; Lehtiö, L.; Schechtman, D.; Meotti, F.C.; et al. The SARS-CoV-2 Nsp3 Macrodomain Reverses PARP9/DTX3L-Dependent ADP-Ribosylation Induced by Interferon Signaling. *J. Biol. Chem.* **2021**, *297*, 101041. [CrossRef] [PubMed]
24. Brosey, C.A.; Houl, J.H.; Katsonis, P.; Balapiti-Modarage, L.P.F.; Bommagani, S.; Arvai, A.; Moiani, D.; Bacolla, A.; Link, T.; Warden, L.S.; et al. Targeting SARS-CoV-2 Nsp3 Macrodomain Structure with Insights from Human Poly(ADP-Ribose) Glycohydrolase (PARG) Structures with Inhibitors. *Prog. Biophys. Mol. Biol.* **2021**, *163*, 171–186. [CrossRef] [PubMed]
25. Bajusz, D.; Wade, W.S.; Satała, G.; Bojarski, A.J.; Ilaš, J.; Ebner, J.; Grebien, F.; Papp, H.; Jakab, F.; Douangamath, A.; et al. Exploring Protein Hotspots by Optimized Fragment Pharmacophores. *Nat. Commun.* **2021**, *12*, 3201. [CrossRef] [PubMed]
26. Schuller, M.; Correy, G.J.; Gahbauer, S.; Fearon, D.; Wu, T.; Díaz, R.E.; Young, I.D.; Carvalho Martins, L.; Smith, D.H.; Schulze-Gahmen, U.; et al. Fragment Binding to the Nsp3 Macrodomain of SARS-CoV-2 Identified through Crystallographic Screening and Computational Docking. *Sci. Adv.* **2021**, *7*, eabf8711. [CrossRef]
27. Ekblad, T.; Verheugd, P.; Lindgren, A.E.; Nyman, T.; Elofsson, M.; Schüler, H. Identification of Poly(ADP-Ribose) Polymerase Macrodomain Inhibitors Using an AlphaScreen Protocol. *SLAS Discov.* **2018**, *23*, 353–362. [CrossRef]
28. Gahbauer, S.; Correy, G.J.; Schuller, M.; Ferla, M.P.; Doruk, Y.U.; Rachman, M.; Wu, T.; Diolaiti, M.; Wang, S.; Neitz, R.J.; et al. Iterative Computational Design and Crystallographic Screening Identifies Potent Inhibitors Targeting the Nsp3 Macrodomain of SARS-CoV-2. *Proc. Natl. Acad. Sci. USA* **2023**, *120*, e2212931120. [CrossRef]
29. Dasovich, M.; Zhuo, J.; Goodman, J.A.; Thomas, A.; McPherson, R.L.; Jayabalan, A.K.; Busa, V.F.; Cheng, S.J.; Murphy, B.A.; Redinger, K.R.; et al. High-Throughput Activity Assay for Screening Inhibitors of the SARS-CoV-2 Mac1 Macrodomain. *ACS Chem. Biol.* **2022**, *17*, 17–23. [CrossRef]
30. Sowa, S.T.; Galera-Prat, A.; Wazir, S.; Alanen, H.I.; Maksimainen, M.M.; Lehtiö, L. A Molecular Toolbox for ADP-Ribosyl Binding Proteins. *Cell Rep. Methods* **2021**, *1*, 100121. [CrossRef]
31. Wazir, S.; Parviainen, T.A.O.; Maksimainen, M.M.; Duong, M.T.H.; Pfannenstiel, J.J.; Cluff, D.; Sowa, S.T.; Galera-Prat, A.; Ferraris, D.V.; Fehr, A.; et al. Discovery of 2-Amide-3-Methylester Thiophenes Inhibiting SARS-CoV-2 ADP-Ribosyl Hydrolysing Macrodomain and Coronavirus Replication. *bioRxiv* **2023**. [CrossRef]
32. Roy, A.; Alhammad, Y.M.; McDonald, P.; Johnson, D.K.; Zhuo, J.; Wazir, S.; Ferraris, D.; Lehtiö, L.; Leung, A.K.L.; Fehr, A.R. Discovery of Compounds That Inhibit SARS-CoV-2 Mac1-ADP-Ribose Binding by High-Throughput Screening. *Antivir. Res.* **2022**, *203*, 105344. [CrossRef] [PubMed]
33. Schuller, M.; Zarganes-Tzitzikas, T.; Bennett, J.; De Cesco, S.; Fearon, D.; von Delft, F.; Fedorov, O.; Brennan, P.E.; Ahel, I. Discovery and Development Strategies for SARS-CoV-2 NSP3 Macrodomain Inhibitors. *Pathogens* **2023**, *12*, 324. [CrossRef] [PubMed]
34. Sherrill, L.M.; Joya, E.E.; Walker, A.M.; Roy, A.; Alhammad, Y.M.; Atobatele, M.; Wazir, S.; Abbas, G.; Keane, P.; Zhuo, J.; et al. Design, Synthesis and Evaluation of Inhibitors of the SARS-CoV-2 Nsp3 Macrodomain. *Bioorg. Med. Chem.* **2022**, *67*, 116788. [CrossRef] [PubMed]

MDPI
St. Alban-Anlage 66
4052 Basel
Switzerland
www.mdpi.com

Pathogens Editorial Office
E-mail: pathogens@mdpi.com
www.mdpi.com/journal/pathogens

www.ingramcontent.com/pod-product-compliance
Lightning Source LLC
Chambersburg PA
CBHW061435210926
51850CB00008B/1518
9783725809868